Theory and Applications of Transport in Porous Media

Volume 26

More information about this series at http://www.springer.com/series/6612

Vyacheslav G. Rumynin

Overland Flow Dynamics and Solute Transport

Springer

Vyacheslav G. Rumynin
The Russian Academy of Sciences
Institute of Environmental Geology
Saint Petersburg, Russia

Saint Petersburg State University
Institute of Earth Sciences
Saint Petersburg, Russia

ISSN 0924-6118 ISSN 2213-6940 (electronic)
Theory and Applications of Transport in Porous Media
ISBN 978-3-319-21800-7 ISBN 978-3-319-21801-4 (eBook)
DOI 10.1007/978-3-319-21801-4

Library of Congress Control Number: xxxxxxxxxx

Springer Cham Heidelberg New York Dordrecht London

Printed on acid-free paper

Springer International Publishing AG Switzerland is part of Springer Science+Business Media (www.springer.com)

Preface

It is generally recognized that overland flow (runoff), surface–subsurface mass transfer, and subsurface water and solute fluxes are key processes governing contaminant transport in the landscape environment. The relevant interdisciplinary studies have been a prime focus of the hydrological society from the past century to the present, resulting in an enormous number of publications dedicated to model development, both conceptual and site-specific. Moreover, in the recent decades, a number of observations were made, evidencing the presence of anomalous phenomena enhancing or restraining water and chemical runoff from contaminated watersheds. However, it is not yet fully understood how both the natural and human-induced mechanisms, controlling these processes, interact and how the temporal and spatial-scale effects control these interactions under different watershed conditions and at different characteristics. Such understanding may help improve the reliability of assessment and prediction of the large-scale human impact on the environment, in particular, for areas contaminated by radioactive fallout from damaged nuclear units, such problems being among the most important applications of this work.

In this context, the purpose of this work is to contribute, marginally at least, to the theoretical framework of the link between overland flow dynamics and water quality, with a special focus on the challenge the author faced in dealing with the ambiguity of existing approaches to conceptualization of some particular transport mechanisms and field conditions. Thus, the main subjects include (1) extension of the theoretical concepts regarding the connection between overland flow dynamics and water quality, with a special focus on the transient system behavior; (2) study of anomalous behavior of the mass transfer accompanying the overland flow, which stems from both the peculiarities of the physicochemical interactions and the overlapping of several transfer mechanisms; and (3) collection of field data required to quantify the parameters and processes controlling the radionuclide transport in the near-surface domains which is closely related to the risk assessment of soil and water contamination through radioactive fallout.

More specifically, this book is aimed to emphasize analytical tools, supported by numerical modeling and illustrative field materials, providing assessment and prediction of contaminant transport in runoff, interacting with the shallow subsurface environment, represented by soil, vadose zone, and phreatic aquifers. The topics discussed here are related to the land surface hydrology and cover a wide range of coupled hydrological processes across a range of scales from hillslope to watershed. Overland dynamics and solute transport are presented and discussed through the application of both physically based models (mostly, using methods from the kinematic wave theory) and the empirical (effective lumped-parameter) approach. Such combination provides a better understanding of the mechanisms of flow and transport and would assist in the development of effective methods to control and predict changes in water components of the environment.

The fundamental problem of studying the formation of surface runoff and its chemistry under anthropogenic pollution of the soil and precipitation contains three major aspects:

First, the identification and description of runoff generation mechanisms providing rain and snowmelt water conversion into water bodies on the landscape

Second, the development of hydrodynamic models, describing water flow over land surface toward an outlet

Third, the development of hydrological models, describing the transfer of contaminants accumulated on the land surface or in the soil profile into water flow and their lateral transport in the form of dissolved species and particulate matter toward an outlet

The variable rainfall conditions, one of the most common features of the synoptic environment, determine the transient effects of rainfall–runoff–infiltration partitioning and chemical response of catchments to excess precipitation. Therefore, special attention is given to the analysis of the coupled transient flow and solute transport with the aim to more precisely formulate the physical and mathematical problem. To simplify the mathematics and reduce the number of required variables and parameters, other lumped runoff and solute transport models are also considered.

Another priority of this book is the focus on the anomalous behavior of mass transfer accompanying the overland flow. Such phenomenon stems from both the specifics of physicochemical interactions (e.g., sorption kinetics and irreversibility) and the overlapping of several transfer mechanisms (infiltration, soil erosion, the flow-focusing or channeling effects of microtopography, etc.). The relevant illustrations are concerned mostly with the model and experimental study of the regional-scale radionuclide transport with runoff induced by radioactive fallout from damaged nuclear reactors or nuclear weapon tests in the atmosphere since 1952.

In the latter respect, special attention is paid to the analysis of the consequences of the Chernobyl 1986 and Fukushima 2011 NPP accidents, supplemented with analysis of the less known Kyshtym 1957 accident, from the viewpoint of fallout radionuclide mobility and retention in the shallow subsurface environment, surface water bodies, and groundwater reservoirs. Monitoring data supported by theoretical findings are used as a basis for coupling NPP accident scenarios (source-term and fallout deposition specification) with a set of hydrological models aimed at prediction of hydrological system response to soil contamination by radioactive materials in the areas of potential influence of existing or designed nuclear power units. To test the methodology, a distributed parameter watershed model of the Beloyarsk NPP location (Middle Urals, RF) was developed and calibrated basing on monitoring data.

This book is organized into seven chapters. The first two Chaps. (1 and 2) consider the runoff generation and water flow dynamics as a mathematical background of the near-surface solute transport (Chaps. 3 and 4) based on distributed parameter approach. Then, in the next two Chaps. (5 and 6), watershed lumped-parameter models for both water flow and solute transport are discussed. The conclusive Chap. 7 illustrates both the applicability of the above risk assessment strategy and the applicability of a selected numerical code for watershed modeling to the investigation of urgent issues related to radioactive fallout after hypothetical accidents at engineered nuclear power units. Monitoring data and data from field-site characterization experiments are also discussed in this chapter.

The author very much appreciates the help of Dr. Leonid Sindalovsky in the implementation of many numerical algorithms and codes considered in the book, as well as the contribution of Dr. Anton Nikulenkov and Elena Vereschagina, who shared their data on regional study of soil and surface water systems in the influence area of the Beloyarsk NPP. The author also appreciates the attention to his work and fruitful discussions with other colleagues – researchers from E.M. Sergeev Institute of Environmental Geology, St. Petersburg Division, RAS, and staff from St. Petersburg State University, Institute of Earth Sciences. Finally, the author deeply thanks Gennady Krichevets for his help in the professional translation of the book and many useful comments from him allowing the author to make certain improvements to the book.

Thus, this book, along with theoretical findings, contains field information, which will facilitate the understanding of near-surface solute transport and the development of a methodology for practical application in watershed hydrology. This book addresses scientists and engineers who are interested in the quantitative approach to studying contaminant transport processes. The book can also be profitably read by students.

St. Petersburg, Russia — Vyacheslav G. Rumynin
March 31, 2015

Contents

Part I
Response Mechanisms of Hydrological Processes in the Near-Surface Environment

To describe the flow and contaminant transport, induced by rainfall or snowmelt, through the landscape media, one is to consider several coupled processes occurring in the near-surface environment represented by soil and vadose zone in contact with a periodically forming movable water domain or with the atmosphere through the surface (Chow et al. 1988; Brutsaert 2005; Shaw et al. 2011). Thus, rain or snow-melting events intensify several mechanisms and processes, including: (a) the formation of a water body on the landscape; (b) the accumulation of pollutants of natural or anthropogenic origin in this layer after their release from the surface or from soil solution; (c) flow of polluted water over the surface; (d) the descending infiltration of a part of this water through the porous medium under the effect of capillary and gravity forces – a process, which controls the flow depth and, accordingly, the degree of water saturation with solutes, as well as the rate of water flow; (e) the development of paths of rapid pollution transport through macropores and fractures in the vadose zone and from depressions on the land surface toward water table, and, finally, (f) lateral contaminant transport through temporary or permanent phreatic horizons (sloping shallow aquifers).

The relevant hydrological analysis shall account for the differences between the space and time scales of the processes in the near-surface domains in contact with aquifer materials. Thus, the time scale for the conditions of runoff formation and solute migration is commonly of the order of hours, rarely a few days, while those of flow and solute transport in aquifers are of the order of months or years. The length of water flow paths in such systems is of the order of hundreds of meters or some kilometers. In terms of the time a water particle spends in it, the soil and vadose zone, with rare exceptions, occupy an intermediate position; however, unlike the systems involving surface and subsurface runoff, the vertical flow paths are much shorter in the sediments above the water table. For example, at such combination of time scales, a description of subsurface flow and solute transport can be based on the mean annual values of groundwater recharge and solute inputs, because the long pollutant residence time in the aquifer smoothes the effect on the solution of transport problem caused by daily and seasonal fluctuations in water flux and concentration functions on the upper aquifer boundary. From the viewpoint of stochastic analysis (Duffy and Gelhar 1985), for systems with large residence times (such as phreatic aquifers), small input correlation scale variations in continuous flow and solute inputs will produce little variation in the outflow characteristics. Under the same condition, the nearsurface domains are much more sensitive to small-scale changes in the flow and solute input characteristics. Therefore, to properly model the behavior of such system, in some practical situations it is important to use rainfall records with high temporal resolution.

The large differences in the time and space scales between the flow of water within the surface and subsurface domains allow a researcher to formalize the interaction between the domains through the transfer of boundary conditions from one domain into another, thus avoiding the solution of fully coupled equations of surface and subsurface flow (Furman 2008). In this part of the work, the relevant decoupling of the hydrological processes in the two domains is based on the raincontrolled infiltration interface approach allowing analytical solution of flow and solute transport problems with the assumption of prescribed infiltration rate or depth.

In this, first, part of the book, the behavior of the near-surface system under rainfall conditions will be analyzed based on an analytical framework at the column and hillslope scales. Mega-scale system's behavior is the subject of the second part of the book.

References

Brutsaert W (2005) Hydrology: An Introduction. Cambridge University Press, Cambridge, UK, p 605

Chow VT, Maidment DR, Mays LW (1988) Applied hydrology. McGraw-Hill, New York, p 572

Duffy CJ, Gelhar LW (1985) A frequency domain approach to water quality modeling in groundwater: theory. Water Resour Res 21:1175–1184

Furman A (2008) Modeling coupled surface–subsurface flow processes. A review. Vadose Zone J 7(2):741–756

Shaw EM, Beven KJ, Chappel NA, Lamb R (2011) Hydrology in Practice. Fourth Edition. Spon Press, London, p 543

Chapter 1
Surface Runoff Generation, Vertical Infiltration and Subsurface Lateral Flow

In this and the following three chapters, we will focus explicitly on the dynamic (transient, short-time-scale) hydrological processes that determine the partitioning of rainfall into runoff and infiltration and control the flow and chemical response of a catchment or its segments to the anthropogenic impact. Two principal components of runoff, surface and subsurface, which differ remarkably in their response time to precipitation or snow-melting events, are considered; however we do not present here a general mathematical framework for coupling the surface and subsurface flow equations, relying instead on an approach based on the transfer of boundary conditions (from one model domain to another). Soil infiltration theory, as discussed here briefly, plays the central role in such approach as well as in the solution of various problems of the surface and subsurface hydrodynamics. With this in view, special attention will be paid to some nonlinear and threshold phenomena in structured (discontinued by macropores and cracks) soils having a major impact on hydrological processes as well.

However, before we pass to the substantive part of the chapter, it is reasonable to discuss the key terminological issues and the description of individual mechanisms that govern the flow of water over the surface and within the shallow subsurface domains.

1.1 Key Definitions

Rain falling onto land surface is accompanied by rainwater partitioning into the surface and subsurface components in different proportions, depending on the rainfall rate, the properties of the cover deposits (their permeability and capillary characteristics), and their initial moisture content. This process can be considered at different space and time scales with different degree of detail or, conversely, generalization of the processes governing it. Thus, many problems can be analyzed at scale 1D of a soil profile. The next scale of hydrological process consideration is

V.G. Rumynin, *Overland Flow Dynamics and Solute Transport*, Theory and Applications of Transport in Porous Media 26, DOI 10.1007/978-3-319-21801-4_1

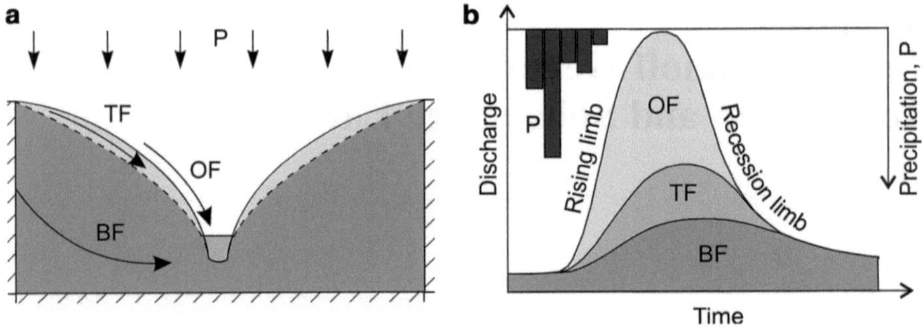

Fig. 1.1 (**a**) A diagram illustrating three major runoff components produced by rainfall: OF overland flow (HOF or/and DOF, see Sect. 1.2), TF throughflow (sometimes in the form of subsurface stormflow, SSF, see Sect. 1.2), and BF base flow; (**b**) corresponding (hypothetical) storm hydrograph. *P* precipitation

the scale of a hillslope (a slope of a river valley or a sloped urban district). Then, of particular importance is the hydrological analysis at the scale of a watershed (catchment) area, a topographic region in which all water drains to a common outlet. According to classical concepts, the entire land surface can be divided into polygons, representing a «matrix» of watersheds of different orders. As a rule, watersheds are associated with stream systems and some of them are identified geographically. In this book, the terms watershed and catchment are used interchangeably without defining the distinctions between them.

The flow of a water layer over the surface and through the pores of soils and sediments that is coming out of the watershed is termed runoff (Fig. 1.1a). There are three components of the runoff from watersheds (Shaw et al. 1994; Dingman 2002; Zhang et al. 2002): (1) surface runoff or overland flow (sometimes termed as direct runoff), (2) subsurface runoff or interflow (throughflow), and (3) groundwater runoff or baseflow. A dynamic form of rapid soil interflow that results from heavy rainfall is associated with subsurface stormflow. Overland flow and soil interflow together are sometimes referred to as quickflow. All three components contribute to the total hydrograph, a plot showing the rate of flow (discharge) versus time (Fig. 1.1b).

The surface runoff is the water that travels over the ground surface driven by gravitational forces in the form of sheet flow (interrill flow), rill and gully flow, towards the stream. It can be generated by different mechanisms discussed below. The water that moves over surface, i.e., surface runoff, rapidly reaches the nearest discharge zones, thus showing a quick response to a rain event or snow melting.

Subsurface runoff or interflow represents the portion of water that moves laterally in the upper part of soil, litter layer covering the soil surface, or in the soil–bedrock interface. Such lateral flow appears when soils have impermeable or semi-permeable layers at shallow depths. Subsurface runoff moves slower than surface runoff.

The water that has been absorbed by soil and has passed through the vadose zone supplements the storage of the topmost aquifer. This process is termed the groundwater recharge. The water flow thus forming in an aquifer is groundwater runoff. It responds to rainfall with a noticeable delay and does not fluctuate rapidly.

The mean annual values of the flow characteristics mentioned above, expressed in terms of volume (L^3, commonly, m^3) or runoff depth (L, commonly, mm), from a unit drainage area, along with precipitation and evapotranspiration, are the main components of water balance. Their values and the ratio between them are determined based on the soil conditions and actual evapotranspiration, as well as landscape–climatic characteristics of the area (Sects. 1.3 and 1.4).

1.2 Surface Runoff Mechanisms: Conversion of Rainfall into Runoff

The development of a functional basis for quantifying the transformation of rainfall to runoff has been a prime target for hydrologists of several generations. Several models (with different degree of physical soundness) are known to describe the transformation of rainfall to overland flow. The models that have the largest recognition and application in the practice of hydrological analysis are (1) the model of infiltration excess runoff (Hortonian overland flow, HOF) and (2) the model of saturation excess runoff (Dunne or saturation overland flow, DOF). Subsurface stormflow (SSF) as a component of runoff has been a subject for much research and discussion over the years as well.

1.2.1 Infiltration Excess Runoff

The physical ideas and the mathematical relationships that form the basis of the infiltration excess model were formulated and developed in the early works (the 1930s) of the well-known American hydrologist Robert Elmer Horton. In his theory, which has become classical, R.E. Horton proceeds from the basic assumption that the surface (slope) runoff forms due to the limited capacity of soil (or rock) to imbibe and pass water that arrives to its surface as rain. This assumption requires the introduction of additional definitions and criteria, enabling quantitative evaluation. To quantify Hortonian overland flow (HOF) generation it is necessary to have a sub-model of infiltration that interacts with the rainfall input (Vieux 2004).

R.E. Horton introduces the concept of soil infiltration capacity, $f = f(t)$, implying the maximal rate at which rainwater can be adsorbed by soil under given conditions. In this case, water is assumed to have unlimited access to the porous surface; therefore, function f is also called potential infiltration (by analogy with potential evaporation). Later the term infiltrability was suggested to replace the infiltration capacity to represent the surface flux under any set of conditions, whatever the rate or pressure at which the water is supplied to the soil (Hillel 2004).

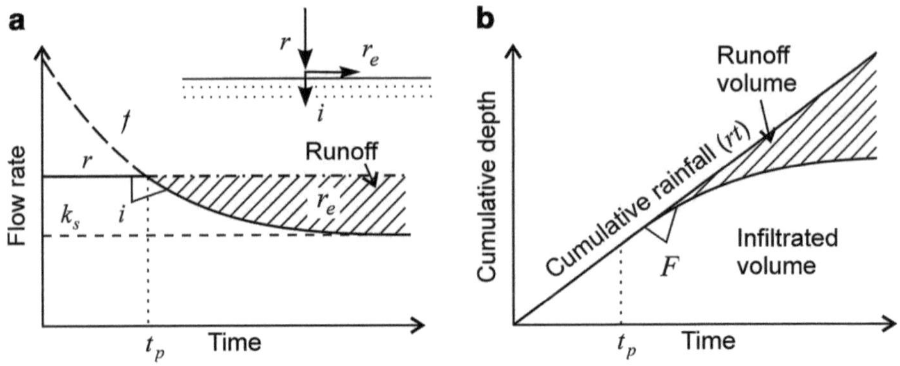

Fig. 1.2 Conceptual model of surface runoff (HOF) generation (uniform rainfall, $r = \text{const}$). (**a**) Main flow rate components and infiltrability vs time, (**b**) cumulative curves and volumes

As seen (Fig. 1.2a), due to the nonlinearity of flow in unsaturated media (soil), f decreases continuously throughout the rainfall period, thus it behaves similar to the decay function. When the rainfall rate, r (the simplest case $r = \text{const}$ is considered), is less than f, all the rain water will be imbibed. When r is greater than f, the rate of water penetration into soil is f, so the rate of water accumulation on the surface will be $r_e = r - f$ (Fig. 1.2a), the so-called rainfall excess, which determines the potential of the process of water flow over the surface (Chap. 2).

Summarizing the above, the following mathematical formula can be suggested:

$$i(t) = \min[r(t), f(t)], \tag{1.1}$$

where $i(t)$ is the actual rate of infiltration [LT^{-1}]. The quantitative description of the process implies determining the moment t_p (Sect. 1.3.2) when surface ponding begins, i.e., when the infiltration capacity f equals r. At the moment $t = t_p$, the surface of soil or sediment shows the maximal moisture content ($\theta \approx \theta_s$), and its permeability equals the hydraulic conductivity, $k \approx k_s$ (Fig. 1.2a); therefore, obviously, HOF is possible when $r > k_s$. After moment t_p, $f = f(t - t_p)$ becomes less than r (i.e., not all water infiltrates), and surface runoff starts forming. In this case, the mathematical description of this process takes the form:

$$i(t) = \begin{cases} r & 0 \le t \le t_p, \\ f(t - t_p) & t > t_p. \end{cases} \tag{1.1a}$$

Because the precipitation rate exceeds the infiltration capacity, there is excess precipitation available for surface runoff

$$r_e(t) = \begin{cases} 0 & 0 \le t \le t_p, \\ r - f(t - t_p) & t > t_p; \end{cases} \tag{1.1b}$$

$r_e(t)$ is the infiltration excess runoff rate (or intensity) [LT^{-1}]. Mathematically, the ponding time, t_p, is the moment when the flux boundary condition changes to a head boundary condition (Sect. 1.3).

The rainfall excess pattern, $r_e(t)$, can be directly routed across the watershed to the stream network. More realistic routing methods are usually based on surface water hydraulics, utilizing, for example, the kinematic wave approach (Chap. 2).

To forecast the actual rates of infiltration and runoff and to determine the characteristic time t_p, one needs the form of the functional relationship $f(t)$ to be established, i.e., an infiltration model is to be chosen; in most cases, preference should be given to relationships $f = f(F)$, which relate infiltration rate with the cumulative infiltration and volume of water F (Fig. 1.2b). Cumulative infiltration curves, $F(t)$, indicate the amount of water entering the soil at any given time or the time required for a given amount of water to infiltrate into soil (Biswas and Mukherjee 1994). Functions $f(t)$ and $f(F)$ can be empirical or a mathematical result of solving moisture transfer problems in an incompletely saturated porous medium (Sect. 1.3). Because of this, the HOF mechanism of surface runoff formation can be described within the framework of rather strict mathematical constructions.

Finally, it should be mentioned that in this chapter the infiltration is considered independent of overland flow dynamics resulting in weak coupling of the two processes.

1.2.2 Saturation Excess Runoff

Currently the majority of authors incline to the opinion that the Hortonian overland flow forms under a limited range of natural conditions, mostly in arid areas, where moisture content is low and the vadose zone thickness is large. Indeed, there are many confirmations of the fact that surface runoff can form in areas with surface permeability high enough for the precipitation rate r never be in excess of the infiltration capacity f. Such behavior of the system can be seen in areas with humid climate, where groundwater table is high and the soils and vadose zone rocks are heterogeneous.

In porous media and under conditions where the difference between rainfall intensity and infiltration rate is not a criterion of overland flow formation, more suitable is the type of surface runoff models referred to as models of saturation excess runoff. These models employ the so-called Dunne assumption that all precipitation enters the soil and that runoff (Dunne overland flow, DOF) occurs due to the soil's inability to absorb any more water (Hydrology Handbook 1996; Todini 2007). They reflect the limited ability of soil and underlying rocks to accumulate water, as is the case, for example, in areas where a thin soil layer is underlain by low-permeability deposits or in areas with shallow groundwater table. When falling onto such areas, precipitation, which has no storage reserve for imbibition, will directly transform into surface runoff. Therefore, the DOF generation is controlled, along with soil hydraulic properties, by two major factors, namely, the geomorphology (the shape and the slope) of the catchment and its subsurface hydrology (Willgoose and Perera 2001). T. Dunne with coauthors (Dunne and Black 1970; Dunne et al. 1975) were among the first who, based on data of detailed hydrological observations and experiments, offered a physical explanation to this phenomenon. That is why, such models, associated with rapid overfilling of the pore storage of soil layers, is often referred to as Dunne-type (DOF).

Thus, in contrast to the infiltration excess (HOF) model, runoff in humid regions with coarse-texture soils is generated by saturation from below by a rising groundwater table (Ogden and Watts 2000) or by discharging sporadic horizontal flow of water (throughflow) within the soil layer (Hydrology Handbook 1996). Many studies have shown that overland flow in areas with humid climate form within relatively small areas (as compared with the total watershed area) with higher water table. Such areas commonly occur at the bases of hill slopes, in river valleys and swales. Such saturated areas, where saturation excess runoff is produced, are referred to as variable source areas (VSAs). Detailed analysis of the rainfall–runoff processes in experimental watersheds shows that up to two thirds of direct runoff can originate from the VSAs occupying only 5–20 % of the watershed area (Boughton 1993; Hydrology Handbook 1996; Ogden and Watts 2000). Moreover, due to the nonlinearity of the saturation excess mechanisms, the amount of runoff produced in watersheds depends on individual rainfall characteristics (intensity and duration).

1.2.3 Subsurface Stormflow Runoff

Many hydrological observations suggest that overland flow may not occur even during high-intensity rainfall. Indeed, in some catchments with well-developed soil cover, lateral flow is concentrated in the lower part of the soil profile underlain by a clay layer or rigid bedrock (Weyman 1970; Sloan and Moor 1984; Zhang et al. 2006). On steep hillslopes, such subsurface stormflow (SSF) could be fast enough to be the main water contributor to downslope drainage area forming stream runoff in a catchment. The rapid vertical movement of stormwater in the soil profile can be explained by the presence of macropores or other soil structural elements that provide preferential flowpaths (Buttle and McDonald 2002; Beckers and Alila 2004; Dusek and Vogel 2014). This type of pathways can take a variety of forms, including movement as a thin saturated layer above unsaturated bedrock; runoff, concentrated in the shallow weathered zone of bedrock; pipeflow at the base of the soil profile; and flow associated with macropores or subhorizontal layers with high permeability, embedded in the soil matrix (McDonnell 1990; Buttle and McDonald 2002; Uchida et al. 2002; Beckers and Alila 2004).

Stormflow generation during heavy rainfall events may play an important role in determining the various hydrological and soil-mechanical phenomena like saturation excess overland flow, changes in the soil moisture, slope stability changes from a stable to an unstable condition, water pollution impacts. Thus, a rise of water table may induce chemical leachate from near-surface layers (Weiler and McDonnell 2006).

In a uniform soil profile, flow response to an increase in hydraulic gradient due to infiltrating water can be described on the base of the hydraulic theory for unconfined groundwater flow in a sloping aquifer using an extended Dupuit–Forcheimer model (Henderson and Wooding 1964; Beven 1981). The model originates from a continuity equation,

$$\phi_n \frac{\partial h}{\partial t} + \frac{\partial q}{\partial x} = w, \tag{1.2}$$

in which the flow discharge (per unit width), q, is derived from Darcy equation as formulated by Boussinesq (Childs 1971)

$$q = -k_s h \left(\frac{\partial h}{\partial x} \cos \varphi - \sin \varphi \right); \tag{1.2a}$$

here x is the coordinate parallel to the impermeable layer underlying the saturated soil; t is the time; φ is the bed slope; h is the hydraulic head; k_s is the saturated hydraulic conductivity; ϕ_n is the effective storage coefficient (drainable porosity); w is the rate of water input to the saturated zone from the unsaturated zone (in other words, the recharge rate, the portion of the precipitation that recharges the water table). The model (1.2) assumes that the streamlines are parallel to the sloping impermeable bed and the hydraulic head h is independent of depth. The model (1.2) was analyzed by many authors (Beven 1981; Brutsaert 1994; Verhoest and Troch 2000; Xiangjun et al. 2006; Harman and Sivapalan 2009), and several approximations describing the water table shape and flow discharge as functions of x and t were obtained. Among these solutions, a solution for the output hydrograph is of main interest.

When φ is large, the derivative ("diffusive") term in (1.2a) vanishes, thus, $q = k_s h \sin \varphi$, and the continuity equation (1.2) takes the form of a linear kinematic wave equation

$$\phi_n \frac{\partial h}{\partial t} + k_s \sin \varphi \frac{\partial h}{\partial x} = w. \tag{1.2b}$$

Equation (1.2b) implies that the hydraulic gradient at any point is equal to the bed slope, φ.

It was shown (Beven 1981) that for a dimensionless ratio $4w \cos \varphi / k_s \sin^2 \varphi$ (representing combinations of slope conditions and input rate), less than 0.75, or

$$\lambda = 4w \cos \varphi / k_s \sin^2 \varphi < 0.75, \tag{1.2c}$$

the kinematic solution is a reasonable representation of the output hydrograph and the subsurface water table. As can be seen from the structure of criterion λ, the kinematic wave approximation becomes increasingly useful with increasing slope angle and increasing saturated hydraulic conductivity (Verhoest and Troch 2000).

This range of conditions for λ is of practical interest. Under steady-state conditions ($\partial h/\partial t = 0$), from solution of Eq. (1.2b) it follows that aquifer thickness, h, is a linear function of x, $h(x) = wx/k_s \sin\varphi$, thus the mean aquifer thickness is $h_{av} = wL/2k_s \sin\varphi$, and criterion (1.2c) can be rewritten in a form:

$$h_{av}/L \tan\varphi < 0.1, \tag{1.2d}$$

where L is the hillslope length. The inequality (1.2d) limits the slope and flow characteristics, which provide conditions for the potential gradient, driving the flux (Eq. 1.2a), to be dominated by the topographic gradient. The estimate (1.2d) does not contradict that of Harman and Sivapalan (2009), where one can find much more thorough analysis of the links between these characteristics and other parameters in hillslope-scale models of subsurface lateral saturated flow.

The kinematic wave Eq. (1.2b) has been studied well (Chaps. 2 and 3), allowing us to give a quantitative description to this type of water flow, generated at a slope with increasing angle under rainstorm conditions. The considered linear model suits the idealized concept of SSF. In particular, this model predicts the hydrological response in the form of a continuous water flow in soil for any storm size, a situation, which is not common in the nature. Meanwhile, the kinematic wave assumption does not accommodate the lower boundary condition (Singh 2002). As a result, it will always overpredict the length of the seepage face at the downhill boundary, since it takes no account of downstream effects in the vicinity of the boundary (Beven 1981). Therefore, the farther the distance from the lower boundary, the more accurate the theory, since the effect of the downstream boundary will decline (Singh 2002).

On the other hand, field studies in hillslope hydrology indicate that the SSF is a threshold phenomenon with respect to storm rainfall input: to initiate downslope drainage effect in soil, the atmospheric precipitation layer accumulated in the subsurface over certain time should have the depth in excess of some critical (threshold) value.

In general, the interaction between hillslope attributes (slope, soil depth and permeability, etc.) and storm size usually causes unexpected behavior of outlet hydrograph. Such behavior, resulting, in particular, in the mentioned above delay in lateral flow initiation, may be explained by the interplay between subsurface topography and the overlaying soil mantle with its varying soil depth distribution as suggested by Hopp and McDonnell (2009) and Graham and McDonnell (2010).

The authors proposed a concept of hydrological connectivity between isolated patches of saturation, distributed along the soil–bedrock interface. Lateral subsurface flow only occurred when well-connected hillslope-scale areas of saturation or near saturation developed at the soil–bedrock interface, or, the other words, the SSF is dominated by a connected preferential flow network located at the soil–bedrock interface (Graham et al. 2010). In the relevant physical model, downslope micrographic impediments in the subsurface can be a barrier for flow, interrupting and/or redirecting flow (Hopp and McDonnell 2009). When precipitation exceeds the threshold dictated by topographic impedance, the subsurface water can be efficiently routed downslope.

This model can be visualized as water motion along the topographic lows of the bedrock surface between water-filled bedrock depressions. During a rainfall event, those depressions form fill-and spill areas within a hillslope, the connectivity between them being determined, in particular, by the average slope angle. At a gentle slope, many isolated fill areas are usually identified across the hillslope but hardly any spill areas. Subsurface stormflow is very small if any. At a steep slope, the rain size and the characteristics of soil–bedrock interface being the same, the fill areas will be very well connected to each other by spill areas. The generation of such "system" is associated with a period of intensive SSF. Following the authors' concepts (Hopp and McDonnell 2009), one may conclude that the subsurface flow generation always positively correlates with slope angle and storm size, and negatively correlates with bedrock permeability and soil depth.

Finally, SSF can be generated due to water table rise from below into shallower, more transmissive layers, which provide the pathways for rapid lateral hillslope flow (Gabrielli et al. 2012; McDonnell 2013).

1.2.4 On a Combination of Runoff Generation Processes

The properties of cover deposits within a catchment can vary widely, and the topographic features of relief and soil morphology vary over space. Therefore, the same drainage area can show a combination of radically different mechanisms of precipitation transformation into surface runoff, and the area distribution of zones of water appearance on the surface can be extremely uneven. Such combination often hinders the adequate reproduction of a real natural process by homogeneous (limiting) models.

Moreover, the mechanisms of runoff generation at the hillslope scale are highly nonlinear, their manifestation depending also on rainfall intensity, rainfall pattern, and antecedent soil moisture conditions. The rainfall rate may exceed the infiltrability for some storms, while for others, the rain may be not so heavy, leading to the saturation of surface soil layer. In such cases, subsurface runoff may dominate in some parts of the catchment where the slope angles of impermeable bases of slopes are large. Thus, the infiltration and saturation excess-generating mechanisms as well as a mechanism responsible for the movement of water in the porous space beneath the surface are not mutually exclusive at a point on a watershed (Smith and Goodrich 2005).

The search for some universal criteria to distinguish between different mechanisms of runoff generation in extremely diverse slope (catchment) environments is a problem, which cannot be solved in a strict formulation. Here, we have to agree with the quotation from McDonnell's work (2013): "... no overarching theory currently exists for runoff generation across all climates, geology, and topography". Further, he writes "This is a problem – both for experimentalists, ... wondering how to diagnose the dominant processes governing water flow ..., and for modelers, who wish to capture the key aspects of runoff behavior at a site, rather than simply imposing a one-size-fits-all model structure at the site".

For the present, qualitative estimates and approaches are rather to be expected here. From this viewpoint, the qualitative analysis of the effect of major environmental controlling factors on the different runoff mechanisms presented in a pioneer work by Dunne (1978) and later works of this and other authors still remains relevant. It is evident that many factors will influence the amount of runoff

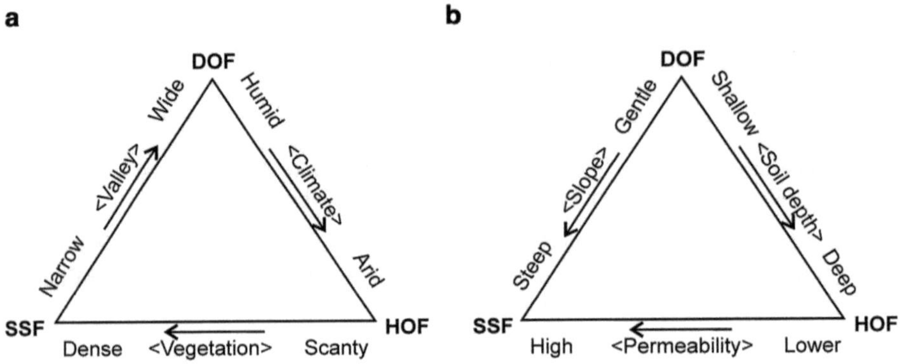

Fig. 1.3 Simplified diagrams of hillslope response in terms of the type of runoff generation (HOF, DOF or SSF) in: (**a**) different landscape–geographic conditions and (**b**) soil and topographic environments

or infiltration of a drainage basin. Conventionally, two groups of such factors can be identified. The former is associated with the climate and landscape–geographic conditions in a drainage basin; and the latter, with its soil and topographic features. The contribution of these attributes to runoff generation determines the near-surface flow regime that can acquire features of Horton overland flow (HOF) or Dunne overland flow (DOF) and subsurface storm flow (SSF). This contribution can be represented schematically by triangle diagrams as shown in Fig. 1.3. These diagrams, which inherited features of the relevant analysis by Dunne and other authors, qualitatively illustrate the occurrence of different runoff generation mechanisms.

Thus, most authors agree with the thesis that the increase in the aridity index shifts the runoff generation from saturation excess (DOF) to infiltration excess process (HOF). Dense vegetation usually increases infiltration into the ground and thus reduces HOF and facilitates the generation of SSF. Subsurface flow mechanisms also dominate in deep narrow valleys with steep sides. In wide river valleys, groundwater table is located near the land surface, thus favoring the generation of DOF (Fig. 1.3a).

There are three major slope attributes, changes in which may considerably influence the runoff generation (Fig. 1.3b): the slope, soil depth, and soil permeability. The diagram in Fig. 1.3b shows HOF generation to be most likely in the case of large enough thickness of the soil unit, represented by low-permeability soils on relatively steep slopes. HOF occurs when the saturated hydraulic conductivity of the topsoil represented by homogeneous sediments is below the rainfall intensity and the subsurface environment is characterized by a deep water table. A high permeability of the soil along with a shallow depth to the impermeable bed or to groundwater table is the condition for generation of DOF, i.e., the latter is more sensitive to topography and soil thickness. If the layer is steep enough and the soil section contains high-permeability layers, it is likely that the total hydrograph will be determined by the contribution of SSF, especially in forested catchments (Whipkey 1965; Sidle et al. 2000; Beckers and Alila 2004; Alaoui et al. 2011). Obviously, SSF dominates when the rainfall intensity is below the saturated hydraulic conductivity of the soil. When SSF convergence promotes a rise of the water table to intersect the surface, DOF is generated (Mirus and Loague 2013).

In addition, other important factors influence the runoff generating process. They have been a subject of intensive studies performed by a number of research groups for the past decades. Some such studies try to find qualitative criteria to predict the conditions for the occurrence of different runoff generation mechanisms (HOF, DOF, or SSF).

Thus, Mirus and Loague (2013) used a comprehensive physics-based model of coupled surface–subsurface flow to simulate rainfall-runoff events to systematically explore the impact of soil hydraulic properties and rainfall characteristics. One of the principal conclusions from the work of reference is that the hydraulic conductivity and rainfall intensity are not the only controls on runoff generation mechanisms. The authors showed how rainfall intensity/depth, subsurface permeability contrasts, characteristic curve shapes, and topography provide important controls on the hydrologic-response dynamics.

Mirus and Loague (2013) suggested two dimensionless terms for the rate of flow and storage, which are supposed, together with the average slope along the modeled catchment hollows, to control the hydrological-response processes. The rate term is the ratio of event-averaged rainfall flow intensity, r, to the saturated hydraulic conductivity, k_s. The storage term is the ratio of the cumulative depth of the rainfall event, R, to the depth, H_u, equivalent to the initial unsaturated storage prior to the rainfall event (in the authors' interpretation, this is the initial volume of unsaturated pores above the soil/bedrock interface in the area of the catchment midpoint). A plot in the dimensionless coordinates, R/H_u (as y-coordinate) – r/k_s (as x-coordinate), facilitates incorporating the important effects of the soil-hydraulic properties, rainfall characteristics, and initial conditions in the modeling study.

Based on such dimensionless format, the general trends in the rainfall-runoff generation processes were determined. Thus, for steep slopes (topography having an incline in excess of 15 %) surface runoff (predominantly in a form of DOF) occurs over a range of low values for the rate (~0.0005–2) and storage ratios (~0.5–10). For catchments with more gentle topography, surface runoff (more probably in a form of HOF) generally occurs for higher rate ratios (>1) and storage ratios within a wider range (~0.03–20). SSF was observed to be generated mostly in soil profiles with steep slopes, when the catchment conditions are characterized by low values of the rate (<0.1) and storage (<0.5) ratios.

The main authors' findings can be also formulated as follows: (1) for gently sloping topography, the variability in rainfall and saturated hydraulic conductivity provide the primary controls on the threshold between the Horton and Dunne mechanisms; (2) the Dunne mechanism is active over a wide range of average slopes, rainfall intensities, and soil hydraulic properties, due to soil layering and topographic convergence; (3) the topographic convergence and average slope exert a very strong control on lateral drainage through permeable soil layers, so they influence the transition between SSF- and DOF-dominated response.

1.3 Infiltrability Models for Non-Structured Soils (Infiltration in an Ideal Soil)

Infiltration occupies a special place among the processes responsible for mass and energy exchange between the surface and subsurface hydrospheres. Infiltration is governed by two driving forces: gravity and capillarity. Under precipitation conditions, according to the mass conservation law, the infiltration (actual inflow rate), $f(t)$, must equal the change in soil storage beneath the surface (Smith 2002; Smith and Goodrich 2005):

$$f(t) = \frac{d}{dt}\int_0^Z (\theta - \theta_i)dz, \tag{1.3}$$

where θ is the soil water content, θ_i is the initial value of θ (assumed to be uniform), Z is the depth below the advance of the wetting zone. Given the infiltration rate, the cumulative depth of water infiltrated into the soil during a given time period, $F(t)$, can be determined from the integral relationship:

$$F(t) = \int_0^t f(t)dt = \int_0^Z (\theta - \theta_i)dz. \tag{1.4}$$

From Eqs. (1.3) and (1.4), it is seen that the infiltration rate is the time derivative of the cumulative infiltration:

$$f(t) = \frac{dF(t)}{dt}. \tag{1.5}$$

In this section, we consider the infiltration process, when a storm is sufficiently intense to create infiltration excess ($r > k_s$). From the mathematical viewpoint, the infiltration can be most rigorously conceptualized as the inflow of rainwater into soil under a flux boundary condition. This type of boundary condition is controlled by both gravity and nonlinear soil water diffusivity. Under the flux boundary condition, water intake by the soil surface is characterized by a change in soil water content over time, $\theta = \theta(t)$, and θ tends to reach the saturation limit, $\theta = \theta_s$. In this, transient, period, the rate of water delivery to the surface is smaller than the soil's infiltrability, f. One may assume that at the end of the transient period, the flux condition switches to a more simple constant-head boundary condition, $\psi_0 = 0$ (or $\theta = \theta_s$). The latter determines the infiltration rate being equal to the infilrability, f, implying the infiltration under unlimited water supply at the surface.

Because of the high nonlinearity of the flow process in the unsaturated media, there is a near equality between the infiltration relations for the inflow rate, f, under rainfall and that under unlimited water supply at the surface (Smith 2002; Smith and Goodrich 2005). This allows one to use models for f (usually in the F domain, Sect. 1.3.2) to predict both the onset of ponding under rainfall, and the behavior of the inflow after that (Smith and Goodrich 2005).

The infiltrability models used in hydrology can be conventionally divided into two categories: empirical models that have no rigorous physical–mathematical basis and models based on relationships derived from partial solutions of equations of moisture transport in an incompletely saturated porous medium.

1.3.1 Empirical Infiltrability Models

Empirical models describing soil infiltrability number dozens. Each model was used to obtain more or less adequate approximation of field data or verified by correlation with the results of numerical solution of vertical moisture transfer in incompletely saturated sediments. The authors of such studies commonly noted that ideal models cannot exist. Each model has its pros and cons (Mishra et al. 2003; Carlier 2007). Thus, some models give good descriptions of imbibition and infiltration of rainwater within a wide time range, though their application is limited to a certain type of soil sections (represented by a limited type of sediments). Other models, on the contrary, work well within limited time intervals, but can be successfully applied under a wide range of natural conditions.

Therefore, for illustration purposes, we will limit ourselves to several, most often cited models. Among them are empirical Kostiakov (1932) and Horton (1940) equations proposed in the early past century. Due to their simplicity, they have been quite often used in the hydrological practice up to now.

Kostiakov formula describes the drop in soil infiltrability by a power time function

$$f(t) = Bt^{-n}, \tag{1.6}$$

where B (> 0) and n are constants ($0 < n < 1$). The formula is applicable for $t \neq 0$.

Integrating (1.6) from 0 to t gives an expression for accumulated infiltration

$$F(t) = \frac{B}{1-n} t^{1-n}. \tag{1.6a}$$

A modification of formula (1.6) is known (Smith 1972):

$$f(t) = \begin{cases} f_c + Bt^{-n}, & t \leq t_p, \\ f_c + B(t - t_p)^{-n}, & t > t_p, \end{cases} \tag{1.6b, c}$$

where f_c is asymptotically stationary imbibition (at large time when the porous medium becomes fully saturated with water), t_p is surface ponding time.

Horton formula uses an exponentially decreasing function

$$f(t) = f_c + (f_0 - f_c)\exp(-kt), \tag{1.7}$$

where k is an exponent factor, f_0 and f_c are the initial and final rates of water imbibition.

Integrating (1.7) yields an expression for accumulated infiltration:

$$F(t) = f_c t + \frac{1}{k}(f_0 - f_c)[1 - \exp(-kt)]. \tag{1.7a}$$

The coefficients in Eqs. (1.6) and (1.7), are evaluated by model calibration against data on precipitation and runoff in individual watersheds. Data of large-scale experiments at specially equipped runoff grounds can also be used for this purpose.

The competitive capacity of formulas (1.6) and (1.7) for the description of infiltration process was studied by comparing the appropriate analytical calculations with results of computation experiments (Carlier 2007). The mathematical simulation was based on a numerical solution of the Richards' equation for Van Genuchten approximation of the retention function $h(\theta)$ and hydraulic conductivity, $k(\theta)$ (see below Sect. 1.5). Such analysis has shown that the infiltrability of typical soils can be best described by Kostiakov formula. The obtained values of parameter n varied from 0.26 (sand) to 0.37 (sandy clay, clay loam), thus suggesting a strong nonlinearity of the process as a whole.

It is worth mentioning that, though formulas (1.6) and (1.7) have no physical basis, they have certain statistical (probabilistic) meaning (Carlier 2007). Thus, Kostiakov formula (1.6) corresponds to a process, which is not a Markov process from probabilistic viewpoint: the Markov property would mean here, in particular, that the expected value of infiltration rate depends on its previous value. The Horton formula (1.7) describes the process, which has another (Markov) statistical nature.

As mentioned above, the results of such studies for the comparison of different empirical formulas for soil infiltrability have been also published in many papers. Generalization of those publications does not yield unambiguous recommendations regarding the choice of specific empirical relationships for various types of landscape–soil conditions.

1.3.2 *Physically Based Infiltrability Models (One-Phase Flow Approach)*

The basic theoretical equation for vertical unsaturated flow is Richards' equation. The equation stems from a simple statement of continuity for one-phase flow system:

$$\frac{\partial\theta}{\partial t}+\frac{\partial w}{\partial z}=0, \tag{1.8}$$

where z is taken as positive downward (or negative up) with ground surface as the datum ($z = 0$) [L]; t is the time [T]; w is the water discharge (positive downwards) [LT^{-1}]; θ is the moisture content of the soil [$-$] . Consider Darcy's law

$$w=-k\frac{\partial h}{\partial z}, \tag{1.8a}$$

where k is the hydraulic conductivity of the soil (can be function of θ or h); $\partial h/\partial z$ is the hydraulic gradient. Because the flow is driven by the gravity (pressure of the overlying water, basically, z) and capillary action (suction head, ψ), the hydraulic head will be the sum:

$$h = \psi - z. \tag{1.8b}$$

Thus, the equation of continuity (1.8) becomes

$$\frac{\partial \theta}{\partial t} = \frac{\partial}{\partial z}\left(k\left(\frac{\partial \psi}{\partial z} - 1\right)\right). \tag{1.9}$$

Depending on the functional type of the nonlinear coefficient k ($k(\theta)$ or $k(h)$), Richards' equation (1.9) can be rewritten in several forms, water-content-based, potential-based, or mixed Richards' equation. For this variety, the dependence $\psi(\theta)$ is of importance.

The Richards' model has been developed implicitly relying on an assumption that the soil air can easily escape the soil. Therefore, the limitation of the Richards' equation is connected with ignoring the flow of soil air during infiltration process. More sophisticated, two-phase flow model, accounting for this phenomenon, can result, in particular, in deviation of the inflow rate from what is predicted by the "Richards' family models" (Sect. 1.4.1.2).

Different simplifications enable a solution to the Richards' equation. The Green–Ampt (1911) and Philip (1957) models are among the best known approximate analytical solutions to Richards' equation that are commonly used in hydrologic theory and practice to study infiltration process, in particular, to describe the soil infiltrability.

1.3.2.1 The Green–Ampt Infiltration Model

When considering the piston-like (with an abrupt interface) motion of the wetting front in a relatively dry soil (Fig. 1.4a), we can try to take into account the combined effect of capillarity and gravity in most simple way following Green and Ampt (1911) concept. Suppose that soil potential is given by the hydraulic head

$$h = h(z,t) = -h_c - z, \tag{1.10}$$

where $h_c = -\psi_c$ is the capillary suction head (capillary pressure head) at wetting front [L]. In the liquid at the interface between the media ($z = 0$), h is determined by the reference atmospheric pressure assumed to equal 0, and the head of the liquid layer covering the surface (Fig. 1.4b):

$$h(0,t) = h_0, \tag{1.10a}$$

where h_0 is the depth of the water layer [L]. The head at the wetting front, $h(z_f,\ t)$, whose position is determined by the current coordinate $z_f(t)$, will be the sum of the

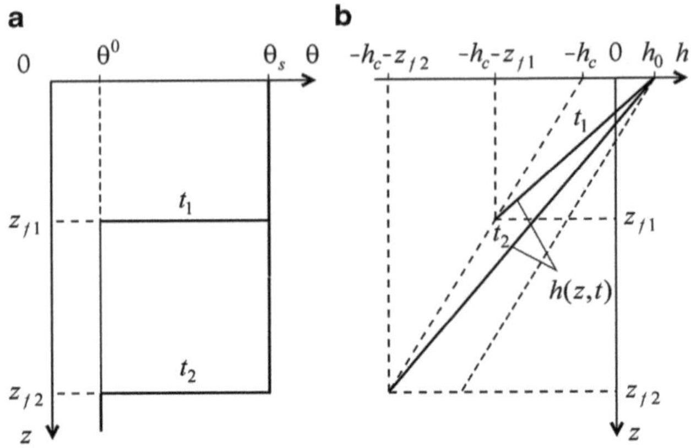

Fig. 1.4 The distribution of (**a**) volumetric moisture content and (**b**) the hydraulic head during piston-like infiltration into semi-infinite unsaturated soil column

capillary head at the wetting front and the head provided by the weight of the water column above:

$$h\left(z_f(t), t\right) = -h_c - z_f. \tag{1.10b}$$

Under the assumption of an abrupt jump of pressure at the advancing front of the transmission zone (saturated soil), the time can be implicitly expressed through the depth to the front below the surface, $z_f(t)$, and thus the time derivative can be excluded from the continuity equation (1.8), which is reduced to an ordinary differential equation:

$$\frac{dw(z)}{dz} = 0, \quad 0 \leq z \leq z_f(t). \tag{1.10c}$$

Such formulation of the boundary value problem requires finding function $z_f(t)$ as the time characteristics of the infiltration process.

Substituting the head gradient from (1.8a) into (1.10c) yields a second-order ordinary differential equation

$$\frac{d^2h(z)}{dz^2} = 0. \tag{1.10d}$$

Its solution with boundary conditions (1.10a) and (1.10b) or in terms of function $h(z)$,

$$h(0) = h_0, \quad h\left(z_f\right) = -h_c - z_f, \tag{1.10e}$$

has a linear form (Fig. 1.4b):

$$h(z) = h_0 - \left(1 + \frac{h_c + h_0}{z_f}\right) z. \tag{1.10f}$$

Now the hydraulic gradient is the same as that defined by Green and Ampt (1911) in their hydrostatic model:

$$\frac{dh}{dz} = -\frac{h_0 + h_c + z_f}{z_f} (< 0). \tag{1.10g}$$

To describe the infiltration process in dynamics, one can define the velocity of the wetting front:

$$\frac{dz_f}{dt} = -\frac{k_s}{\theta_s - \theta^0} \frac{dh}{dz} = \frac{k_s}{\Delta\theta}\left(1 + \frac{h_0 + h_c}{z_f}\right), \tag{1.10h}$$

where $\Delta\theta = \theta_s - \theta^0$ can be considered as the initial saturation deficit (available water storage). Integrating the ordinary differential Eq. 1.10h yields the following formula

$$t = \frac{\Delta\theta}{k_s}\left[z_f - (h_0 + h_c)\ln\left(1 + \frac{z_f}{h_0 + h_c}\right)\right], \tag{1.11}$$

known as the Green–Ampt solution. This solution, which was published in 1911 (Green and Ampt 1911), is among the fundamental solutions that have been in wide use for the analysis of water infiltration from the surface and for the assessment of moisture transfer in the soil and unsaturated zone.

Assuming

$$f = k_s(z_f + h_c + h_0)/z_f, \tag{1.11a}$$

the Green–Ampt model (1.11) can be adopted to calculate the infiltration rate ($f \equiv w$)

$$t = \frac{\Delta\theta(h_0 + h_c)}{k_s}\left[\frac{k_s}{f - k_s} - \ln\left(\frac{f}{f - k_s}\right)\right], \quad f > k_s. \tag{1.12}$$

Solution (1.12) implies an abrupt drop in the specific discharge f over time at the initial stages of infiltration and its asymptotic approximation to the limiting value $f = k(\theta_s) \equiv k_s$ during long stages (Chow et al. 1988; Biswas and Mukherjee 1994;

Smith 2002). Thus, the solution (1.11), (1.11a) and (1.12) has two asymptotics and appropriate expressions for the inflow rate, *f*:

– at early time, when capillary forces dominate over gravitational ones,

$$t = \frac{\Delta\theta}{2k_s(h_0 + h_c)} z_f^2 \quad \text{and} \quad f = \Delta\theta \frac{dz_f}{dt} = \frac{1}{2}\frac{S}{\sqrt{t}}; \tag{1.13}$$

– at sufficiently large time, when the motion of wetting front in the unsaturated zone is determined by the weight of the above water column:

$$t = \frac{\Delta\theta}{k_s} z_f \quad \text{and} \quad f = \Delta\theta \frac{dz_f}{dt} = k_s. \tag{1.13a}$$

In Eq. (1.13), the characteristics S quantifies the capillary forces affecting water movement in the soil, and can be conventionally associated with a sorptivity, a term which appeared much later than the Green–Ampt solution (Philip 1957), in a form of relationship

$$S^2 = 2k_s(h_0 + h_c)\Delta\theta. \tag{1.13b}$$

In the classical interpretation, S depends also on the soil's moisture status and surface boundary condition (Sect. 1.3.2.2).

Calculations of $f(t)$ using (1.12) and (1.13) require an estimate of the effective capillary pressure head, h_c, at the wetting front, which is a parameter that can vary significantly across the wetting front (Wang et al. 1997) depending on the moisture content, θ, and some other measurable physical characteristics of the soil material. There have been several attempts to relate h_c to soil water relations (Neuman 1976; Aggelides and Youngs 1978; Springer and Cundy 1987; Kao and Hunt 1996; Wang et al. 1997; Williams et al. 1998; Smith 2002).

The most general approach is believed to be that based on the use of water retention (h versus θ) or water tension versus relative hydraulic conductivity (h versus k_r) curves. Thus, h_c can be defined by the relative conductivity, $k_r = k(h)/k_s$, weighted average value of the capillary pressure across the wetting retention curve as follows (Springer and Cundy 1987)

$$h_c = \int_{h_i}^{h_s} k_r(h)dh, \tag{1.13c}$$

where $h_i = h(\theta^0)$; $h_s = h(\theta_s)$; θ^0 is the initial moisture content before infiltration began; θ_s is the saturated moisture content. In this manner, formulas for calculating h_c can be obtained for the basic soil water retention curves (e.g., Brooks–Corey and van Genuchten). An approximate expression for h_c can be also obtained from the phenomenological capillary model of porous medium (Kao and Hunt 1996).

After the original model was presented by Green and Ampt, many studies were aimed to modify this model to extend its application domain and prediction capacity. For example, a solution was obtained for a constant-intensity rainfall input condition (Mein and Larson 1973), as well as for a sloping surface (Chen and Young 2006). Of particular interest is the derivation of explicit equations for infiltration (Schmed 1990; Salvucci and Entekhabi 1994; Smith 2002; Craig et al. 2010). The relevant formulae are of particular importance for application in storm runoff modeling and prediction.

Indeed, Eq. (1.12) is not always convenient to use in the study of infiltration. Therefore, it is of interest to carry out an additional study of the Green–Ampt solution to obtain analytical relationships describing the behavior of function $f(t)$ in the explicit form.

Thus, the expression (1.12) can be introduced in the following dimensionless form (Salvucci and Entekhabi 1994)

$$\bar{t} = \frac{1}{\bar{f} - 1} - \ln\left(\frac{\bar{f}}{\bar{f} - 1}\right), \quad \bar{f} = \frac{f}{k_s}, \quad \bar{t} = \frac{t}{\chi}, \quad \chi = \frac{S^2}{2k_s^2}. \tag{1.14}$$

The solution (1.14) can be differentiated with respect to $\bar{t}$, yielding, upon the introduction of a new dimensionless variable $\tau = \bar{t}/(1 + \bar{t})$, the ordinary differential equation

$$\bar{f}(\bar{f} - 1)^2 + (1 - \tau)^2 \frac{d\bar{f}}{d\tau} = 0. \tag{1.15}$$

Next, it was proposed (Salvucci and Entekhabi 1994) to approximate $\bar{f}(\tau)$ by a power series, which finally results in an approximate solution of the Eq. (1.15):

$$\bar{f} \approx \frac{\sqrt{2}}{2}\tau^{-1/2} + \frac{2}{3} - \frac{\sqrt{2}}{6}\tau^{1/2} + \frac{1 - \sqrt{2}}{3}\tau. \tag{1.16}$$

For dimensionless time $\bar{t} << 1$ ($\tau \to 0$), the number of terms in the right-hand part of Eq. (1.16) can be reduced to two, allowing the following approximate equality to be written:

$$f(t) \approx \frac{1}{2}\frac{S}{\sqrt{t}} + \frac{2}{3}k_s. \quad S^2 = 2k_s(h_0 + h_c)\Delta\theta. \tag{1.16a}$$

The Eq. (1.16) can be integrated, allowing one to obtain (according to Eq. 1.4) an expression for the cumulative infiltration, $F(t)$ (Salvucci and Entekhabi 1994).

The Green–Ampt solution (1.11) can be rewritten in a form facilitating its application in the rainfall–infiltration analysis ($h_0 << h_c$) (Mein and Larson 1973; Chow et al. 1988; Smith 2002; Todd and Mays 2005):

$$t(F) = \frac{1}{k_s}\left[F - \Delta\theta h_c \ln\left(1 + \frac{F}{\Delta\theta h_c}\right)\right], \tag{1.17}$$

where F is the cumulative depth of infiltration (1.4), $F = \Delta\theta z_f$.

The flux at the surface is equal to the infiltration rate, $f = -k(\theta_s)dh/dz > 0$, or

$$f(F) = k_s\left(1 + \frac{\Delta\theta h_c}{F}\right). \tag{1.18}$$

In the infiltration theory, Eq. (1.18) is known as infiltrability-depth approximation (IDA) (Smith 2002). The IDA means that cumulative infiltration serves as a surrogate for time and may be treated as a state variable. The relation $f(F)$ (1.18), rather than $f(t)$, is valuable for estimating the ponding time and infiltration for a variable rainfall pattern.

At some moment $t = t_p$ after precipitation began, the infiltration rate becomes equal to the precipitation rate, $f = r$, and water starts ponding the surface. The infiltration depth at that moment is given by $F_p = rt_p$, and the infiltration rate is determined by $f = r$. Substituting these equalities in Eq. (1.18) yields:

$$F_p = \frac{\Delta\theta h_c}{r/k_s - 1}(r/k_s > 1), \tag{1.19}$$

in other words, the infiltrated amount F_p (see also comments in Sect. 1.3.2.2) is equal to that at $f = r$ in the case where the boundary condition is "ponded" from the start (Smith 2002). The time of ponding is given by

$$t_p = F_p/r. \tag{1.19a}$$

The infiltration equation developed by Green and Ampt (1911) was extended to the conditions of non-immediate ponding (Mein and Larson 1973; Chow et al. 1988). In terms of $F(t)$, the solution is:

$$F(t) = \begin{cases} rt & 0 \le t < t_p, \\ F(t_p) + \Delta\theta h_f \ln\left[\dfrac{F(t) + \Delta\theta h_c}{F(t_p) + \Delta\theta h_c}\right] + k_s(t - t_p) & t \ge t_p. \end{cases} \tag{1.20}$$

The infiltration rate may be calculated as

$$i(t) = \begin{cases} r & 0 \le t \le t_p, \\ k_s\left[1 + \dfrac{h_c\Delta\theta}{F(t)}\right] & t > t_p. \end{cases} \tag{1.21}$$

Thus, Eqs. (1.20) and (1.21) are appropriate to describe the infiltration process after ponding.

1.3.2.2 The Philip Two-Term Infiltration Model

This model is an approximate analytical solution of Richards' equation written in the form of an infinite series solution (valid for finite t) for the case of infiltration in a semi-infinite soil domain at a constant-head boundary (Philip 1957, 1987). For practical purpose, the series solution can be restricted to the first two terms

$$f(t) = \frac{dF}{dt} = \frac{1}{2}\frac{S}{\sqrt{t}} + Ak_s, \tag{1.22}$$

$$F(t) = S\sqrt{t} + Ak_s t, \tag{1.23}$$

where S is the soil water sorptivity [$LT^{-1/2}$]; Ak_s is a gravity factor [LT^{-1}]. The term sorptivity, introduced by John Philip, defines the ability of soil to absorb or desorb liquid by capillarity.

The formal comparison of Eq. (1.22) with the asymptotics (1.13), (1.16a) introduced earlier shows that Philip model and Green–Ampt model have a common combined parameter (Charbeneau 2006):

$$S = \sqrt{2k_s(h_0 + h_c)\Delta\theta}, \tag{1.23a}$$

which, however, does not have the original physical meaning as sorptivity in the Philip concept.

The solution (1.22) and (1.23) is valid at constant head boundary condition, $\theta(0,t) = \theta_s$, and for uniform initial condition, $\theta(z,0) = \theta^0$. Most authors recommend using the value of coefficient A (in the gravity term in Eqs. (1.22) and (1.23)) between 1/3 and 2/3 for not very long periods. One may note also that the form of the Philip two-term Eq. (1.22) is very similar to that of modified Kostiakov equation (1.6b). In fact, the infiltrability equation (1.6b) with $n = 1/2$, $f_c = Ak_s$, and $B = S/2$ is essentially the same equation.

Analysis of solution (1.22), and (1.23) shows that in the course of time (tentatively, at $t > t_g = (3S/2k_s)^2$), the effect of the first term, which accounts for the role of capillary-adsorption forces, becomes negligible and the rate of imbibition is determined by soil hydraulic conductivity, i.e., by the second term of the Eq. (1.22), which accounts for the gravity movement of moisture.

The obtained relationships allow us, by analogy with the Green–Ampt problem, to write the generalized expressions determining the cumulative infiltration:

$$F(t) = \begin{cases} S\sqrt{t} + Ak_s t & 0 \le t < t_g, \\ F(t_g) + Ak_s(t - t_g) & t \ge t_g. \end{cases} \tag{1.24a, b}$$

It can be seen that the infiltration time is divided into two periods: at times less than t_g, the moisture front propagation velocity is governed by both the gravity and capillary mechanisms, while the gravity moisture transport dominates for time greater than t_g.

The Philip infiltration model (1.22) and (1.23) can be recast in the format of the ADA (Sect. 1.3.2.1) by eliminating t between these equations:

$$f(F) = Ak_s + \frac{Ak_s S}{\sqrt{S^2 + 4Ak_s F} - S}. \tag{1.25}$$

For the Philip two-term infiltration model, the time of surface ponding at $r > Ak_s$ can be calculated as:

$$t_p = \frac{S^2(r - Ak_s/2)}{2(r - Ak_s)^2}, \tag{1.25a}$$

where r is the intensity of rainfall. This expression is obtained by setting $f = r$ in Eq. (1.25) and equating the cumulative infiltration, $F = F_p$, to rt_p (1.19a) – Fig. 1.5a.

The infiltration rate under ponded conditions is given by $f(t)$ as well, but with the time origin shifted by the amount (Fig. 1.5b):

$$t_c = t_p - t_e, \tag{1.25b}$$

where t_e is an equivalent time which can be obtained by equating $f(t = t_e)$ from Eq. (1.22) to r,

$$t_e = \frac{S^2}{4(r - Ak_s)^2}. \tag{1.25c}$$

Note, that $f(t)$ in the period $t \le t_e$ represents only a potential value based on unlimited water supply at the surface at that time, rather than the actual infiltration rate (Smith 2002). Thus, taking into account (1.25b),

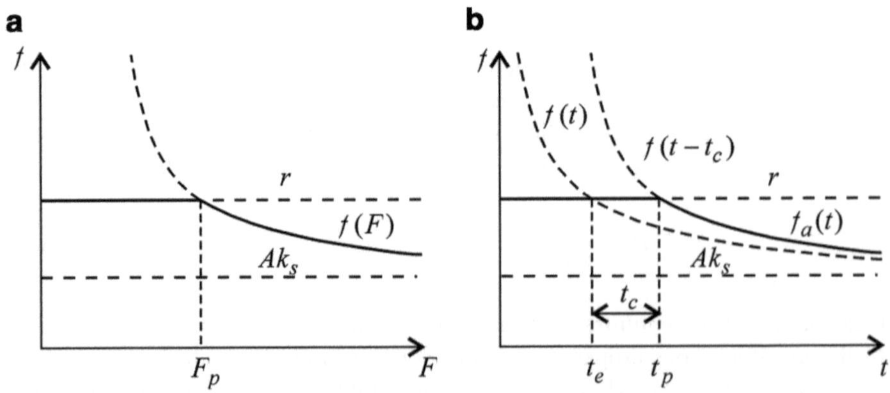

Fig. 1.5 Infiltrability as a function of cumulative depth, F (**a**), and time, t (**b**)

$$t_c = \frac{S^2}{4r(r - Ak_s)}, \tag{1.25d}$$

the resulting formula for the infiltration rate can be written as:

$$i(t) = \begin{cases} r, & t \leq t_p, \\ (1/2)S(t - t_c)^{-1/2} + Ak_s, & t > t_p. \end{cases} \tag{1.25e}$$

The cumulative infiltration, F, can be obtained via integration of $f(t)$ according to Eq. (1.4) (Kim et al. 1996).

1.3.3 Two-Phase Flow Approach

Unfortunately, the classical approaches considered above are theoretically limited in their ability to accurately model: (1) the beginning of infiltration, because they predict initially infinite flow velocity, i.e., $f(t)$ (both Green–Ampt and Philip models); (2) the large-time behavior of the $f(t)$ function, because they assume the transmission zone to be fully saturated (Green–Ampt model). These all would lead to an overestimation of the actual infiltration capacity of soils.

The main limitation of the traditional models is connected with ignoring the participation of soil gas (air) in the infiltration process and dynamic capillary pressure phenomenon. The effect of the soil air flow on the process of water infiltration is associated with such phenomena as *air compression*, *air counterflow*, and *air entrapment* in soil. Thus, during infiltration, air can be compressed in the soil below the wetting front (especially if the water table is shallow), resulting in reduction in infiltration. Compressed air can make its way out to the atmosphere through the transmission zone exhibiting the so-called counterflow phenomenon. Because air flows upward, in the opposite direction to the infiltrating liquid, the water content in the transmission zone (Green–Ampt model) decreases and the effective hydraulic conductivity and infiltration rate decrease as well. The same effects can be expected from the influence of air entrapment limiting the saturation of the transmission zone by infiltrating water from the ponded surface, and, as a consequence, the degree of saturation and the infiltration rate may be significantly reduced as well.

Many successful attempts have been made to adapt the Green–Ampt model to account for air counterflow, air compression and dynamic capillary pressure using rather simple physically based approaches or varying soil properties (Wang et al. 1997; Pellichero et al. 2012). Those, as well as many other studies, showed that the agreement between the Green–Ampt model predictions and observations can be artificially improved.

However, the most rigorous solutions of infiltration problem, explaining why experimental data could not be described accurately with the classical Green–Ampt or Philip approaches, were obtained in a two-phase formulation, accounting for the flow of air in soil, within the transmission zone and ahead of the wetting front (Morel-Seytoux and Khanji 1974, 1975; Touma and Vauclin 1986; Szymkiewicz 2013). Those solutions predict the behavior of functions $f(t)$ and $F(t)$ and yield asymptotics other than formulas (1.12), (1.13) and (1.17). In particular, it was shown that air compression ahead of the wetting front is the major cause of the decrease in the infiltration rate, $f(t)$ (Touma and Vauclin 1986). Moreover, due to air counterflow from ahead of the wetting front, this rate decreases continuously with time (Wang et al. 1997) instead of reaching a steady state constant infiltration rate predicted by the asymptotic formula (1.13a).

In light of the above-mentioned facts, two approaches, Green–Ampt and two-phase flow, to calculating two infiltration characteristics, $\overline{f}(\overline{t})$ and $\overline{F}(\overline{t})$, were compared (Fig. 1.6). Using TOUGH2 as a solver for a system of differential equations describing two-phase flow in unsaturated zone (Pruess 1991, 2004) we formulated a numerical problem which is identical to the original Green and Ampt setup.

The parametric basis includes the relationships between relative permeability, k_{rw} (water), k_{ra} (air), and effective saturation, S_{ew} for a water–air system through a simple power law function (Pinder and Gray 2008; Szymkiewicz 2013)

$$k_{rw} = S_{ew}^n, \quad k_{ra} = (1 - S_{ew})^n, \quad k_{r(w,a)} = k_{w,a}(S_{ew})/k_s, \tag{1.26}$$

where $S_{ew} = (S_w - S_{wr})/(S_{w\max} - S_{wr})$ is the effective saturation with water; S_{wr} is the water residual saturation; $1 - S_{ew} = S_{ea}$ is the effective saturation with air; $k_{w,a}$ is the unsaturated hydraulic (phase w and a) conductivity; k_s is the hydraulic conductivity at saturation; n is an exponent. For the abrupt wetting front, a step-vise capillary function can be represented as follows

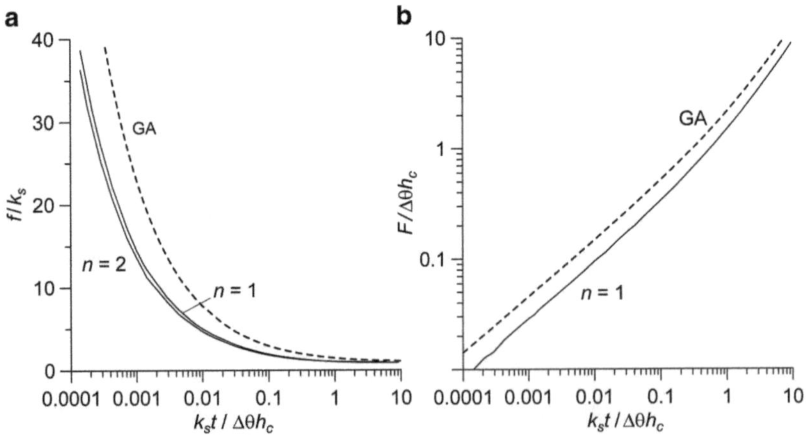

Fig. 1.6 Dimensionless time dependence of infiltrability (**a**) and cumulative infiltration (**b**) for zero water ponding on the soil surface ($h_0 = 0$). *Solid curves* correspond to the two-phase infiltration model; *dashed curve* (GA) is the Green and Ampt solution (1.12) and (1.17)

$$h_c = \begin{cases} h_{c(\max)}, & \text{if } S_w \le S_{wr}, \\ 0, & \text{if } S_w \ge S_{w\max}; \end{cases} \tag{1.26a}$$

in the interval $[S_{wr},\ 1]$, h_c changes linearly; $S_{w\max} = 1$.

Although the interpretation of the very initial stage of the numerical modeling is a matter of uncertainty, it seems, similar to the analytical one-phase infiltration models discussed above, the two-phase flow approach exhibits an infinite limit at $t = 0$ (Fig. 1.6a). As seen, within all specified time range, the Green–Ampt curves are located above the two-phase flow curves, both in plot $\overline{f}(\overline{t}) = f/k_s$ (Fig. 1.6a) and in plot $\overline{F}(\overline{t}) = F/\Delta\theta h_c$ (Fig. 1.6b), and hence the Green–Ampt model overestimates the infiltration capacity of the soil. The major feature in the evolution of the Green–Ampt infiltrability curve over the time is $f(t)$ tending to the hydraulic conductivity at saturation, k_s, at late time infiltration. The limit of $f(t)$ that has been reached in the two-phase flow simulation is slightly less. This effect was observed earlier by Morel-Seytoux and Khanji (1974) during their analysis of two-phase flow in unsaturated media. Curves in Fig. 1.6a show that infiltrability accounting for two-phase infiltration is weakly sensitive to exponent, n.

Though the classical Green–Ampt approach has some shortcomings, its applicability to a wide variety of initial, boundary, and soil profile conditions makes it still one of the most popular in hydrology, since the data obtained with it meet the needs of practice. The essential advantage of the model is that it allows one to obtain the physical parameters for the model from soil properties estimated using rather traditional approaches and techniques.

1.4 Influence of Macropores and Surface-Exposed Fractures on Infiltration, Runoff Generation and Lateral Preferential Flow

Soils are generally structured media where the mineral and organic matter components are organized into aggregates that vary in size, porosity, pore size, and continuity. Soils, also having a multitude of irregular opening voids, represented by bio-pores, fractures in clayey soils or soil aggregates, exhibit heterogeneous properties. These soil voids are commonly referred to as macropores in spite of the fact that some soils may have discontinuities in the shape of fractures/cracks. Therefore, for convenience, in the further analysis, we will often not differentiate these two types of the structural soil discontinuities forming voids in soil material.

The conductivity through the macropores and fractures is several orders of magnitude larger than the conductivity of the intact micropore matrix. Therefore, these soil discontinuities can have a large influence on both the vertical and lateral flow of water at hillslope scale, providing different effects related to the excessive surface ponding and soil profile saturation.

The recognition of the importance of macropores and fractures in both water and air flows in soils plays a central role in advanced hydrological analysis of hillslope and watershed systems (Beven and Germann 1982; Šimůnek et al. 2003; Jarvis 2007; Nieber et al. 2006; Köhne et al. 2009; Beven and Germann 2013). The quoted researches and many others have shown that the problem has two aspects: first, the partitioning of infiltrating water between matrix and macropore or fracture domains during fallout resulting in the partitioning of rainfall to surface runoff and infiltration, and second, the generation of macropore runoff in a soil profile (subsurface runoff) in a form of preferential flow. Such vertical and lateral preferential flow, triggered by rainfall events, becomes the major mechanism providing a rapid solute migration through the soil and unsaturated zone towards surface streams and groundwater (McGrath et al. 2010).

The most accurate mathematical formulation of the preferential flow phenomenon is based on the so-called dual-porosity and -permeability concept (Šimůnek et al. 2003; Jarvis 1998; 2007; Köhne et al. 2009). In more general setup, one should find the solution of coupled Richards' equations describing nonequilibrium flow in the soil matrix and macropore systems. The first-order term can be used to describe water transfer from voids to the soil matrix. Water exchange between the two domains can be simulated also in a direct way without a need to resort to the first-order approximation.

Compared to the predominantly vertical preferential flow at smaller scales, overland flow and lateral preferential flow appear as important rapid runoff components on hillslopes (Köhne et al. 2009). Therefore, in this section, the influence of preferential flow on soil systems behavior under rainfall conditions will be analyzed at both the column scale (based on a simplified analytical framework) and the hillslope scale (conceptually).

1.4.1 *The Influence of Natural Voids in Soil on Infiltration and Runoff Generation*

In well-structured soils, flow through continuous voids (fractures, cracks or macropores) may result in a deeper and faster penetration of rainfall and solutes than what is predicted by uniform displacement (Beven 1981; McDonnell 1990). One may also expect that the kinetics of capillary imbibition of water from voids to the surrounding matrix can affect the soil infiltration capacity. This phenomenon is of special interest for this section as it demonstrates the possible influence of surface and subsurface flow interaction on runoff generation.

1.4.1.1 On Conceptual Approaches to the Development of Flow Models

It is commonly agreed that downward water movement in well-structured media (both macroporous soils and fractured soils/rocks) in the end of a prolonged precipitation following the wettest antecedent conditions is provided by macropores or/and surface-exposed fractures, and overland flow commences only when a certain threshold determined by a combination of parameters of the dual-capacity system is exceeded. In this analysis and further studies (Sect. 2.4), to be specific, we will focus on an analytical framework for describing the rainfall response behavior of fractured soils based on: (1) a physically-based approach to coupling models with distributed parameters for overland flow and preferential flow infiltration (Beven and Germann 1982; Ruan and Illangasekare 1998; Novak et al. 2002; McGrath et al. 2010), and (2) a lumped-parameter approach (dual-storage-based model) adopted from works by Kohler and Struthers with co-workers (Kohler et al. 2003; Struthers et al. 2007).

The discussed physically-based approach (Sect. 1.4.1.2) describes a flow through soil discontinuities which is initiated when the soil surface is ponded. Water entering the fractures or macropores is imbibed by unsaturated soil matrix, which controls the rate of water transfer from surface to the lower part of the soil profile. The vertical movement of such water package may result in the generation of lateral flow at the interface between the soil and bedrock, which is a special focus of this work.

The hydrological response of a hillslope or a catchment area, represented by structured soils, to rainfall application in the form of surface runoff can be divided into several stages corresponding to different types of boundary conditions for the unsaturated soil surface (Novak et al. 2002; Beckers and Alila 2004).

1. At the beginning of the rainfall, water is preferentially imbibed by soil (porous) matrix exposed to the surface, while direct infiltration into macropores can be neglected since they contribute very little to the total surface area (Beven and Germann 1982). As long as the soil remains unsaturated, the infiltration rate through soil matrix, i_m, is equal to rainfall rate, r, similar to that we observe for the case of homogeneous soil:

$$i_m = r, \quad h = 0. \tag{1.27a}$$

2. Later, as the soil matrix becomes unable to infiltrate (as before) more water than it is supplied by precipitation, because of its limited infiltration capacity, a “subcritical surface layer of water” is being formed on the surface:

$$i_m + \frac{dh}{dt} = r, \quad h < h_s. \tag{1.27b}$$

3. When the critical depth, h_s, is reached, excess rainwater flows over the surface, filling up soil macropores (fractures, cracks), with the infiltration rate i_f, and the following relationship for rainfall excess, r_e, is valid:

$$r_e = r - i_m - i_f, \quad h \geq h_s. \tag{1.27c}$$

The boundary condition (1.27a) permits water to build up on the surface (Novak et al. 2002), giving rise to overland flow (Luce and Cundy 1992) and preferential flow.

Contrary to the above approach, where the interaction between voids and matrix domains has a continuous nonequilibrium character, a lumped-parameter model (Sect. 1.4.1.3) relies on an assumption that a fracture system is allowed to participate in infiltration in the form of preferential flow when the matrix is fully saturated or the infiltration capacity is exceeded, i.e. the preferential flow through voids in porous matrix is triggered when the ability of the latter to store rainwater is exceeded.

1.4.1.2 Application of the Green–Ampt Concept to Preferential Flow Analysis

The interrelation between overland flow and the infiltration into the macropores/ fractures can be analyzed based on simplified models of vertical flow in a periodical macropore system represented by cylindrical pores or fractures surrounded by much less permeable porous material. As downward flow in such systems is initiated when the surface is ponded, the upper boundary condition can be specified as a pressure boundary. If the soil column is deep enough, the effect of the lower boundary can be neglected.

The classical problem setup takes into account two main flow-driven forces, i.e., the gravity and capillarity. The full description of transient water movement in structured soils and vadose zone represented by fractured-porous rocks is usually based on the numerical solution of the Richards' equation (Šimůnek et al. 2003; Šimůnek and van Genuchten 2008). Under some simplified assumptions relating the position of the fracture liquid front as a function of time, it was shown that the flow problem solutions can be obtiained in the form of integro-differential equations (Nitao and Buscheck 1991). On the other hand, the kinematic wave theory is well suited to the problem of flow (and solute transport) in the two overlapping regions as it was suggested by Germann (1985) and Germann and Beven (1985), who used a kinematic wave equation to describe gravitational movement of water in a macropore domain, where capillarity may be neglected.

There are two main conceptual approaches to account for preferential flow in structured soil (Kohler et al. 2003): (1) continuous dual-domain (or multi-domain) approach, and (2) discrete (network or structure-based network) approach. The first approach implies that flow is conceptually separated in different, but spatially overlapping and interacting continua (domains), one consisting of macroporous or fracture network, the other, of the porous matrix (Šimůnek et al. 2003; Jarvis 1998; 2007). In the second approach, the voids are taken into account explicitly, as discrete fine-scale elements with defined shape and physical properties deterministically or stochastically implemented in homogeneous matrix (Wienhöfer and Zehe 2014).

In this work, we utilize the first approach because it provides a more rational and common means allowing us to study the flow problems remaining in the same mathematical framework, based on a representative elemental volume. To be specific, but without loss of generality, we treat the soil and vadose-zone material as a uniform fractured porous continuum assuming that the fracture network is surrounded by a homogeneous low-permeability matrix. Both domains represent a single overlapping system. In such preferential-flow problem formulation, the term "fractures" can synonymously be used for "macropores" as well.

Similar to Sect. 1.3.2.1, where vertical piston-like (*one-phase*) flow in homogeneous soil under Green–Ampt assumption is considered (Eq. 1.10c), we will start with continuity equation including a source term, i.e., transfer rate function, $\Gamma_w(t)$,

$$\frac{dw}{dz} + \Gamma_w(t - \Omega(z)) = 0, \tag{1.28}$$

where w is the specific discharge [LT^{-1}]; $\Omega(z)$ is the arrival time of the fracture front at the depth z from the inlet boundary [T]. Under the assumption that the structured media act as a uniform dual domain continuum, source-tem can be postulated as follows:

$$\Gamma_w(t) = S_b i_m(t), \tag{1.28a}$$

$i_m(t)$ is the matrix imbibition rate [LT^{-1}]; S_b is the specific surface of the fracture domain which is defined as the total area of fractures' surface per volume of the surrounding soil matrix [L^{-1}].

The correspondence of the model of two overlapping domains, characterized by S_b and ϕ_f (fracture porosity), to the model of flow in a single vertical fracture (with half-aperture b and matrix block half-width a) in contact with porous matrix (Nitao and Buscheck 1991) can be achieved when the following relationships holds: $S_b = a^{-1}$, $w = u\phi_f$ (u is the flow velocity in the fracture). An analytical representation of the imbibition function, $i_m(t)$, can be obtained from a solution of a supplementary boundary-value problem related to imbibition in a porous block of an idealized geometry shape at a certain time scale.

Assuming the Darcy's law in the fracture domain (Eq. (1.8a), where $k = k_{fs}$ is the saturated hydraulic conductivity of fractures), and continuity equation (1.28), we come to the ordinary differential equation:

$$\frac{d^2h}{dz^2} = \frac{S_b}{k_{fs}} i_m(t - \Omega(z)). \tag{1.29}$$

The solution of this equation with boundary conditions (1.10a) and (1.10b where $h_c \equiv h_{fc}$), with the kinematic identity

$$\frac{dz_f(t)}{dt} = \frac{w(z_f(t), t)}{\phi_f} \tag{1.29a}$$

taken into account, leads to an integro-differential equation (Nitao and Buscheck 1991):

$$z_f(t)\frac{dz_f(t)}{dt} = \frac{k_{fs}}{\phi_f}\left(z_f(t) + h_{fc} + h_0\right) - \frac{S_b}{\phi_f}\int_0^t i_m(t-\xi)z_f(\xi)\frac{dz_f(\xi)}{d\xi}d\xi, \qquad (1.30)$$

which can be provisionally referred to as extended Green–Ampt equation: if $i_m = 0$ we come to Eq. 1.10h, describing the movement of wetting front in a homogeneous medium ($k_s \equiv k_{fs}$, $\Delta\theta \equiv \phi_f$, $h_c \equiv h_{fc}$). The saturation of the fracture domain is assumed to drop from 1 to 0 (no residual water in fractures).

The infiltrability of the cracked soil material under ponding conditions, $f(t)$, can be found from the equation of water balance in fractures

$$w(0,t)F = \phi_f F\frac{dz_f}{dt} + S_b F\int_0^{z_f} i_m(t - \Omega(z))dz \qquad (1.31)$$

(F is the surface area), where the left-hand part of (1.31) is the total flux (inlet flux), and the right-hand part is the sum of the flux at the fracture front and the imbibition flux into the matrix. The substitution of $z = z_f(\xi)$ in the integral yields

$$f(t) = w(0,t) = \phi_f\frac{dz_f}{dt} + S_b\int_0^t i_m(t-\xi)\frac{dz_f(\xi)}{d\xi}d\xi. \qquad (1.32)$$

The solution of Eqs. (1.30) and (1.32) involves some mathematical problems. The asymptotic behavior of the function was examined by Nitao and Buscheck (1991). In particular, they showed that at early time, the inflow rate is controlled by the capillary fracture pressure h_{fc} and the boundary pressure h_0, and the behavior of $f(t)$ is consistent with that of a homogeneous imbibing medium, $f \sim t^{-1/2}$. For sufficiently large time, the inlet flow rate approaches the values in a relatively narrow bend around k_{fs} to $\pi k_{fs}/2$ (see below).

The behavior of function $z_f(t)$ and other characteristics of the system can also be studied using approximate solutions. One such solution, obtained by the method of characteristics for kinematic wave equation, is given below. For one-dimensional vertical water flow in unsaturated fractured-porous media, a transfer rate function is included in the continuity equation as follows

$$\frac{\partial\theta_f}{\partial t} + \frac{\partial w_f}{\partial z} + S_b i_m(t) = 0, \qquad (1.33)$$

where $\theta_f = \theta_f(z,t)$ is the soil water content in fracture network; the specific discharge, $w_f = w_f(z,t)$, is determined using the one-dimensional form of Darcy's law (Bear 1972), assuming a unit hydraulic gradient (i.e., the role of suction pressure on the wetting front in fractures is insignificant):

$$w_f = k(S_f), \quad S_f = \theta_f/\phi_f; \qquad (1.34)$$

we assume that the residual water content in the fracture domain is zero; S_f is the effective (relative) saturation of the fracture domain. Equation (1.33) can be seen to be similar to infiltration equation for homogenous soils (1.8); however, it contains a source term responsible for the matrix imbibition of water from fractures.

Using a simple linear relationship

$$k(S_f) = k_{fs} S_f, \tag{1.34a}$$

to present the unsaturated hydraulic conductivity, $k(S_f)$, one may obtain from (1.33):

$$\frac{\partial S_f}{\partial t} + \overline{k}\frac{\partial S_f}{\partial z} + \frac{S_b}{\phi_f} i_m(t) = 0, \tag{1.35}$$

where $\overline{k} = k_{fs}/\phi_f$.

Let us assume (Rumynin 2011) that the early stage of capillary-controlled water imbibition by porous matrix is described by Philip's asymptotic solution for "horizontal infiltration" (Philip 1957), thus, the imbibition rate is determined by relationship

$$i_m(t) = \frac{S(\theta_{ms},\ \theta_m^0)}{2\sqrt{t}}, \tag{1.36}$$

where $S(\theta_{ms},\ \theta_m^0) = \sqrt{2k_{ms}h_{mc}(\theta_{ms} - \theta_m^0)}$ is the water sorptivity [$LT^{-1/2}$] (Eq. 1.23a, Sect. 1.3.2.2); k_{ms} is the saturated hydraulic conductivity of porous matrix [LT^{-1}]; h_{mc} is the capillary head at the moistening front [L]; θ_s is moisture content at complete saturation of the porous matrix; θ_m^0 is the initial moisture content (at $t = 0$).

The imbibition function (1.36) can be rewritten in a more tradition form:

$$i_m = \Delta\theta_m \sqrt{D_{we}/\pi t}, \tag{1.37}$$

where $D_{we} = (\pi/4)S^2/\Delta\theta_m$ (the coefficient of effective matrix diffusivity) coincides with the estimate in (Philip 1955); $\Delta\theta_m = \theta_{ms} - \theta_m^0$ is the initial matrix saturation deficit. Then, the mass transfer term in (1.37) can be written as

$$i_m = \Delta\theta_m D_{we}/l(t), \quad l^2(t) = \pi D_{we} t, \tag{1.37a}$$

where $l(t)$ is moisture penetration depth into the porous matrix [L]; $\Delta\theta_m/l(t)$ is the current moisture gradient [L^{-1}].

Thus, matrix imbibition can be modeled as a linear function of the inverse square root of time ($1/\sqrt{t}$). The model (1.37) has some limitation because of the assumption that the matrix features unlimited capacity. With this point in view, Eq. 1.37 is approximately true for $t << t_a$ (Nitao and Buscheck 1991), where $t_a = \pi a^2/D_{we} = \pi S_b^2/D_{we}$. For practical estimates, we can assume $t < 0.1\, t_a$.

We represent the system of Eqs. 1.35, 1.37 for the early stage of the infiltration as

$$\frac{\partial S_f}{\partial t} + \bar{k}\frac{\partial S_f}{\partial z} = -\frac{\sigma S_b D_{we}}{l(z,t)}, \quad 0 < S_f \le 1, \tag{1.38a}$$

$$\frac{\partial l^2(z,t)}{\partial t} = \begin{cases} \pi D_{we}, & 0 < S_f \le 1, \\ 0, & S_f = 0; \end{cases} \tag{1.38b}$$

here $\sigma = \left(\theta_{ms} - \theta_m^0\right)/\phi_f$.

For the further analysis, we hypothesize that, at the very beginning of precipitation event, the rainfall rate quickly exceeds the infiltration capacity of the soil matrix on its surface, and water from the ponded surface flows into macropores/fractures. Thus, the one-dimensional infiltration process will be considered under a prescribed head condition (Basha 1999), and, as a reasonable approximation, one may assume that the soil surface is wetted to near saturation. The initial moisture content is considered uniform throughout the soil profile. The above said can be summarized as follows:

$$S_f(0,\ t) = 1, \quad S_f(z,\ 0) = 0, \quad l(z,0) = 0, \quad \theta_m(z,\ 0) = \theta_m^0. \tag{1.38c}$$

The solution of the system of partial differential equations (1.38a, 1.38b) can be associated with a problem related to the determination of two unknown movable boundaries of saturation, i.e., wetting front in the fractures, $t = \Omega(z)$, and wetting front in the rock matrix, $t = \ell(z)$ (Kosterin and Selin 2000). The time coordinate $\Omega(z)$ corresponds to the arrival time of the wetting front at the depth z from the upper entry boundary. Clearly, the wetting front $t = \Omega(z)$ should move with a decreasing velocity and a lag behind the hypothetical migration front $t_0 = z\phi_f/k_{fs}$ of solution not imbibed by porous blocks.

The solution of the second equation (Eq. 1.38b) of the system, which serves as an auxiliary equation in the search for solution of the problem as a whole, can be obviously written as

$$l^2(z,t) = \pi D_{we}[t - \Omega(z)], \quad t > \Omega(z). \tag{1.39}$$

The expression in brackets indicates that the fluid flow from fractures into blocks forms only after the front, whose time coordinate is $t = \Omega(z)$, has reached the point z.

The solution of the governing partial differential equation (1.38a) is sought for by the method of characteristics, i.e., this equation is replaced by its equivalent in the form of a system of ordinary differential equations:

$$\frac{dt}{1} = \frac{dz}{\bar{k}} = \frac{dS_f}{-\sigma S_b D_{we}/l(z,t)}. \tag{1.40}$$

The effective saturation changes along characteristics according to the law determined by the solution of the second equation of the system (1.40) written as

$$\frac{dS_f}{dz} = -\frac{N}{l(z,t)}, \quad N = \sigma S_b D_{we}/\bar{k}. \tag{1.41}$$

Since the saturation varies from 1 to 0 along any characteristic, we can write the integral identity

$$N\int_0^z dz'/l(z') = 1, \tag{1.41a}$$

where z is the coordinate of the saturation front, governed by the trajectory of the leading saturation front $t = \Omega(z)$, and

$$l(z') = \sqrt{\pi D_{we}}\sqrt{\Omega(z) - \Omega(z') - (z - z')/\overline{k}}. \tag{1.42}$$

Kosterin and Selin (2000) introduced a new function, $y(z) = \Omega(z) - z/\overline{k}$, reducing the integral identity (1.41a) to Abel's integral equation

$$\int_0^z \frac{dz'}{\sqrt{y(z) - y(z')}} = \frac{\sqrt{\pi D_{we}}}{N}, \tag{1.43}$$

whose solution has the form

$$y = \left(\pi N/2\sqrt{\pi D_{we}}\right)^2 z^2. \tag{1.44}$$

Thus, we come to the equation

$$\Omega(z) = \frac{z}{\overline{k}} + \frac{\pi}{4}\frac{\sigma^2\lambda_m}{\overline{k}^2}z^2, \quad \lambda_m = S_b^2 D_{we}, \tag{1.45}$$

describing the movement of the leading front of saturation (the time taking the front to reach depth z from the surface, i.e., the fracture network entrance); λ_m is the imbibition coefficient characterizing the capillary-driven transfer. The solution (1.45) has the following dimensionless form

$$\sigma^2\tau = \sigma^2\eta + \frac{\pi}{4}\left(\sigma^2\eta\right)^2, \tag{1.46}$$

where $\eta = \lambda_m \phi_f z/k_{fs}$, $\tau = \lambda_m t$, $\lambda_m = S_b^2 D_{we}$.

This solution is in good agreement with the solution obtained earlier (Rumynin 2011) using the mass-balance integral approach (Barenblatt et al. 1990)

$$\sigma^2\eta = \frac{2}{\sqrt{\pi}}\sigma\sqrt{\tau} + \exp\left(\sigma^2\tau\right)\operatorname{erfc}\left(\sigma\sqrt{\tau}\right) - 1, \tag{1.47}$$

as illustrated by the plot in Fig. 1.7. Similar results were obtained by other methods by Castaing (1991), Rangel-German and Kovscek (2001).

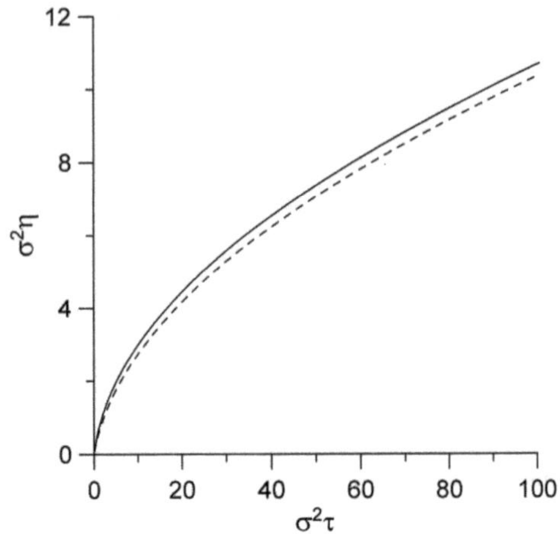

Fig. 1.7 Relationship between the dimensionless depth of wetting front from the surface and dimensionless time. The *solid curve* is the solution (1.46); the *dashed curve*, the solution (1.47)

According to (1.45), the equation for determining the current position of fracture wetting front has the form:

$$z_f(t) = \frac{2\bar{k}}{\pi\sigma^2\lambda_m}\left(\sqrt{1+\pi\sigma^2\lambda_m t}-1\right), \tag{1.48}$$

and the velocity of front motion is

$$\frac{dz_f(t)}{dt} = \frac{\bar{k}}{\sqrt{1+\pi\sigma^2\lambda_m t}}. \tag{1.49}$$

Now, the solution for infiltrability (1.32) becomes:

$$f(t) = \frac{\phi_f\bar{k}}{\sqrt{1+\pi\sigma^2\lambda_m t}} + \Delta\theta_m\sqrt{\frac{\lambda_m}{\pi}}\int_0^t \frac{\bar{k}d\xi}{\sqrt{(t-\xi)(1+\pi\sigma^2\lambda_m\xi)}}. \tag{1.50}$$

The integral in (1.50) has a finite analytical representation; therefore, the solution (1.50) becomes:

$$f(t) = k_{fs}\left(1+\frac{1}{\sqrt{1+\pi\sigma^2\lambda_m t}}-\frac{2}{\pi}\arctan\frac{1}{\sqrt{\pi\sigma^2\lambda_m t}}\right). \tag{1.51}$$

The solution (1.51) and its graphical representation (Fig. 1.8) quantify the effect of the fracture and matrix interaction on infiltration rate implying that the capillarity

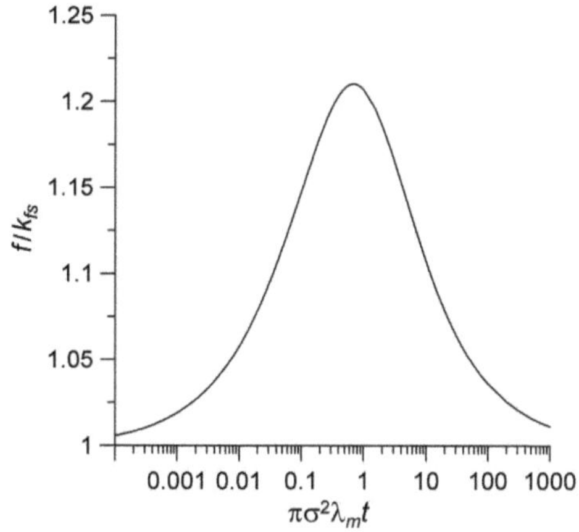

Fig. 1.8 Relative infiltrability, f/k_{fs}, as a function of dimensionless time, $\pi\sigma^2\lambda_m t$

of fracture flow is not significant. The plot in Fig. 1.8 has a maximum at around $\pi\sigma^2\lambda_m t = 3/4$ showing that infiltration velocity exceeds the hydraulic conductivity at saturation by about 20 %. As seen, at the early time, matrix imbibition causes changes in the pressure distribution in the fracture domain, increasing the effective weight of liquid column, hence an increase in the vertical flow velocity near the inflow boundary; therefore, the ratio f/k_{fs} tends to grow over time. Next, gravitation forces start to dominate over the lateral capillary forces at the fracture–matrix interface; therefore, infiltration rate decreases, tending to the Darcy's velocity, determined by the unit hydraulic gradient and hydraulic conductivity, $f \approx k_{fs}$, i.e., the inlet velocity is independent of matrix capillary properties. Such system behavior is in agreement with numerical solution of the integro-differential equations (1.30), (1.32) related to unsaturated flow in a single fracture in contact with porous matrix (Nitao and Buscheck 1991). It was shown that the inflow rate exhibits extreme behavior with a maximum around $1.4k_{fs}$. Both results demonstrate that the process of infiltration is weakly sensitive to matrix imbibition properties, so the capillary forces in the fractures can be neglected for practical purposes.

Finally, we consider the solution of the problem in a wider formulation, accounting for the *two-phase flow* nature of the infiltration process in structured medium, where gravity, capillary forces, and matrix imbibition act simultaneously. Vertical flow through a single high-permeability fracture in contact with a lower permeability matrix is considered. The simulation was performed with the program package TOUGH2/EOS3, allowing the solution of two-phase flow problems (Pruess 1991; 2004).

A pressure-head boundary condition is maintained at the inlet boundary. The basic parameters of the numerical model are as follows: $k_{fs} = 1\ \text{md}^{-1}$, $b = 1 \cdot 10^{-3}$ m (1 mm), $k_m = 1 \cdot 10^{-4}\ \text{md}^{-1}$, $\phi_m = 0.1$, $\theta_{ms} = 0.1$, $\theta_m^0 = 0.01$, $\Delta\theta = 0.09$, $h_{fc} = 0.1$ m, $h_0 = 0$.

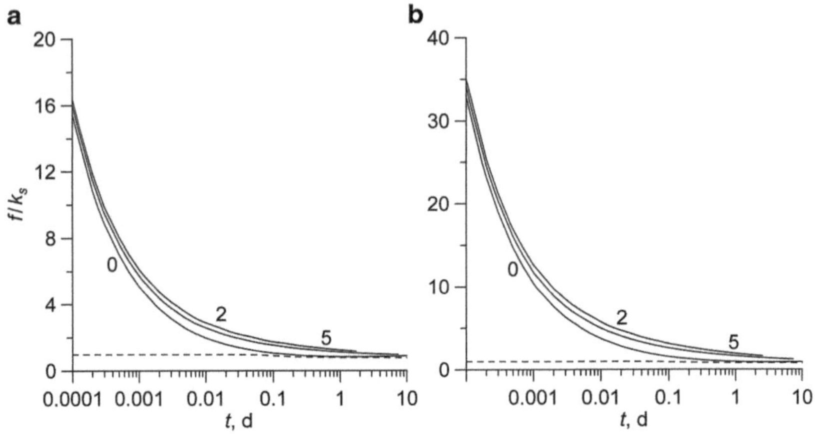

Fig. 1.9 Dependence of the relative infiltrability, $\overline{f}(t)$, on the initial capillary head in the matrix, h_{mc} (figures at the curves, m): (**a**) $h_{fc} = 0.1$ m, (**b**) $h_{fc} = 0.5$ m. The *dashed curves* are the modeling results for $h_{fc} = 0$

The size of porous matrix in horizontal direction (a) was assumed to be 1 m, thus within the simulation time (10 days), the imbibition front will not reach the outer boundary of the model domain, i.e., the matrix behaves like a surrounding with unlimited capacity.

Plots in Fig. 1.9 show that at the beginning of infiltration, while the wetting front is localized near the inflow boundary, the infiltration rate appreciably exceeds the Darcy velocity at the unit hydraulic gradient due to capillary forces in the fracture. At that time, the effect of the matrix is less significant. This effect, however, becomes more considerable over time, resulting in a higher rate of water infiltration through the inlet boundaries compared with the case $h_{mc} = 0$. Soil infiltrability can be seen to be 30–50 % greater than the infiltrability, determined by fracture capillarity ($h_{mc} = 0$). Over time, the gravity becomes the dominant force in the fracture flow, resulting in a decrease in the inflow rate, which tends to values less than the saturation conductivity. This value, coincides with the final segment of curves calculated at $h_{fc} = h_{mc} = 0$.

Overall, the simulation has confirmed the conclusion that the matrix imbibition has a minor effect on infiltrability from the practical viewpoint.

1.4.1.3 A Lumped-Parameter Model for the Rain-Triggered Near-Surface Flows

According to the lumped-parameter approach, the preferential flow through soil and vadose zone discontinuities, e.g., fractures/cracks and macropores, forms when the ability of the intact matrix to store and drain infiltrating water is exceeded (i.e., a saturation excess mechanism). Such preferential flow may serve a trigger for surface runoff when storage and drainage capacities of the fracture or macropore domain are also exceeded (Struthers et al. 2007). As before, to be specific, we will treat the near-surface medium as a uniform fractured porous continuum, and the term "fractures" can synonymously be used for "macropores".

By employing an equivalent medium representation of soil matrix and fractures (Fig. 1.10), we may describe the dual-domain system by a set of similar parameters (the subscripts m and f are for matrix and fracture domains, respectively): the average moisture content, θ_m, and θ_f (both are functions of time); field capacity, θ_{mc} and θ_{fc}; and moisture content at saturation, θ_{ms} and θ_{fs}. The governing equations for flow in each domain can be represented separately as

$$dS_m/dt = i_m - w_m, \tag{1.52a}$$

$$dS_f/dt = i_f - w_f, \tag{1.52b}$$

where $S_m = S_m(t)$ and $S_f = S_f(t)$ are the "zero"-dimensional moisture store in the matrix and fracture domains, respectively, which can be defined as:

$$S_m = (1 - \omega)\theta_m(t)z_0, \quad S_f = \omega\theta_f(t)z_0; \tag{1.53}$$

i_m, i_f and w_m, w_f are source functions representing the inflow (associated with infiltration) and outflow rates (associated with recharge), respectively; ω is the volumetric proportion of fractures; z_0 is the soil depth.

Equations (1.52a) and (1.52b) are not coupled by any transfer functions, but their solutions are dependent upon prescribed threshold values of rainfall rate and moisture content in the model domains (Struthers et al. 2007). Thus, the change in rainfall partitioning between the two domains and the relevant source functions, i_m and i_f, is controlled by three distinct thresholds.

1. For the initial period, prior to matrix saturation, it is assumed that all rainfall volume enters the pore matrix domain through direct contact at the surface and instantaneous imbibition (via the capillary suction) of moisture from fractures (Sect. 1.4.1.1), thus

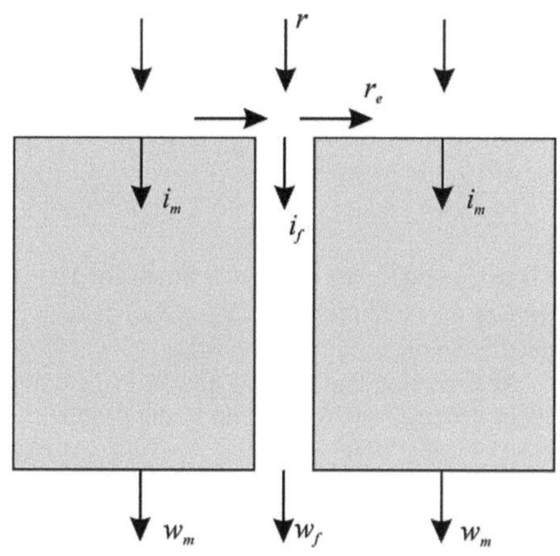

Fig. 1.10 A conceptual diagram showing main flow components in a fracture–matrix system under rainfall conditions

$$i_m = r, \quad i_f = 0, \tag{1.54a}$$

that is, fracture infiltration is absent during this period which is limited ($t \leq t_m$) by the storage and drainage capacities of the matrix domain. In this context, t_m is the fracture flow triggering time.

2. The second period starts ($t > t_m$) when the matrix domain becomes saturated and effective rainfall, represented by the difference

$$i_f = r - i_m(S_{ms}), \quad i_m = w_m(S_{ms}), \tag{1.54b}$$

infiltrates the fracture network. Similar to the above, the vertical flow process is limited by the store and drainage capacities of the fracture domain.

3. The third, final period, associated with surface ponding, begins ($t \geq t_p$) when the fracture domain moisture content reaches the saturation limit, $\theta_f = \theta_{fs} = 1$. The source functions are defined as

$$i_f = w_f(S_{fs}), \quad i_m = w_m(S_{ms}), \tag{1.54c}$$

and runoff with the rate

$$r_e = r - w_m(S_{ms}) - w_f(S_{fs}) \tag{1.55}$$

is generated.

To solve Eqs. 1.52a, 1.52b and 1.53, the drainage rates, w_m and w_f, are to be defined in terms of moisture system specification. It can be assumed (Struthers et al. 2007) that water discharge occurs only at storage above field capacity for each domain, at a rate given by

$$w_m(t) = \frac{S_m(t) - S_{mc}}{\tau_m}, \quad w_f(t) = \frac{S_f(t) - S_{fc}}{\tau_f}, \tag{1.56}$$

where τ_m and τ_f are characteristics of drainage response time for the matrix and fracture domains.

General solutions of two ordinary differential equations (1.52a), (1.52b) are similar:

$$S_m = \tau_m i_m + S_{mc} + C_m \exp(-t/\tau_m), \tag{1.57a}$$

$$S_f = \tau_f i_f + S_{fc} + C_f \exp(-t/\tau_f), \tag{1.57b}$$

where C_m and C_f are constants, which are determined by the initial conditions. For the matrix domain: $S_m = S_m^0$ at $t = 0$ (if $S_m^0 > S_{mc}$) or $S_m = S_{mc}$ at $t = (S_{mc} - S_m^0)/r$ (if $S_m^0 \leq S_{mc}$); for the fracture domain: $S_f = 0$ at $t = t_m$.

Recognizing that w_m (w_f) and S_m (S_f) are connected with each other through (1.56), one obtains the following solution to the problem:

$$w_m = \begin{cases} \left[r\tau_m + (S_m^0 - r\tau_m - S_{mc})\exp(-t/\tau_m)\right]\tau_m^{-1}, & S_m^0 > S_{mc}, \\ r[1 - \exp(-(t - t_i)/\tau_m)], & S_m^0 \leq S_{mc}; \end{cases}$$

$$w_f = (r - w_{ms})\left[1 - \exp(-(t - t_m)/\tau_f)\right]; \tag{1.58}$$

$w_{ms} = (S_{ms} - S_{mc})/\tau_m, t_i = (S_{mc} - S_m^0)/r$. Based on the solutions (1.57a) and (1.57b), time intervals t_m and t_s ($S_m(t_f) = S_{ms}$ and $S_f(t_p) = S_{fs}$) can be determined as follows:

$$t_m = \begin{cases} \tau_m \ln \dfrac{r\tau_m + S_{mc} - S_m^0}{r\tau_m + S_{mc} - S_{ms}}, & \text{if } S_m^0 > S_{mc}, \\ \dfrac{S_{mc} - S_m^0}{r} + \tau_m \ln \dfrac{r\tau_m}{r\tau_m + S_{mc} - S_{ms}}, & \text{if } S_m^0 \le S_{mc}; \end{cases} \quad (1.59a)$$

$$t_p = t_m + \tau_f \ln\left(1 - \frac{w_{fs}}{(r - w_{ms})}\right)^{-1}, \quad w_{fs} = z_0\omega/\tau_f. \quad (1.59b)$$

Obviously, surface runoff determined by Eq. 1.30 requires rainfall duration, T, to exceed fracture flow triggering time, t_m, and fracture saturation time (the second term in 1.59b). In addition, it is seen that this criterion depends on the given antecedent moisture conditions.

Assuming that, after the rain has ceased, the land surface rapidly (theoretically, instantaneously) gets free of water layer ($t > T$), we can use equations (1.52) and (1.56) to obtain an equation describing the dewatering of the porous matrix and fractures due to the drainage effect:

$$w_m = w_{ms}\exp[-(t - T)/\tau_m], \quad w_f = w_{fs}\exp[-(t - T)/\tau_f]. \quad (1.60)$$

The response behavior of a fractured soil in terms of the flux component generation by a rectangular rainfall pulse is illustrated by an example plot in Fig. 1.11. Parameter values are given in the capture.

The plots in Fig. 1.11 demonstrate that the fractured soil properties selected for this example are sufficient for the fracture flow to be activated by a given rainfall, and that the time of runoff generation for the given drainage response time (τ_m, τ_f) depends upon the initial (antecedent) moisture conditions (θ_m^0).

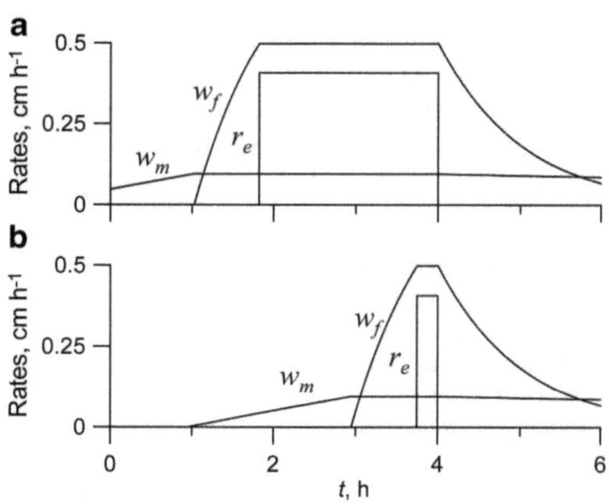

Fig. 1.11 Influence of antecedent moisture condition on the flow rates, w_m, w_f, and r_e. (**a**) $\theta_m^0 = 0.4$, (**b**) $\theta_m^0 = 0.2$. Rainfall intensity $r = 10$ mmh^{-1}, rainfall duration $T = 4$ h. Other system parameters are: $\theta_{ms} = 0.5$, $\theta_{mc} = 0.3$, $\tau_m = 20$ h, $\tau_f = 1$ h, $\omega = 0.05$, $z_0 = 10$ cm, $\theta_{fs} = 1$, $\theta_{fc} = 0$

1.4.2 Subsurface Macropore Runoff in a Soil Profile (Hillslope Scale)

In this section, we will focus briefly on factors determining the contribution of macropores to lateral subsurface flow, namely, to stormwater runoff at hillslope scale (Sects. 1.2.3 and 1.2.4). This type of preferential flow provides a quick runoff response in streams. Runoff generation from hillslopes in the presence of variously oriented macropores has been studied in details by Noguchi et al. (1999), Sidle et al. (2000, 2001), Freer et al. (2002) and others. In more recent publications, macropores are not treated to be continuous throughout the soil profile of the hillslope, thus the relevant downward flow cannot be attributed to individual macropores (Nieber et al. 2006; Andersen et al. 2009; Nieber and Sidle 2010). It was proved that preferential flow in the macropore networks is dynamic and that the extent of their hydrologic activity is influenced strongly by antecedent moisture conditions (Sidle et al. 2001; Nieber et al. 2006). The initiation of preferential flow may require certain thresholds related to the extent of soil saturation or saturation at the soil–bedrock interface to be exceeded (Freer et al. 2002).

Individual macropores that make up preferential flow networks are typically very short (smaller than approximately 0.5 m in length); however, they are able to link over relatively long slope distances due to various mechanisms (Sidle et al. 2001). To quantify the linkage between initially unconnected macropores and the formation of dynamical network structures, several conceptual models of preferential flow have been developed.

Modeling results and field observations demonstrate that that the connectivity of macropore networks and the resistance to overall through flow depend on soil wetness, the depth of storm water, and hillslope geometry (Fig. 1.12). The increase in preferential flow is attributed to an expansion of macropore networks in time and space. Such expansion may occur through a series of complex mechanisms as antecedent moisture increases (Sidle et al. 2000). The relationship between total rainfall depth and total preferential flow volume is highly nonlinear and threshold-like at each site (Uchida et al. 2005). The macropores become effective on large scales mainly during storm events when the extent of wetness of soil increases dramatically and the effects of capillary tension decrease (Sidle et al. 2000; 2001; Beckers and Alila; Uchida et al. 2005; Nieber and Sidle 2010).

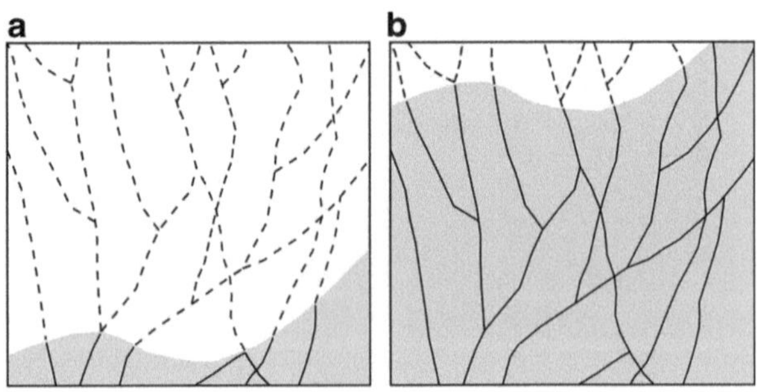

Fig. 1.12 The extension of the macropore network with an increase in saturation of the soil network. (**a**) Dry conditions, (**b**) wet conditions. The *shadowed area* – soil with increased moisture content

In general, the analysis of available publications has shown that the preferential flow is a significant component of the subsurface water balance, and the connectivity of the preferential flow network is an important factor governing the subsurface runoff and thus controlling solute transport. However, a certain rainfall threshold is required to trigger rapid preferential flow near the surface or in the deeper part of the soil profile. The inclusion of the preferential flow in modeling analysis usually showed clearly better results than modeling without preferential flow.

At the same time, it is obvious that each hillslope is unique and may have more or less developed preferential flow networks that are activated only under certain fallout and antecedent moisture conditions. For example, some field studies show that greater rain event frequency promotes preferential flow and transport when storms are smaller in magnitude (McGrath et al. 2010). Other studies show quite different tendency (Andersen et al. 2009). Therefore, to model hillslope behavior properly, it is important to be able to carry out high-density field survey at hillslope scale supported by high resolution of rainfall records.

1.5 Non-Infiltration Types of Rainfall Losses

Apart from rainfall loss due to infiltration, which has been discussed in the previous section, there are three processes that can affect runoff generation and flow path dynamics: (1) the interception of rainfall by plants and soil vegetation, (2) the retention and storage of water in surface depressions, and (3) evapotranspiration. The former two processes are commonly called the surface retention loss or initial (prior to any surface runoff) abstraction. The latter process mostly affects runoff implicitly, because moisture conditions of the soil mantle and its contribution to watershed water budget become important only for long-term forecasting. Evaporation is also considered as a part of the total surface retention loss.

1.5.1 Interception

Interception is defined as the process of evaporation from intercepted rainfall, or, quantitatively, the amount of daily rainfall that evaporates, thus, the process has a typical timescale of 1 day. Such surface retention losses for various surface cover conditions and land-use can vary considerably, from a hundredth to a tenth of inch (per day) reaching, e.g., up to 50 % of the total precipitation falling as rain on forests in temperate humid latitudes (Savenije 2004; Gerrits et al. 2010). Interception is an important water flux component, especially in semi-arid catchments, where a few intense rainfall events may generate much of the season's runoff (Lange and Leinbundgut 2003; Love et al. 2010) mostly in a form of infiltration excess overland flow. Therefore, interception causing less water to be available for infiltration may play a controlling role in runoff generation determining the water balance trends and water cycle of catchment areas. Obviously, from this viewpoint, interception may be responsible for the spatial and temporal (seasonal) distribution of infiltration and, consequently, groundwater recharge.

As many hydrological processes, interception is considered as a threshold phenomenon depending on time scale, rainfall intensity and some evaporation characteristics. Usually interception is modeled as a threshold process at a daily time scale (De Groen 2002; Savenije 2004; Love et al. 2010):

$$I_d = \min(P_d;\ D_d),$$

where I_d is the daily interception; P_d is the daily rainfall; and D_d is the daily interception threshold. The value of D_d may vary over seasons and depends on meteorological conditions. According to the threshold concept, only the amount of rainfall that exceeds the threshold takes part in subsequent processes such as infiltration and surface runoff. If some rainfall was intercepted on the previous day, and the amount of intercepted rainfall was more than could be evaporated on that day, some moisture will remain in interception storage until the next day (Love et al. 2010).

Using special statistical procedures and time series analysis, the daily interception can be upscaled to monthly or annual interception (De Groen and Savenije 2006). For expert evaluations, one may use tabular information or empirical relationships represented in the reference literature (Gash 1979; Van Dijk and Bruijnzeel 2001).

1.5.2 Depression Storage

On the surface of a natural catchment, overland flow retention is determined by the gradual filling of micro-relief surface depressions with rainwater. Depression storage contributes considerably to the initial rainfall abstraction showing some threshold effect (Luce and Cundy 1992). Depressions represented by negative landforms vary in size from place to place across a catchment, resulting in that small depressions are almost immediately filled up at the beginning of a rainfall, whereas the larger ones remain only partially filled with water during the whole rainfall event. The phenomenon may proceed at varying rates depending upon the physical conditions of a catchment.

To assess the effects of water surface storage on overland flow, a simplified lumped model was offered by Linsley et al. (1949):

$$V = S_d\left(1 - \exp\left(-\frac{P - F}{S_d}\right)\right), \tag{1.61}$$

where V is the volume of water stored (mm); S_d is the maximum depression storage capacity (mm); $P - F$ is the rainfall that reaches the surface (mm) minus infiltration depth (mm). Then the supply rate, v, to the depression storage becomes:

$$v = \frac{dV}{dt} = \frac{S_d}{S_d}\exp\left(-\frac{P-F}{S_d}\right)\frac{d(P-F)}{dt} = (r-f)\exp\left(-\frac{P-F}{S_d}\right), \tag{1.62}$$

where r is the rainfall rate, f is the infiltration capacity. From here, the rainfall excess rate is defined as

$$r_e = r - (f + v) = (r-f)\left(1 - \exp\left(-\frac{P-F}{S_d}\right)\right). \tag{1.63}$$

At the beginning of the rainfall, when $F \approx P$, the exponent in Eq. (1.63) is close to 1, and the excess fallout available for surface runoff is negligible, while, as the fallout continues the exponent in Eq. (1.36) quickly approaches zero, and the rate r_e takes a standard form (Eq. 1.1b).

1.5.3 Evapotranspiration

Evapotranspiration (actual, ET), is a major component of the hydrological cycle, significantly affecting the hydrological response of watersheds. It accounts for the water that plants extract through their roots from the soil. ET depends on a highly complex set of processes and local conditions, such as plant type, as well as climate and soil characteristics, in particular, soil moisture content. Many approaches are known to enable the evaluation of this component of the hydrological cycle.

One of the ways is to determine ET through a supplementary assessment of a potential evapotranspiration (PET). PET is water demand (maximum possible evapotranspiration when there is plenty of water available in the soil), while ET is actual water use (which depends on how much water is really available). Many simple and rather sophisticated models for PET rate prediction are known; they were reviewed in a number of publications (Dunn and Mackay 1995; Allen et al. 1998; Dingman 2002). The most commonly used are the Penman equation, and Thornthwaite's, Turc's, and Blaney-Criddle formulas. Some authors believe that the evaporation has much less spatial and temporal variability than rainfall, and the average monthly evaporation can be used as a surrogate for daily evaporation without any significant loss of accuracy in the modeling of runoff (Chapman 2003).

We can hypothesize that the actual evapotranspiration rate (denoted as ET [LT^{-1}]) increases linearly from zero when the soil water content reaches the wilting point, θ_{wp} (soil is nearly dry), to the potential rate (denoted as PET [LT^{-1}]) at critical moisture content, θ_{fc}, corresponding to field capacity. Following this logic, we come to the threshold type of equations:

$$ET = 0 \text{ when } \theta < \theta_{wp}, \tag{1.64a}$$

$$ET = \bar{\theta}PET \text{ when } \theta_{wp} \le \theta < \theta_{fc}, \tag{1.64b}$$

$$ET = PET \text{ when } \theta \ge \theta_{fc}; \tag{1.64c}$$

the dimensionless moisture content in (1.64b) is determined by $\overline{\theta} = (\theta - \theta_{wp})/(\theta_{fc} - \theta_{wp})$. This type of ET behavior (Eqs. 1.64a, 1.64b and 1.64c) is illustrated by Dingman (1994). The structure of ET function as represented by Eq. (1.64) has been incorporated in many well-known water balance models (Bergström 1992; Kling and Gupta 2009; see also Sect. 5.4.2). Some additional comments can be found in a review by Xu and Singh (2004).

References

Aggelides S, Youngs EG (1978) The dependence of the parameters in the Green and Ampt infiltration equation on the initial water content in draining and wetting states. Water Resour Res 14(5):857–862

Alaoui A, Caduff U, Gerke HH et al (2011) Preferential flow effects on infiltration and runoff in grassland and forest soils. Vadose Zone J 10:367–377

Allen RG, Pereira LS, Raes D et al (1998) Crop evapotranspiration, guidelines for computing crop water requirements. FAO Irrigation and Drainage Paper 56, Food and Agriculture Organization of the United Nations, Rome

Andersen AE, Weiler M, Alila Y (2009) Subsurface flow velocities in a hillslope with lateral preferential flow. Water Resour Res. doi:10.1029/2008WR007121

Barenblatt GI, Entov VM, Ryzhik VM (1990) Fluid flow in natural reservoirs. Kluwer, Dordrecht

Basha HA (1999) One-dimensional nonlinear steady infiltration. Water Resour Res 35:1697–1704

Bear J (1972) Dynamics of fluids in porous media. Dover Publ Inc, New York

Beckers J, Alila Y (2004) A model of rapid preferential hillslope runoff contributions to peak flow generation in a temperate rain forest watershed. Water Resour Res 40(3), W03501

Bergström S (1992) The HBV model – its structure and applications. SMHI RH

Beven KJ (1981) Kinematic subsurface stormflow. Water Resour Res 17(5):1419–1424

Beven KJ, Germann PF (1982) Macropores and water flow in soils. Water Resour Res 18(5):1311–1325

Beven KJ, Germann PF (2013) Macropores and water flow in soils revisited. Water Resour Res 49:1–22

Biswas TD, Mukherjee SK (1994) Textbook of soil sciences. Tata McGraw-Hill Publishing Company Limited, New Delhi, p 430

Boughton WC (1993) A hydrograph-based model for estimating the water yield of ungauged catchments. Proc Hydrol Water Resour Symp Newcastle Inst Engs Aust. Nat Conf Publ 93, 14:317–324

Brutsaert W (1994) The unit response of groundwater outflow from a hillslope. Water Resour Res 30(10):2759–2763

Buttle JM, McDonald DJ (2002) Coupled vertical and lateral preferential flow on a forested slope. Water Resour Res 38(5). doi:10.1029/2001WR000773

Carlier E (2007) A probabilistic investigation of infiltration in the vadose zone a proposal for a new formula for infiltration rate. Hydrol Process 27:2845–2849

Castaing R (1991) Un modele simple pour la migration de radionucléides par transport colloidal dans un milieu fracture. J Hydrol 125:55–92

Chapman TG (2003) Estimation of evaporation in rainfall-runoff models. In: Proceedings MODSIM 2003 international congress on modelling and simulation, modelling and simulation society of Australia, vol 1, pp 148–153

Charbeneau RJ (2006) Groundwater hydraulics and pollutant transport. Waveland Press, Long Grove, p 593

Chen L, Young MH (2006) Green-Ampt infiltration model for sloping surfaces. Water Resour Res 42(7). doi: 10.1029/2005WR004468
Childs EC (1971) Drainage of groundwater resting on a sloping bed. Water Resour Res 7(5): 1256–1263
Chow VT, Maidment DR, Mays LW (1988) Applied hydrology. McGraw-Hill, New York, p 572
Craig JR, Liu G, Soulis ED (2010) Runoff–infiltration partitioning using an upscaled Green–Ampt solution. Hydrol Process. doi:10.1002/hyp.7601
Dingman SL (1994) Physical hydrology. Macmillan Publishing Company, New York, p 575
Dingman SL (2002) Physical hydrology. Prentice-Hall Inc, Upper Saddle River
De Groen MM (2002) Modelling interception and transpiration at monthly time steps; introducing daily variability through Markov chains. PhD thesis, IHE-Delft, Swets and Zeitlinger, Lisse, The Netherlands, p 211
De Groen MM, Savenije HHG (2006) A monthly interception equation based on the statistical characteristics of daily rainfall. Water Resour Res 42, W12417. doi:10.1029/2006WR005013
Dunn SM, Mackay R (1995) Spatial variation in evapotranspiration and the influence of land use on catchment hydrology. J Hydrol 171:49–73
Dunne T (1978) Field studies of hillslope flow processes. In: Kirkby MJ (ed) Hillslope hydrology. Wiley, Chichester/New York, pp 227–294
Dunne T, Moore TR, Taylor CH (1975) Recognition and prediction of runoff-producing zones in humid regions. Hydrol Sci Bull 20(3):305–327
Dunne T, Black RD (1970) Partial area contributions to storm runoff in a small New England watershed. Water Resour Res 6(5):1296–1311. doi:10.1029/WR006i005p01296
Dusek J, Vogel T (2014) Modeling subsurface hillslope runoff dominated by preferential flow: One- vs. two dimensional approximation. Vadose Zone J 13. doi:10.2136/vzj2013.05.0082
Freer J, McDonnell JJ, Beven KJ et al (2002) The role of bedrock topography on subsurface storm flow. Water Resour Res 38(12):1269. doi:10.1029/2001WR000872
Gabrielli C, McDonnell JJ, Jarvis T (2012) The role of bedrock groundwater in rainfall-runoff response at hillslope and catchment scales. J Hydrol 450–451:117–133
Gash JHC (1979) An analytical model of rainfall interception by forest. Quart J R Meteorol Soc 105:43–55
Germann P (1985) Kinematic wave approach to infiltration and drainage into and from soil macropores. Trans Am Soc Agric Eng 28(3):745–749
Germann P, Beven K (1985) Kinematic wave approximation to infiltration into soils with sorbing macropores. Water Resour Res 21:990–996
Gerrits AMJ, Pfister L, Savenije HHG (2010) Spatial and temporal variability of canopy and forest floor interception in a beech forest. Hydrol Process 24:3011–3025
Graham CB, McDonnell JJ (2010) Hillslope threshold response to rainfall: development and use of a macroscale model. J Hydrol 393(1–2):77–93
Graham CB, Woods RA, McDonnell JJ (2010) Hillslope threshold response to rainfall: a field based forensic approach. J Hydrol 393(1–2):65–76
Green WH, Ampt G (1911) Studies of soil physics, part I – the flow of air and water through soils. J Agric Sci 4:1–24
Harman C, Sivapalan M (2009) A similarity framework to assess controls on shallow subsurface. Water Resour Res 45(1), W01417. doi:10.1029/2008WR007067
Henderson FM, Wooding RA (1964) Overland flow and groundwater flow from a steady rainfall of finite duration. J Geophys Res 69(8):1531–1540
Hillel D (2004) Introduction to environmental soil physics. Academic, Amsterdam, p 494
Hopp L, McDonnell JJ (2009) Connectivity at the hillslope scale: identifying interactions between storm size, bedrock permeability, slope angle and soil depth. J Hydrol 376:378–391
Horton RE (1940) An approach toward a physical interpretation of infiltration capacity. Soil Sci Soc Am Proc 5:399–417
Hydrology handbook (1996) 2nd ed American Society of Civil Engineers. New York, p 784

Jarvis NJ (1998) Modelling the impact of preferential flow on non-point source pollution. In: Selim HH, Ma L (eds) Physical non-equilibrium in soils: modelling and application. Ann Arbor Press, Chelsea, pp 195–221
Jarvis NJ (2007) A review of non-equilibrium water flow and solute transport in soil macropores: principles, controlling factors and consequences for water quality. Eur J Soil Sci 58:523–546
Kao CS, Hunt JR (1996) Prediction of wetting front movement during one-dimensional infiltration into soil. Water Resour Res 32:55–64
Kim CP, Stricker JNM, Torfs PJJF (1996) An analytical framework for the water budget of the unsaturated zone. Water Resour Res 32(12):3475–3484
Kling H, Gupta H (2009) On the development of regionalization relationships for lumped watershed models: the impact of ignoring sub-basin scale variability. J Hydrol 373:337–351
Kohler A, Abbaspour KC, Fritsch M (2003) Using simple bucket models to analyze solute export to subsurface drains by preferential flow. Vadose Zone J 2:68–75
Köhne JM, Köhne S, Šimůnek J (2009) A review of model applications for structured soils: (a) Water flow and tracer transport. J Contam Hydrol 104:4–35
Kosterin AV, Selin VI (2000) Liquid hydrocarbons migration in the unsaturated zone presented by fractured-porous rocks. Prob Nucl Sci Technol Math Model Phys Proc 2:53–57 (In Russian)
Kostiakov AN (1932) On the dynamics of the coefficient of water percolation in soils and on the necessity for studying it from a dynamic point of view for purposes of amelioration. In: Trans, 6th Comm Int Soc Soil Sci, Russian Part A: 17–21
Lange J, Leinbundgut C (2003) Surface runoff and sediment dynamics in arid and semi-arid regions. In: Simmers I (ed) International contributions to hydrogeology 238: Understanding water in a dry environment: hydrological processes in arid and semi-arid zones. Balkema, Rotterdam, pp 114–150
Linsley RK, Kholer MA, Paulhus JLH (1949) Applied hydrology. McGraw Hill, New York
Love D, Uhlenbrook S, Corzo-Perez G et al (2010) Rainfall–interception–evaporation–runoff relationships in a semi-arid catchment, northern Limpopo basin, Zimbabwe. Hydrol Sci J 55(5):687–703
Luce CH, Cundy TW (1992) Modification of the kinematic wave-Philip infiltration overland flow model. Water Resour Res 28(4):1179–1186
McDonnell JJ (1990) A rationale for old water discharge through macropores in a steep, humid catchment. Water Resour Res 26(11):2821–2832
McDonnell JJ (2013) Are all runoff processes the same? Hydrol Process 27:4103–4111
McGrath GS, Hinz C, Sivapalan M et al (2010) Identifying a rainfall event threshold triggering herbicide leaching by preferential flow. Water Resour Res. doi:10.1029/2008WR007506
Mein RG, Larson CL (1973) Modeling infiltration during a steady rain. Water Resour Res 9(2):384–394
Mirus B, Loague K (2013) How runoff begins (and ends): characterizing hydrologic response at the catchment scale. Water Resour Res 49(5):2987–3006. doi:10.1002/wrcr.20218
Mishra SK, Tyagil JV, Singh VP (2003) Comparison of infiltration models. Hydrol Process 17:2629–2652
Morel-Seytoux HJ, Khanji J (1975) Equation of infiltration with compression and counterflow effects. J Hydrol Sci 20:505–517
Morel-Seytoux HJ, Khanji J (1974) Derivation of an equation of infiltration. Water Resour Res 10:795–800
Neuman SP (1976) Wetting front pressure head in the infiltration model of Green and Ampt. Water Resour Res 12:564–566
Nieber JL, Sidle RC (2010) How do disconnected macropores in sloping soils facilitate preferential flow? Hydrol Process 24:1582–1594
Nieber JL, Steenhuis TS, Walter T (2006) Enhancement of seepage and lateral preferential flow by biopores on hillslopes. Biologia Bratislava 61(Suppl 19):225–228
Nitao JJ, Buscheck TA (1991) Infiltration of a liquid front in an unsaturated, fractured porous medium. Water Resour Res 27(8):2099–2112. doi:10.1029/91WR01369

Noguchi S, Tsuboyama Y, Sidle RC et al (1999) Morphological characteristics of macropores and the distribution of preferential flow pathways in a forested slope segment. J Soil Sci Soc Am 63(5):1413–1423

Novak V, Simunek J, van Genuchten MT (2002) Infiltration into a swelling, cracked clay soil. J Hydrol Hydromech 50(1):3–19

Ogden FL, Watts BA (2000) Saturated area formation on nonconvergent hillslope topography with shallow soils: a numerical investigation. Water Resour Res 36(7):1795–1804

Pellichero E, Glantz R, Burns M et al (2012) Dynamic capillary pressure during water infiltration: Experiments and Green-Ampt modeling. Water Resour Res 48. doi:10.1029/2011WR011541

Philip JR (1955) Numerical solution of equations of the diffusion type with diffusivity concentration-dependent. Trans Faraday Soc 51:885–892

Philip JR (1957) The theory of infiltration: 4 Sorptivity and algebraic infiltration equations. Soil Sci 8:257–264

Philip JR (1987) The infiltration joining problem. Water Resour Res 23:2239–2245

Pinder GP, Gray WG (2008) Essentials of multiphase flow in porous media. Wiley, Hoboken, p 374

Pruess K (1991) EOS7 An equation-of-state module for the TOUGH2 simulator for two-phase flow of saline water and air Earth Science Division, Lawrence Berkeley Laboratory. Report N LBL-31114, Berkeley

Pruess K (2004) The TOUGH codes – a family of simulation tool for multiphase flow and transport processes in permeable media. Vadose Zone J 3:738–746

Rangel-German ER, Kovscek AR (2001) Experimental and analytical study of multidimensional imbibition in fractured porous media. Technical report, Stanford University, Stanford, CA, USA

Ruan H, Illangasekare TH (1998) A model to couple overland flow and infiltration into macroporous vadose zone. J Hydrol 210:116–127

Rumynin VG (2011) Subsurface solute transport models and case histories (with applications to radionuclide migration), vol 25, Theory and applications of transport in porous media. Springer Science + Business Media BV, Dordrecht, p 815

Salvucci GD, Entekhabi D (1994) Explicit expression for Green-Ampt (delta function diffusivity) infiltration rate and cumulative storage. Water Resour Res 30:2661–2663

Savenije HHG (2004) The importance of interception and why we should delete the term evapotranspiration from our vocabulary. Hydrol Process 18(8):1507–1511

Schmed BH (1990) Derivation of an explicit equation for infiltration on the basis of the Mein-Larson Model. J Hydrol Sci 35,2,4:197–208

Shaw EM, Beven KJ, Chappel NA, Lamb R (1994) Hydrology in practice, 3rd edn. Chapmam and Hall, London, p 569

Sidle RC, Noguchi S, Tsuboyama Y et al (2001) A conceptual model of preferential flow systems in forested hillslopes: evidence of self-organization. Hydrol Proc 15:1675–1692

Sidle RC, Tsuboyama Y, Noguchi S et al (2000) Storm flow generation in steep forested headwaters: a linked hydrogeomorphic paradigm. Hydrol Process 14:369–385

Šimůnek J, Jarvis NJ, van Genuchten MT et al (2003) Review and comparison of models describing non-equilibrium and preferential flow and transport in the vadose zone. J Hydrol 272:14–35

Šimůnek J, van Genuchten MT (2008) Modeling nonequilibrium flow and transport processes using HYDRUS. Vadose Zone J 7(2)

Singh VP (2002) Kinematic wave solutions for pollutant transport by runoff over an impervious plane, with instantaneous or finite-period mixing. Hydrol Process 16:1831–1863

Sloan PG, Moor ID (1984) Modeling subsurface stormflow on steeply sloping forested watersheds. Water Resour Res 20(12):1815–1822

Smith RE (1972) The infiltration envelope: results from a theoretical infiltrometer. J Hydrol 17:1–21

Smith RE (2002) Infiltration theory for hydrologic applications. With Smettem KRJ, Broadbridge P, Woolhiser DA. American Geophysical Union. Water Resour Monograph Series, vol 15, Washington, DC, p 215
Smith RE, Goodrich DC (2005) Rainfall excess overland flow. In: Anderson MG (ed) Encyclopedia of hydrological science. Wiley, Chichester, pp 1708–1718
Springer E, Cundy TW (1987) Field-scale evaluation of infiltration parameters from soil texture for hydrologic analysis. Water Resour Res 23(2):325–334
Struthers I, Sivapalan M, Hinz C (2007) Conceptual examination of climate-soil controls upon rainfall partitioning in a pen-fractured soil: Single storm response. Adv Water Resour 30:505–517
Szymkiewicz A (2013) Modeling water flow in unsaturated porous media. Springer, Dordrecht, p 250
Todd DK, Mays LW (2005) Groundwater hydrology. Wiley, Arizona
Todini E (2007) Hydrological catchment modeling: past, present and future. Hydrol Earth Syst Sci 11(1):468–482
Touma J, Vauclin M (1986) Experimental and numerical analysis of two-phase infiltration in a partially saturated soil. Transp Porous Media 1:27–55
Uchida T, Kosugi K, Mizuyama T (2002) Effects of pipe flow and bedrock groundwater on runoff generation in a steep headwater catchment in Ashiu, central Japan. Water Resour Res 38(7):1119. doi:10.1029/2001WR000261
Uchida T, Tromp-van Meerveld I, McDonnell JJ (2005) The role of lateral pipe flow in hillslope runoff response: an intercomparison of non-linear hillslope response. J Hydrol 311:117–133
Van Dijk AIJM, Bruijnzeel LA (2001) Modelling rainfall interception by vegetation of variable density using an adapted analytical model. Part 1. Model description. J Hydrol 247(3):230–238
Verhoest NEC, Troch PA (2000) Some analytical solutions of the linearized Boussinesq equation with recharge for a sloping aquifer. Water Resour Res 36(3):793–800
Vieux BE (2004) Distributed hydrologic modeling using GIS. Kluwer Academic Publishers, Dordrecht, p 289
Wang Z, Feyen J, van Genuchten MT (1997) Two-phase flow infiltration equations accounting for air entrapment effects. Water Resour Res 33(12):2759–2767
Weiler M, McDonnell JJ (2006) Testing nutrient flushing hypotheses at the hillslope scale: a virtual experiment approach. J Hydrol 319:339–356
Weyman DR (1970) Throughflow on hillslopes and its relation to the stream hydrograph. Int Assoc Sci Hydrol Bull 15(2):25–33
Whipkey RZ (1965) Subsurface stormflow from forested slopes. International Association of Scientific Hydrology. Bulletin 10(2):74–85
Wienhöfer J, Zehe E (2014) Predicting subsurface stormflow response of a forested hillslope –the role of connected flow paths. Hydrol Earth Syst Sci 18:121–138. doi:10.5194/hess-18-121[--]2014
Willgoose GR, Perera H (2001) A simple model of saturation excess runoff generation based on geomorphology, steady state soil moisture. Water Resour Res 37(1):1471–55
Williams JR, Ouyang Y et al (1998) Estimation of infiltration rate in the vadose zone: Application of selected mathematical models. Environmental Protection Agency, USA, Report EPA/600/R-97/128d
Xiangjun T, Zhenghui X, Shenglei Z et al (2006) A subsurface runoff parameterization with water storage and recharge based on the Boussinesq-storage equation for a land surface model. Sci China Ser D Earth Sci 49(6):622–631
Xu C-Y, Singh VP (2004) Review on regional water resources assessment models under stationary and changing climate. Water Resources Management 18. Kluwer Academic Publishers, Dordrecht, pp 591–612
Zhang GP, Savenije HHG, Fenicia F et al (2006) Modelling subsurface storm flow with the Representative Elementary Watershed (REW) approach: application to the Alzette River Basin. Hydrol Earth Syst Sci 10:937–955. www.hydrol-earth-syst-sci.net/10/937/2006/
Zhang L, Walker GR, Dawes WR (2002) Water balance modelling: concepts and applications. In: McVicar TR, Li Rui, Walker J, Fitzpatrick RW, Liu Changming (eds) Regional water and soil assessment for managing sustainable agriculture in China and Australia. ACIAR Monograph N 84, pp 31–47

Chapter 2
Rainfall-Induced Runoff and Subsurface Stormflow at the Hillslope Scale

Surface runoff (or *overland flow*), which is generated by the precipitation that falls within a drainage area (catchment, watershed), is governed by several factors and processes, including rainfall rate and duration, the characteristics of infiltration (capillary imbibition and gravity-driven) and the temperature regime of soil, landscape surface characteristics, vegetation type, and some others. Hillslopes are regarded as a basic element of catchments, therefore the mathematical and physical description of the hydrological processes that occur at the hillslope scale is the first step to designing more general hydrological models describing hydrological response at catchment/watershed scale.

Surface runoff can be also generated by snowmelt. To predict snowmelt-induced runoff, the frozen subsurface domains are to be mathematically described using coupled models of subsurface heat and mass transfer. This description is a separate research problem, which is out of the scope of this work. However, in some simplified settings, snowmelt-induced runoff and rainfall runoff can be treated in a similar manner.

When runoff flows over the ground surface, it can pick up soil contaminants (fertilizers, pesticides, heavy metals, radionuclides, petroleum, and others) to become a part of a nonpoint pollution source. The process of contaminant detachment is accompanied by *erosion* of the soil and other materials, which increases water turbidity. Runoff maintains transport of contaminants in the form of dissolved solutes or/and colloidal/suspended matter (in adsorbed state) from drainage areas toward the discharge zones of streams into river network or surface water reservoirs (either natural or artificial). Rainwater also accumulates in surface depressions, which may focus infiltration over a relatively small area.

Such water flow and solute transport processes can be described by rigorous mathematical procedures based on systems of differential equations of continuity, mass conservation, and momentum balance (physically-based models with distributed parameters), or by models based on functions and parameters averaged over time and space (lumped parameter models).

V.G. Rumynin, *Overland Flow Dynamics and Solute Transport*, Theory and Applications of Transport in Porous Media 26, DOI 10.1007/978-3-319-21801-4_2

In this chapter, we discuss a widespread approach, which is based on 1D equation of kinematic wave and used to describe surface runoff at the hillslope scale. Such physically based model, dealing with the so-called *sheet-flow,* can more easily be used in the prediction of the depth of water flowing over a sloped surface, which is the main characteristic of the hydrological process. However, such reduction of dimensionality may lead to misrepresentation of important effects, which are controlled by the spatial pattern of surface water flow, such as lateral flow convergence/divergence and water storage in surface depressions. Thus, as the runoff rate increases downslope, the flow converges into micro-scale channels called *rills,* which gradually develop until they form large-scale channels called *gullies* (Julien and Simons 1985). In this work, we focus exclusively on sheet-flow (interrill flow) processes.

The derivation of a catchment unit hydrograph caused by a storm rainfall is often related to two flow components, i.e., surface (overland flow) and subsurface (saturated *subsurface stormwater flow*). Saturated flow in hillslope soil is considered to be initiated when the wetting front from infiltration of rainwater arrives at the lower boundary of a permeable soil layer or shallow water table. Such lateral movement of soil water can occur at any stage of rainfall, which precedes or accompanies surface runoff generation. Therefore, it is logical to consider in this chapter the time-dependent response of this type as a supplemental flow component in a hillslope hydrograph based on the 1D approach, as well.

Despite the obvious limits of 1D models with simplified hydrology, they demonstrate the effects of dominant mechanisms, which are in common in all watershed systems. This chapter, which gives an overview of this theoretical framework, is based on well-known approaches to the description of the dynamics of near-surface flows. This will help develop mathematical models for the description of solute transport in surface and subsurface runoff, which is the subject of the following chapters.

2.1 Kinematic Flow Approximation to Runoff on Idealized Hillslopes and Basic Characteristic Solutions

A rigorous enough description of overland flow is provided by shallow-water equations for a layer with horizontal velocity components u_x and u_y (along the x and y coordinates, respectively), vertically averaged within a layer of thickness h, the so-called Saint-Venant equations, including equations of continuity and the conservation of momentum (Chow 1959). This system is also referred to as the system of equations of dynamic wave. Under certain conditions, this system degenerates into particular equations of diffusion or kinematic wave. Most researchers agree in that the best applicable approximation used to solve overland flow problem is the *kinematic wave approach* as a special case of the diffusion approximation (Eagleson 1970; Govindaraju et al. 1988, 1990). The suitability and

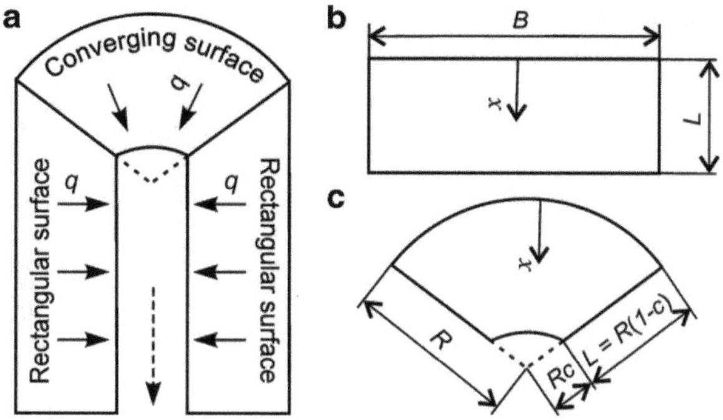

Fig. 2.1 (**a**) Fragmentation geometry of watershed landscape (*left*- and *right*-bank and upstream sub-basins), (**b**) rectangular plane, (**c**) converging surface

success of the kinematic wave approach in the description of overland flow on simple surfaces have been proved by many authors (Woolhiser and Liggett 1967; Parlange et al. 1981; Rose et al. 1983; Govindaraju et al. 1988, 1990; Singh 1997). Analytical solutions have been developed mostly using the *method of characteristics* for idealized flow patterns, and initial and boundary conditions.

The kinematic wave equation, as a form for representation of unsteady-flow continuity equation for the overland flow, can be written as

$$\frac{\partial A}{\partial t} + \nabla Q - F_w = 0, \tag{2.1}$$

where A is the cross section area of the overland flow [L^2]; Q is the total discharge [L^3T^{-1}]; F_w is a source/sink function characterizing lateral inflow or outflow owing to rainfall and/or infiltration [L^2T^{-1}].

For a conceptual representation of a watershed geometry (Fig. 2.1a), two simplified hillslope configurations can be suggested, i.e., a rectangular plane (Fig. 2.1b) and a converging area (Fig. 2.1c).

For flow on a rectangular plane with a constant width, B, the volumetric flux (overland flow discharge) per unit width, $q = q(x, t)$, is defined as $q = Q/B$, the cross-section area equals $A = Bh$, and the source/sink term is $F_w = B(r - i)$. Now, the continuity Eq. (2.1) takes the form:

$$\frac{\partial h}{\partial t} + \frac{\partial q}{\partial x} = (r - i), \tag{2.2}$$

where $h = h(x, t)$ is the current water depth [L]; the difference $(r - i)$, termed as r_e, is the rainfall excess, i.e., the portion of the rainfall that ponds on the surface during

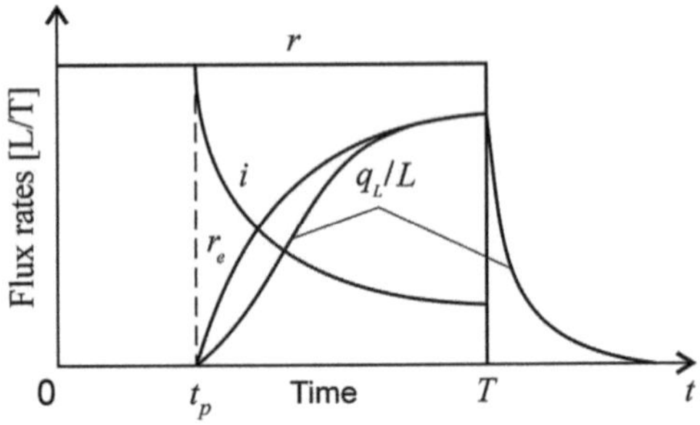

Fig. 2.2 Components of drainage flux in a hillslope

the period when the rainfall rate, r [LT^{-1}], exceeds the infiltration capacity, or infiltration rate, i [LT^{-1}]; all these flow components can be time-dependent.

For flow on a converging surface, the width B can be assumed to vary linearly with $(R - x)$: $B = \theta(R - x)$, and $\theta = -dB/dx$. Then the continuity equation has a one-dimensional form as well:

$$\frac{\partial h}{\partial t} + \frac{\partial q}{\partial x} = (r - i) + \frac{q}{R - x}. \tag{2.2a}$$

Geometric interfacing of the two hillslope configurations requires a relationship $L = R(1 - c)$ to hold as shown in Fig. 2.1, where c is the degree of convergence for the converging surface. The values of c vary from 0 to 1, depending on the watershed geometry. If c approaches 1, this means that the converging representation transforms into a rectangular-plane one.

The dynamics of different flow components in Eqs. (2.2) and (2.2a) is illustrated by Fig. 2.2 where the input function, r, corresponds to a rainfall event with duration T. It can be seen that, once the surface is ponded, $t > t_p$, the infiltration rate, i, drops, but rainfall excess, r_e, increases. The specific flux at the outlet of the slope, q_L/L, first lags behind the curve r_e because of the *inertial* character of flow formation on the slope.

In reality, even on slopes that may appear ideal, the spatial pattern of runoff is rather complex, because of the effect of microrelief, which contributes to the formation of preferential flow paths (runoff in rills). Because of this, function h is an effective characteristic of the process, and it would be more correct to call it effective surface water depth.

Thus, we have one continuity equation (Eqs. 2.2 or 2.2a) containing two unknowns, q and h. To close the equation set we must combine this equation with the equation of motion in a form of flow-resistance relationship

$$q = \alpha h^n, \tag{2.3}$$

which uniquely relates water flux and surface water depth; the relationship (2.3) in the power form ($n > 1$) is of fundamental character; here α is a parameter that accounts for the effect of friction forces, commonly, $\alpha = S_0^{1/2}/m$ [$L^{2-n}T^{-1}$]; S_0 is surface slope [LL^{-1}], m is Manning's roughness coefficient; this is an empirically derived coefficient, which is dependent on many factors, including surface roughness and sinuosity; its dimension depends on the exponent n (flow regime characteristic); in the case of turbulent flow, $n = 5/3$ (so that m [$TL^{-1/3}$] and α [$L^{1/3}\,T^{-1}$]), in the case of laminar flow, $n = 3$; some researchers suggest an intermediate value $n = 2$ to be used (a flow with mixed/transient hydrodynamic regime). Thus, the overland flow discharge, q, is determined, other conditions being the same, by the slope and roughness of the land surface.

The derivative $\partial q/\partial h$ has units of velocity and is called the kinematic wave celerity, c_k. The relationship for c_k, following mathematically from the law of motion (Eq. 2.3), is

$$c_k = \frac{\partial q}{\partial h} = \alpha n h^{n-1}. \tag{2.3a}$$

Another form of kinematic wave celerity can be given by

$$c_k = \frac{\alpha n h^n}{h} = n\frac{q}{h} = nu. \tag{2.3b}$$

As can be seen, the kinematic wave celerity is n times greater than the flow velocity, $u = q/h$.

Substituting (2.3) into (2.2) and (2.2a), we obtain equations

$$\frac{\partial h}{\partial t} + \alpha n h^{n-1}\frac{\partial h}{\partial x} = r - i, \tag{2.4}$$

$$\frac{\partial h}{\partial t} + \alpha n h^{n-1}\frac{\partial h}{\partial x} = (r - i) + \frac{\alpha h^n}{R - x}, \tag{2.4a}$$

referred to as *kinematic wave model* for two simplified geometric configurations of hillslope surface (Fig. 2.1, b, c).

2.2 Overland Flow Over Impermeable Surface

When the fallout duration is limited to interval T, the function r is step-wise

$$r(x,t) = r > 0, \quad 0 \le t < T; \quad r(x,t) = 0, \quad t > T, \tag{2.5}$$

and the initial and boundary conditions for function h can be written as:

$$h(x,0) = 0, \quad 0 \le x \le L; \quad h(0,t) = 0, \quad t \ge 0, \tag{2.6}$$

$$h(x,0) = 0, \quad 0 \le x \le R(1-c); \quad h(0,t) = 0, \quad t \ge 0 \tag{2.6a}$$

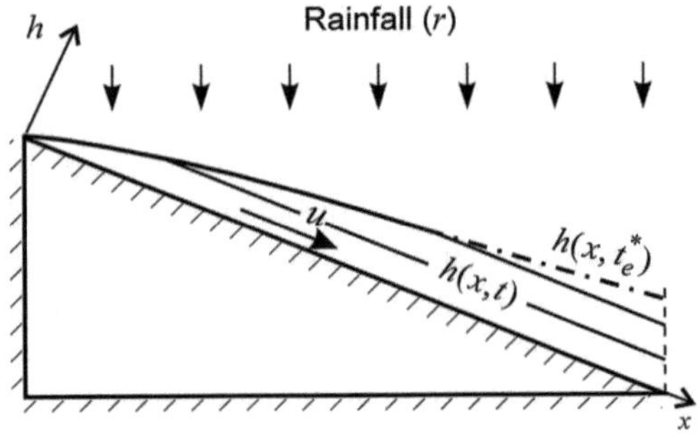

Fig. 2.3 A conceptual representation of overland flow formation on a homogeneous impermeable surface. $h(x, t)$ – water depth profile at $t < t_e^*$; $h(x, t_e^*)$ – equilibrium water depth profile $(t \geq t_e^*)$

for overland flow over a rectangular and converging surfaces, respectively. As one may see, the kinematic wave approximation needs no downstream boundary condition because the characteristics move in the forward direction only (Govindaraju et al. 1990), and therefore it takes no account of downstream effects in the vicinity of the outlet boundary (Beven 1981; Singh 2002a). These boundary constraints are quite common in surface flow hydrology, and they are applicable to a large class of problems. Some limitations of the simplified boundary conditions (2.6) are discussed in papers by Govindaraju with co-authors (1988, 1990).

In this section, we consider partial solutions of (2.4), (2.4a) with the flow assumed to form on an ideal impermeable ($i = 0$) plane with a slope $S_0 = H/L$ (Fig. 2.3). With $i = 0$, Eqs. (2.4) and (2.4a) take simpler forms

$$\frac{\partial h}{\partial t} + \alpha n h^{n-1} \frac{\partial h}{\partial x} = r, \tag{2.7}$$

$$\frac{\partial h}{\partial t} + \alpha n h^{n-1} \frac{\partial h}{\partial x} = r + \frac{\alpha h^n}{R - x}. \tag{2.7a}$$

The features of overland flow dynamics, which are reflected in the character of water depth function, h, on the surface are given in Fig. 2.4. As can be seen from the plots, (1) the hydrograph generated by rainfall of restricted duration, T, features different spatial and temporal behavior of function h; it can be (a) not stationary, but constant within some, long enough, segment along the direction of flow, $x, h = h(t)$; (b) stationary, but increasing along the flow (from the water divide line $x = 0$ toward its discharge zone $x = L$), $h = h(x)$; (c) varying over both time and space, $h = h(x, t)$; (2) depending on rainfall duration, T, the profile of the hydrodynamic flow (hydrodynamic wave) can be (a) equilibrium (when, at $t \leq T$, the flow depth in the section $x = L$ reaches its maximal possible value h, corresponding to steady-state flow), or (b) partial equilibrium (when the rainfall or rainfall excess ends, $t = T$, before the hydrodynamic wave that originates at the top of the sloped plane reaches the outlet of the plane).

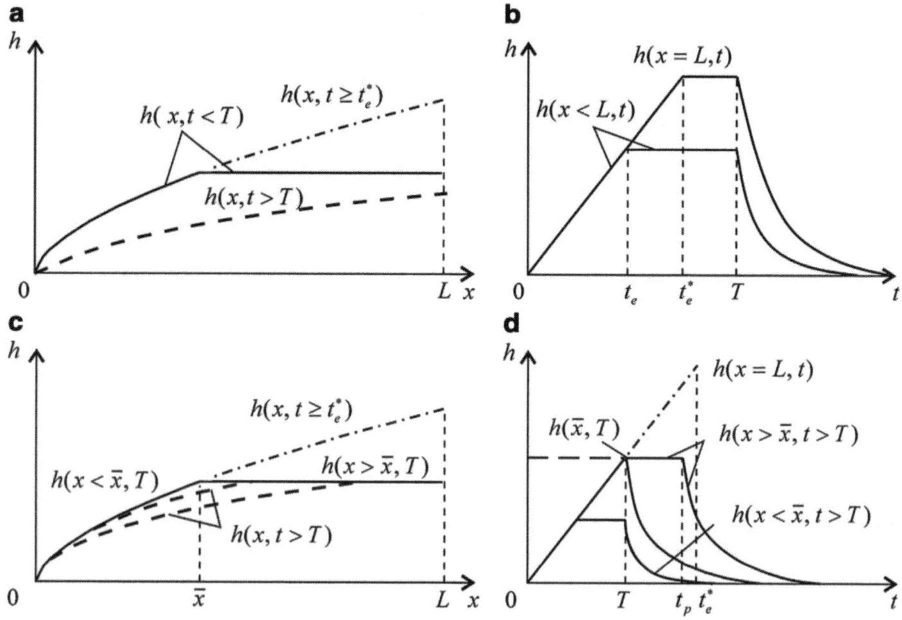

Fig. 2.4 Spatial (1D) and temporal distributions of water depth at the surface, $h(x)$ and $h(t)$, for the cases of formation of equilibrium (**a**, **b**) and partial equilibrium (**c**, **d**) hydrographs at limited rainfall duration T

2.2.1 Basic Characteristic Solutions

The Eq. (2.7) accompanied by the initial and boundary conditions (2.6) corresponds to a model of overland flow on a rectangular geometry hillslope (Fig. 2.1a, b). It can be solved by the method of characteristics, which converts the partial differential equation to a system of ordinary differential equations (Eagleson 1970; Singh 1996)

$$\frac{dt}{1} = \frac{dx}{\alpha n h^{n-1}} = \frac{dh}{r}, \tag{2.8}$$

expressing the absolute time variation of water depth along the characteristic curves. More precisely, characteristic curves $t(x)$, which are defined by the loci of points in the time–space plane (Fig. 2.5), are particular solutions of the first equations in the system (2.8). The second equation in the systems (2.8) is for water depth and the distance along characteristics $t(x)$.

The Eq. (2.7a) corresponding to a model of overland flow on a converging surface (Fig. 2.1a, c) has also a form of representation through a system of ordinary differential equations

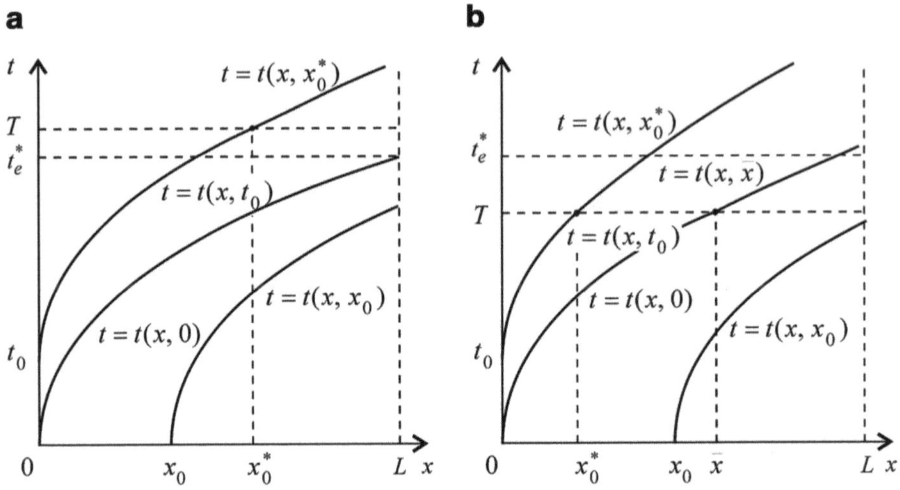

Fig. 2.5 Characteristic lines for (**a**) equilibrium water depth profile and (**b**) partial equilibrium water depth profile at limited fallout duration, T

$$\frac{dt}{1} = \frac{dx}{\alpha n h^{n-1}} = \frac{dh}{r + \alpha h^n/(R - x)}. \tag{2.8a}$$

There are two major domains related to rising and falling stages of the overland flow (sometimes termed the rising and recession limbs of the surface flow hydrograph).

Rising stage of the overland flow (rising hydrograph) Let the displacement of an arbitrary point, x_0, on the free surface of liquid flow moving downward along the sloped *rectangular plane* since the moment of precipitation start, $t = 0$, be $x - x_0$ (Fig. 2.5a). Within this time, $t(x, x_0)$, the flow depth, h, increases from 0 to the value $h(x, x_0)$, which can be obtained by the integration of the second equation in the system (2.8):

$$h(x, x_0) = \left(\frac{r}{\alpha}(x - x_0)\right)^{1/n}. \tag{2.9}$$

Substituting (2.9) into the first Eq. (2.8) and integrating it from 0 to $t = t(x, x_0)$, we obtain the characteristic:

$$t(x, x_0) = r^{-1}\left[\frac{r}{\alpha}(x - x_0)\right]^{1/n}. \tag{2.10}$$

In other words, function h varies along characteristic (2.10) according to (2.9). The function $t(x, 0)$ is the main flow characteristic curve, which originates at the top of the slope at the start of rainfall. It is also termed the equilibrium characteristic because it denotes the time at which the flow reaches steady state under constant rainfall (see below).

Now, combining (2.9) and (2.10), we obtain the relationship:

$$h(t) = rt, \tag{2.11}$$

which implies a linear dependence of the depth h on the process time, t (Fig. 2.4a, b); the applicability domain of (2.11) is determined by the condition of hydrodynamic equilibrium of water layer on the slope and the period of rainfall, as follows from the analysis below.

Let us assume that the duration of rainfall, T, is long enough for the hydrodynamic wave to reach the discharge zone of the overland flow $x = L$, i.e., $T > t_e^*$, where t_e^* is the formation time of *equilibrium profile* of flow (Fig. 2.5a).

According to (2.11), water depth in any point (any cross-section of the flow) increases with constant velocity $dh/dt = r$, which does not depend on x. Such regime, with flow rate varying in time but constant along the direction x, can be seen in any flow section until the moment $t = t_e$ when the boundary wave, which originates from water divide, reaches this section, i.e.,

$$t_e = t(x, 0) = r^{-1}\left(\frac{rx}{\alpha}\right)^{1/n} \tag{2.12}$$

(see Eq. 2.10 at $x = x_0 = 0$). Time t_e is termed the *time of concentration*, which is the equilibrium time required for an impervious slope surface to reach steady state under constant (excess) rainfall intensity. For the time when equilibrium is reached all over the slope with a length L, the time of concentration is

$$t_e^* = t(L, 0) = r^{-1}\left(\frac{rL}{\alpha}\right)^{1/n} \equiv \frac{h(L, 0)}{r}. \tag{2.13}$$

Since the moment $t = t_e$, in point, x, as well as in the upstream domain, $(0, x)$, the flow is steady-state, but its rate increases in the direction from water divide toward the discharge domain

$$q = q(x) = rx. \tag{2.14}$$

The equation of characteristic for any point $t = t_0 > 0$ takes the form

$$t = t_0 + r^{-1}\left(\frac{rx}{\alpha}\right)^{1/n}, \tag{2.15}$$

and the flow surface in this case is described by a power function

$$h(x) = \left(\frac{rx}{\alpha}\right)^{1/n}. \tag{2.15a}$$

Formalizing the above reasoning, we can write particular solutions of the problem, which correspond to the transient period of wave formation ($t < t_e$) and the steady state phase of the process ($t_e \leq t \leq T$), in the generalized form:

$$h = \begin{cases} h(t) = rt, & 0 \leq t \leq t_e \leq T, \\ h(x) = rt_e, & t_e \leq t \leq T. \end{cases} \tag{2.16}$$

In a manner similar to that described above, the method of characteristics can be used to solve Eq. 2.7a or its differential-system analogue, Eq. 2.8a, considered in the context of overland flow on a *converging surface* (Fig. 2.1a, c). However, before searching for the solution, it seems reasonable to re-write the equations in dimensionless form using the depth of flow in the equilibrium state $h_L = h(L)$ (2.15a) and the time to equilibrium $t_e = t(L, 0)$ (2.12) for flow on the rectangular plane as normalizing constants (Singh and Woolhiser 1976; Shokoohi and Saghafian 2012): $h_* = h/h(L), t_* = t/t_e, x_* = x/R(1-c)$. With these dimensionless variables substituted into (2.7a), we obtain the following equation:

$$\frac{\partial h_*}{\partial t_*} + nh_*^{n-1}\frac{\partial h_*}{\partial x_*} = 1 + \frac{(1-c)h_*^n}{1-(1-c)x_*}. \tag{2.17}$$

Note, that α no longer exists in the dimensionless model (2.17). In its turn, the system of ordinary differential equations (2.8a) is transformed to become:

$$\frac{dx_*}{dt_*} = nh_*^{n-1}, \tag{2.17a}$$

$$\frac{dh_*}{dt_*} = 1 + \frac{(1-c)h_*^n}{1-(1-c)x_*}. \tag{2.17b}$$

A solution of this problem by the method of characteristics has been derived earlier (Singh and Woolhiser 1976; Sherman and Singh 1976; Campbell et al. 1984). Therefore, only some basic relationships will be given here. Among them is one of the characteristics describing the response time of watersheds, namely the time of concentration. To evaluate the time of concentration requires the solution of Eqs. (2.17a) and (2.17b) when the characteristic passing through the origin $x = 0$ $(x^* = 0)$ intersects the downstream boundary at $x = L = R(1-c)$ $(x^* = 1)$ as shown in Fig. 2.1c. Upon dividing Eq. 2.17a by Eq. 2.17b we obtain:

$$\frac{dh_*}{dx_*} = \frac{1}{nh_*^{n-1}} + \frac{1}{n}\frac{(1-c)h_*}{1-(1-c)x_*}. \tag{2.17c}$$

Integration of Eq. 2.17c yields a steady-state profile (Singh and Woolhiser 1976; Agiralioglu 1981; Shokoohi and Saghafian 2012):

$$h_* = \left[\frac{1-c}{1-(1-c)x_*} \int_0^{x_*} \left(\frac{1}{1-c} - x_* \right) dx_* \right]^{1/n} = \left[\frac{x_*(2 - x_*(1-c))}{2(1 - x_*(1-c))} \right]^{1/n}. \quad (2.17d)$$

Substituting $x^* = 1$ into (2.17d), we find the dimensionless equilibrium value

$$(h_*)_e = ((1+c)/2c)^{1/n}. \quad (2.17e)$$

Further, from Eq. 2.17a it follows

$$t_* = \frac{1}{n} \int_0^{x_*} h_*^{1-n} dx_*. \quad (2.17f)$$

To determine the dimensionless characteristics $(t_*)_e$, i.e., the ratio of the time of concentration for a converging surface, t_{ec}, to the time of concentration for a rectangular surface, t_e, one can substitute $h_*(x_*)$ from (2.17d) into (2.17f), assuming the upper integration limit to be 1:

$$(t_*)_e \equiv \frac{t_{ec}}{t_e} = \frac{1}{n} 2^{(n-1)/n} \int_0^1 \left[\frac{1 - x_*(1-c)}{x_*(2 - x_*(1-c))} \right]^{(n-1)/n} dx_*. \quad (2.17g)$$

Thus, assuming n to vary from 3/2 to 3 and c, from 0 to 1, from numerical computations based on the integral solution (2.17g), a range for dimensionless characteristics $(t_*)_e$ has been obtained (Agiralioglu 1984) as follows: $(t_*)_e \approx 0.85$ ($c = 0$) – 1 ($c = 1$). The converging representation is seen to yield shorter time of concentration than the rectangular one does under the same conditions. This is what one would expect from the fundamental principles of unsteady free-surface flow (Agiralioglu 1984). Rainwater concentrates in converging directions on the converging surface. Therefore, the flow velocity is relatively greater than that for a rectangular plane surface, depending on the degree of convergence. The dimensionless time of concentration, $(t_*)_e$, decreases with decreasing c.

Some other examples related to the study of the effects of hillslope geometry and topography on the time of concentration considering the degree of flow convergence as well as the curvature of slope profile can be found in many other publications (Agiralioglu 1981, 1984, 1988; Shokoohi and Saghafian 2012; Sabzevari et al. 2013).

Falling Stage of the Overland Flow (Recession Hydrograph) After the cessation of the rain, $t > T > t_e^*$ (the equilibrium profile has formed before the rain ceased), in the recession period, water depth, h, becomes a decreasing function of two coordinates, $h = h(x,t)$. The characteristic solution of the problem of flow on a *rectangular plane* is given by equations (Singh 2002a):

$$h(x, x_0^*) = \left(\frac{rx_0^*}{\alpha}\right)^{1/n}, \tag{2.18}$$

$$t(x, x_0^*) = T + \left(\frac{x - x_0^*}{\alpha n}\right)\left(\frac{\alpha}{rx_0^*}\right)^{(n-1)/n}, \tag{2.19}$$

where x_0^* is a parameter, which represents the point of intersection of the characteristic $t = t(x, t_0)$ with the line $t = T$ on the plot of characteristics (Fig. 2.5a).

Eliminating x_0^* from (2.18) and (2.19), we obtain the following transcendent equation

$$x = \alpha h^{n-1}\left[\frac{h}{r} + n(t - T)\right], \quad t > T > t_e^*, \tag{2.20}$$

describing the decrease in water depth at the recession period. Clearly, at this process stage, flow discharge is also a function, varying over time and space in the entire model domain, $q = q(x,t)$.

Now let us consider a *partially equilibrium profile* of overland flow, which corresponds to the case where rainfall duration is less than the time within which full equilibrium profile will form, i.e., $T < t_e^*$. As can be seen (Figs. 2.4c, d and 2.5b), a particular section is $h(\bar{x})$ ($\bar{x} = \alpha T^n r^{n-1}$), which divides the domain into two zones (upper and lower), which differ in terms of the hydrodynamic regime of the flow at the stage of process recession ($t > T$). For sections $h(x < \bar{x})$, the cessation of precipitation implies an immediate drop in water flow rate and a decrease in its layer thickness on the surface. In sections $h(x > \bar{x})$, the flow rate will remain constant for some time ($T - t_p$) after the cessation of precipitation. The features mentioned above are reflected in the generalized solution:

$$h(t) = rt, \quad 0 \le t \le T \le t_e^*, \tag{2.21a}$$

$$h(x) = rT, \quad T < t \le t_p, \tag{2.21b}$$

$$x = \alpha h^{n-1}[h/r + n(t - T)], \quad t > t_p, \tag{2.21c}$$

where

$$t_p = T\left[1 + \frac{1}{n}\left(\left(\frac{t_e}{T}\right)^n - 1\right)\right]. \tag{2.21d}$$

The latter relationship follows from the characteristics solution of the problem (Singh 2002a)

$$h(x, x_0^*) = rT, \tag{2.21e}$$

$$t(x, x_0^*) = T + \left(\frac{x - x_0^*}{\alpha n}\right)(rT)^{1-n}. \tag{2.21f}$$

Thus, theoretically, the solution (2.21a), (2.21b), and (2.21c) implies that, after the cessation of a short-time rainfall ($T < t_e^*$), hillslope hydrograph should feature a plateau-like interval $q(x = L, T < t < t_p) = \text{const.}$

The solutions of the problems considered above can be expressed in terms of overland flow discharge function. In the case of equilibrium profile, the following dimensionless solutions correspond to the above solutions (2.16) and (2.20)

$$\begin{aligned} &\overline{q} = \tau^n \quad 0 \le \tau < 1, \ r \ge 0, \\ &\overline{q} = 1 \quad 1 \le \tau < \tau_0, \ r \ge 0, \\ &\overline{q}\left[1 + n\overline{q}^{-1/n}(\tau - \tau_0)\right] = 1, \ \tau \ge \tau_0, \ r = 0; \end{aligned} \tag{2.22}$$

here $\overline{q} = q/q_e, q_e = rx, \tau = rt/h_0 \equiv t/t_e, \tau_0 = rT/h_0; h_0 = (rx/\alpha)^{1/n}$. It is seen that for initially dry condition and constant rainfall, runoff discharge, $q(t)$, monotonically increases from 0 (for $t = 0$) to $\alpha(rt_e)^n$ (for $t = t_e < T$). The period $t_e < t < T$ is characterized by a constant overland discharge, $q = q_e = rx$. Starting with $t = T$, $q(t)$ reduces quite slowly to 0.

To quantify the nonequilibrium profile (2.21a), (2.21c), we have expressions:

$$\begin{aligned} &\overline{q} = \tau^n \quad 0 \le \tau < \tau_0, \ r \ge 0, \\ &\overline{q} = \tau_0^n \quad \tau_0 \le \tau < \tau_p, \ r = 0, \\ &\overline{q}\left[1 + n\overline{q}^{-1/n}(\tau - \tau_p)\right] = 1, \ \tau \ge \tau_p, \ r = 0, \end{aligned} \tag{2.22a}$$

where $\tau_p = \tau_0\left[1 + (\tau_0^{-n} - 1)/n\right]$. The second equation in the system (2.22a) indicates that, after the cessation of precipitation, the flow discharge remains constant for some time ($T - t_p$).

Finally, the analytical description of the problem of flow on a *converging surface* (Fig. 2.1c) in the recession period (in the absence of rainfall) requires consideration of Eq. 2.7a in which $r = 0$. This equation is transformed into a dimensionless form as follows:

$$\frac{\partial h_*}{\partial t_*} + nh_*^{n-1}\frac{\partial h_*}{\partial x_*} = \frac{(1-c)h_*^n}{1-(1-c)x_*}. \tag{2.23}$$

As above, Eq. (2.23) can be restated to become a system of ordinary differential equations:

$$\frac{dh_*}{dt_*} = \frac{(1-c)h_*^n}{1-(1-c)x_*}, \tag{2.23a}$$

$$\frac{dx_*}{dt_*} = nh_*^{n-1}. \tag{2.23b}$$

Equation (2.23a) with the initial condition $h_* = h_*^0$ at $x_* = x_*^0$ has a solution

$$h_* = h_*^0 \left[\frac{1-(1-c)x_*^0}{1-(1-c)x_*}\right]^{1/n}, \tag{2.23c}$$

where h_*^0 is the initial depth which can be calculated from the steady-state profile (2.17d) assuming $x_* = x_*^0$:

$$h_*^0 = \left[\frac{x_*^0\left(2 - x_*^0(1-c)\right)}{2\left(1 - x_*^0(1-c)\right)}\right]^{1/n}. \tag{2.23d}$$

Then the differential equation for the characteristic curve (2.23b) becomes:

$$\frac{dx_*}{dt_*} = n\left(h_*^0\right)^{n-1}\left[\frac{1-(1-c)x_*^0}{1-(1-c)x_*}\right]^{(n-1)/n}. \tag{2.23e}$$

After separation of variables, the Eq. (2.23e) can be integrated in the intervals $[t_*^0 - t_*]$, $[x_*^0 - x_*]$ to give the desired solution:

$$t_* = T_* + A\left(x_*^0\right)\left[\left(1-(1-c)x_*^0\right)^{(2n-1)/n} - \left(1-(1-c)x_*\right)^{(2n-1)/n}\right], \tag{2.23f}$$

where

$$A\left(x_*^0\right) = \frac{2^{(n-1)/n}}{(1-c)(2n-1)}\left[x_*^0\left(2-(1-c)x_*^0\right)\right]^{(1-n)/n}, \tag{2.23g}$$

$T_* = T/t_e$ is the dimensionless rainfall duration.

2.2.2 A Particular Example and Its Comparison with Numerical Modeling Data

The slope extends between its left, uppermost, side, represented by a no-flow boundary, and its right, lowermost, side, represented by a free (no resistance) face (Fig. 2.3). The length of the sloped surface is $L = 2000$ m, and its slope has a gradient $S_0 = 0.01$ m/m. The exponent in the formula for the overland flow discharge (2.3) is $n = 5/3$. The Manning's coefficient, which characterizes the roughness of the surface, is $m = 0.05$ s/m$^{1/3}$. In that case, $\alpha = 2$ m$^{1/3}$/s. The rainfall rate $r = 5\cdot10^{-6}$ m/s (18 mm/h). Given $T = T_1 = 15000$ s (4.16 h), $T = T_2 = 7000$ s (1.94 h), we come to two characteristic regimes for water flow profile on the slope surface, i.e., an equilibrium regime (at $T_1 > t_e^*$) and a partially equilibrium regime (at $T_2 < t_e^*$); here $t_e^* = 8325.5$ s (2.31 h) (see Eq. 2.13)

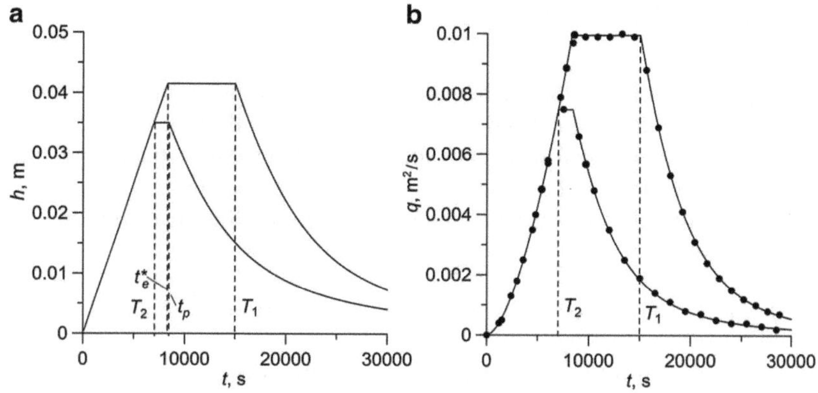

Fig. 2.6 (**a**) Rising and recession limbs of water depth, $h(t)$, and (**b**) the hydrograph, $q(t)$, at the slope outlet. The *full lines* are the result of analytic calculation; *dots* on plot (**b**) show a numerical solution (GSSHA code)

and $t_p = 8406.7$ s (2.36 h) (Eq. 2.21d). Taking into account the test nature of the considered task, to make it more impressive, values r and T_i were increased in comparison with the mean statistical background values; the selected r and T_i are closer to the values corresponding to heavy-storm periods. The same reason underlay the selection of a rather large hillslope length, L.

Increments in water depth at the slope outlet, derived from solutions (2.16), (2.20), (2.21a), (2.21b), and (2.21c), are given in Fig. 2.6a. The specific flux, q, can be readily evaluated using (2.3), see Fig. 2.6b. The analytical curves in Fig. 2.6b are in good agreement with numerical modeling results, obtained with the help of a numerical code GSSHA (Downer and Ogden 2004). The spatial step of the model along the x-axis is 10 m, and the time step is 2 s.

2.3 Coupled Overland Flow and Infiltration Models

Any rainfall and overland flow are accompanied by water infiltration into the soil under the effect of gravity, capillarity, and absorption (Sect. 1.3). An adaptation of the rainfall-runoff mathematical models to such watershed conditions requires considering one more hydrological component, namely infiltration. The dynamic relationship between rainfall, overland flow, and infiltration is one of the central issues in hydrology.

2.3.1 A Qualitative Analysis

It seems quite common in the mathematical problem formulation when infiltration, i, is independently determined and subtracted from rainfall, r; the residual is termed rainfall excess, r_e, which forms input for the overland flow model. The simplest approximation for the rainfall excess rate is to conceptualize it as occurring only when the rainfall rate is greater than the infiltration rate (Sherman and Singh 1976; Stone et al. 1992):

$$r_e(t) = r(t) - f(t) \quad \text{for} \quad r(t) > f(t) \equiv i(t), \quad r_e(t) = 0 \quad \text{otherwise}, \tag{2.24}$$

where $f(t)$ is the soil infiltrability (see also Eq. 1.2b). Thus, when $r(t)$ exceeds $f(t)$, the infiltrability determines the actual infiltration rate $i(t) = f(t)$, and the excess of water, $r(t) - f(t)$, will run off.

The advantage of the model (2.24) is that the infiltration may be computed independently of the overland flow computation. Thus, the infiltration rate has been incorporated in the continuity Eq. (2.2) using Green–Ampt (Baiamonte and Agnese 2010), Horton (Leu and Liu 1988; De Lima and van Der Molen 1988; Guo 1998), Philip (Luce and Cundy 1992), Smith–Parlange (Giráldez and Woolhiser 1996), U.S. SCS (Hjelmfelt 1978, see Sect. 5.3) and some other approximations of the soil infiltrability $f(t)$ (Sect. 1.3.2). The disadvantage is that the infiltration is not computed during the time when water is still flowing on the surface and the rainfall rate is less than the infiltration capacity or when rainfall ceases (Stone et al. 1992), therefore (2.24) always overestimates the volume of runoff during the recession stage of the hydrograph.

As to the analytical description of the process as a whole, the researchers prefer the characteristic method adopted for solving the kinematic wave equations. Commonly identified are the rising and falling stages of the overland flow (associated with the rising and recession limbs of the hydrograph) resulting in the following definition of rainfall excess:

$$0 \le t \le T, \;\; r_e(t) \ge 0; \;\; t > T, \;\; r_e(t) < 0. \tag{2.24a}$$

Recession of the flow process (at $t > T$) results in the formation of a free-surface overland flow boundary condition that provides specific features for solving the initial or boundary problem for Eq. (2.2) or (2.2a).

The characteristic curves in the $x - t$ plane may be grouped in several domains depending on their origin (Wooding 1965; Giráldez and Woolhiser 1996;

Baiamonte and Agnese 2010). Integration of ordinary differential equations related to the basic kinematic wave Eq. (2.4) results in closed-form analytical solutions for integer values of n: $n=2$ for Smith and Parlange (Giráldez and Woolhiser 1996) and Green-Ampt (Baiamonte and Agnese 2010) approximations; $n=3$ for Philip (Luce and Cundy 1992) and Horton (Leu and Liu 1988) approximations. Semi-analytical or numerical methods are used for fractional numbers n.

Thus, Fig. 2.7 illustrates the behavior of the outlet discharge, $q = q(L, \ t)$, affected by the Green–Ampt infiltrability (Eq. 1.12) at basic characteristics of hillslope geometry, overland flow and rainfall (Eqs. 2.2 and 2.24) as follows: $\alpha = 2\ \mathrm{m}^{1/3}/\mathrm{s}$; $n = 5/3$; $L = 1600$ m (outlet); the initial saturation deficit, $\Delta\theta = 0.05$ ($\theta_s = 0.1$, $\theta^0 = 0.05$); $r = 18$ mm/h $= 5{\cdot}10^{-6}$ m/s; $T = 277$ min. The initial and boundary conditions for a rectangular pulse of rainfall (Eqs. 2.5) are represented by Eqs. (2.6). The effect of two governing parameters is taken into account: hydraulic conductivity, k_s, at fixed suction pressure, h_c, and, conversely, h_c at fixed k_s.

In the chosen parameter range, the hydrograph curves, $q(L, t)$, are very sensitive to soil infiltrability. It is clearly seen that hillslope response in the form of overland flow occurs when current time exceeds the time of ponding, t_p, determined by (1.19), (1.19a). In Fig. 2.7a, one can also see that the falling limbs of the hydrographs become steeper with rising k_s values. On the other hand, at this stage of hydrograph formation, the attenuation rate of function q shows a weak sensitivity to changes in suction pressure (Fig. 2.7b).

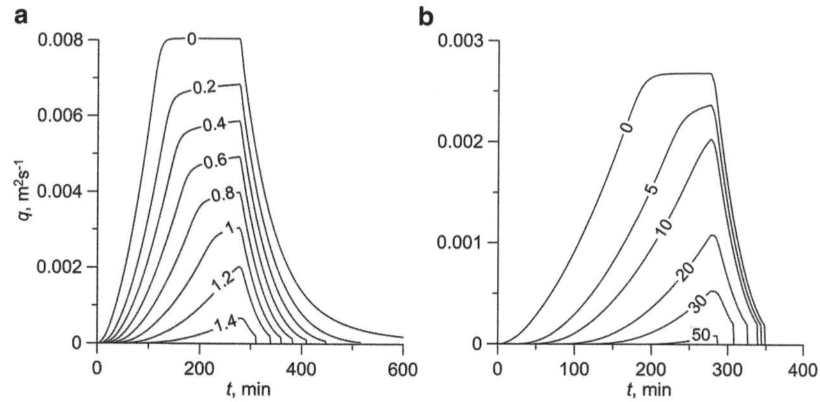

Fig. 2.7 Influence of Green–Ampt transient infiltration on the outlet hydrograph (GSSHA code). (**a**) $h_c = 10$ cm, numbers at the curves are hydraulic conductivity, k_s (cm/h); (**b**) $k_s = 1.2$ cm/h, numbers at the curves are suction potential, h_c (cm)

2.3.2 General Solution

A fully analytical solution to kinematic flow for an arbitrary time-dependent excess rainfall rate, $r_e(t)$, was developed by Parlange and Sander with co-authors (Parlange et al. 1981; Sander et al 1990, 2009; Sander and Parlange 2000). They started from the kinematic wave equation

$$\frac{\partial h}{\partial t} + \alpha n h^{n-1} \frac{\partial h}{\partial x} = r_e(t), \tag{2.25}$$

subject to the initial and boundary conditions (2.6) and (2.24a).

The partial differential equation (2.25) corresponds to a pair of ordinary differential equations:

$$\frac{dh}{dt} = r_e(t), \quad \text{and} \quad \frac{dx}{dt} = \alpha n h^{n-1}. \tag{2.25a}$$

The solution (Sander et al. 1990; Sander and Parlange 2000; Sander et al. 2009) was derived and presented for two domains: (1) $0 \le t \le T, r_e(t) \ge 0$, and (2) $t > T$, $r_e(t) < 0$. In a particular case, T can be associated with the length of the rainfall period; the current time is measured starting from the ponding time.

In the first domain, $0 \le t \le T$, for the initial and boundary conditions (2.6), solutions of (2.25a) are given parametrically (Parlange et al. 1981; Sander et al. 2009):

$$h(t) = \int_{t_0}^{t} r_e(\varsigma)d\varsigma, \quad \text{and} \quad x = \alpha n \int_{t_0}^{t} \left(\int_{t_0}^{\xi} r_e(\varsigma)d\varsigma \right)^{n-1} d\xi, \quad x \le x_e, \tag{2.26}$$

with the parameter t_0, $0 \le t_0 \le t$; $t_0 = t$ corresponding to the boundary condition $h(0,t) = 0$; the initial condition is satisfied by $t_0 = t = 0$. For $x > x_e = x(t_0 = 0)$, or

$$x > x_e = \alpha n \int_{0}^{t} \left(\int_{0}^{\xi} r_e(\varsigma)d\varsigma \right)^{n-1} d\xi, \tag{2.27}$$

h is independent of x and given by the first equation in (2.26).

In the second domain, $t > T$, a free surface forms and begins to move downslope (Sander et al. 1990, 2009) from $x = 0$ so that the boundary condition $h = 0$ occurs at $x = x_d(t)$ where

$$x_d = \alpha n \int_{t_1}^{t} \left(\int_{t_1}^{\xi} r_e(\varsigma)d\varsigma \right)^{n-1} d\xi, \tag{2.28}$$

with $t_1 \leq T \leq t$ and t_1 defined from the first equation in (2.26) as

$$\int_{t_1}^{t} r_e(\varsigma)d\varsigma = 0. \tag{2.29}$$

Equations (2.28) and (2.29) give the time dependence of the free surface edge for $t > T$. The solution for $x > x_d$ is still given by (2.26) but with t_0 restricted to the range $0 \leq t_0 \leq t_1$. The drying time, t_d, can be found from (2.29) and (2.25a) (Sander and Parlange 2000):

$$\int_{0}^{t_d} r_e(\varsigma)d\varsigma = \int_{0}^{t^*} r_e(\varsigma)d\varsigma + \int_{t^*}^{t_d} r_e(\varsigma)d\varsigma. \tag{2.30}$$

Under an idealistic condition, the infiltration rate is assumed to be fixed (time-independent), $i =$const, and therefore: $r_e = r - i \geq 0$ for $0 \leq t \leq T$, and $r_e = -i < 0$ for $t > T$. Obviously, for analyzing the rising overland flow hydrograph, one may use the previously discussed models (Sect. 2.2). Rivlin and Wallach (1995) and Singh (2002b) showed that the movement of the edge of the free surface is described by the characteristic equation:

$$t(x_d) = T + \frac{1}{i}\left[\frac{ix_d}{\alpha}\left(\frac{r-i}{r}\right)\right]^{1/n}. \tag{2.31}$$

In this period, $r = 0$ and, for $x > x_d$, water depth $h = h(x, t)$ is determined by the equation

$$t - T = -\frac{h}{i} + \frac{1}{i}\left[\frac{(r-i)(\alpha h^n + ix)}{\alpha r}\right]^{1/n}, \tag{2.32}$$

which is of transcendent character.

2.4 On an Analytical Approach to Coupled Surface and Subsurface Hydrological Processes

As it was discussed in the previous chapter (Sect. 1.2), the derivation of a catchment unit hydrograph from storm rainfall is often related to two flow components, i.e., surface (Horton overland flow, HOF, or/and Dunne overland flow, DOF) and subsurface (saturated subsurface stormwater flow, SSF). Saturated flow in a hill-slope soil is considered to be initiated when the wetting front from infiltration of rainwater reaches the soil–bedrock interface or a saturated layer above this interface containing the pre-event water. Such lateral movement of water near the base of the soil profile can take a variety of forms (Sect. 1.2.3) and can occur at any stage of rainfall that precedes or accompanies surface runoff generation. Sometimes, extension of the saturated zone due to infiltration of rain water from the surface can generate a variable transient source area (Sect. 1.2.2), neighboring to the hillslope base, which becomes responsible, in turn, for forming the DOF (Freeze 1972a, b; Govindaraju and Kavvas 1991). The relative magnitude of each flow component depends on the interrelation between the rainfall and soil properties as well as antecedent soil profile conditions (Smith and Hebbert 1983; Weiler and McDonnell 2006).

There are several coupled transient mechanisms and processes that make finding closed-form analytical solutions of the problem rather complex and unique: (a) water intake at the soil surface and runoff generation; (b) propagation of the wetting front from the upper interface downwards; (c) vertical drainage of moisture from the upper pore space to the capillary zone located above the water table; (d) lateral saturation flow above the soil–bedrock interface whose water can potentially discharge from saturated zone to the land surface, ponding this surface near the slope base. Even numerical description of such interdependent processes is considered to be among the most complex hydrological problems (Freeze 1972a, b; Govindaraju and Kavvas 1991; Weill et al. 2009).

In the scope of this work, the integrated surface/subsurface flow and transport processes are considered from the viewpoint of conceptual development ignoring many details making analytical description impossible. On the other hand, we concentrate our efforts to create models able to predict rapid response of the subsurface system to rainfall variability rather than stepless and long-term behavior of the subsurface hydrographs. Thus, we focus on the mathematical problem setup that accounts for preferential flow effects (Jarvis 2007; Gerke 2014). Rapid flow along preferential pathways in structured media (both macroporous soils and fractured rocks) is controlled by three major factors: soil structure, matrix potential, antecedent soil moisture, and the input flow pattern and rate. To describe rapid flow through soil and unsaturated (vadose) zone, the term bypass flow is used as well (Gerke 2014).

In this section, as earlier (Sect. 1.4.1), we consider the soil and vadose zone material as a uniform fractured porous continuum, rather than discrete media, assuming that the fracture network is surrounded by a homogeneous low-permeability matrix (dual domain approach). Both domains represent a single overlapping system. In such preferential-flow problem formulation, the term "fractures" can synonymously be used for "macropores" as well. Then, the flow process is investigated to obtain analytical solutions for transient recharge to the water table, with the boundary between the unsaturated and saturated zones represented by subsurface stormflow. Two main approaches will be used to generate recharge function, $w(t)$, which is a source term in the continuity equation for unconfined flow over a steep sloping base (1.2): (1) dual-porosity approach in terms of nonequilibrium models with distributed parameters, allowing flow velocities and water contents during infiltration to be computed (based on the extended Green–Ampt model, Sect. 1.4.1.2); (2) lumped-parameter approach, implying that flow in macropores is triggered when the matrix water content reaches a certain threshold, normally, the saturation limits (Sect. 1.4.1.3).

2.4.1 Subsurface Lateral Flow Equation and its General Solution

A continuity equation for unconfined water flow in porous medium underlined by a steeply sloping impermeable layer has a form of linear differential equation (1.2b) which can be rewritten in the standard form of kinematic wave equation:

$$\phi_n \frac{\partial h}{\partial t} + \frac{\partial q}{\partial x} = w(t), \tag{2.33}$$

where $h = h(x, t)$ is the saturated flow thickness (related to the elevation of the groundwater table) [L]; $q = q(x, t)$ is the subsurface water discharge per unit width [L^2T^{-1}],

$$q = \alpha_s h, \quad \alpha_s = k_s \sin \varphi; \tag{2.33a}$$

$w(t)$ is the rate of recharge induced by vertical downward water movement (infiltration) from the surface [LT^{-1}]; k_s is the hydraulic conductivity [LT^{-1}]; ϕ_n is drainage porosity; φ is the slope.

The model based on a linear kinematic wave formulation, (2.33), (2.33a), has been studied intensively for the past decades, predominantly, assuming a rectangular input rate, $w(t)$, corresponding to a recharge of constant intensity and limited duration, T. Moreover, Fan and Bras (1998), Troch et al. (2002) showed how variable width, slope angle, and the soil mantle can be included in the model framework for the computation of subsurface flow through hillslopes of arbitrary geometry (in plan and profile, including convergent and divergent planforms) using a hillslope storage dynamics theory. Some analytical approaches have been developed also for the computation of return flow or saturation excess overland flow (DOF) induced by the rise of water table, which may intersect the ground surface (Fan and Bras 1998).

The model (2.33) has obvious limitations in its ability to describe the flow on gentle slopes. It can be expected that the model should give a reasonable estimate of water flow on steep hillslopes covered with weathered rock material or highly permeable unstructured soil as it is typical of mountain forested areas. Also in the formulation represented by Eq. (2.33), it is implicitly assumed that flow processes in the two adjacent domains, saturated and unsaturated, are uncoupled; that is, the rise of water table characterizing lateral water flow dynamics cannot affect the vertical flow in the above unsaturated zone, which is responsible for the source term, $w(t)$, in Eq. (2.33).

Equation (2.33) can be restated to be a system of ordinary differential equations:

$$\frac{dt}{\phi_n} = \frac{dx}{\alpha_s} = \frac{dh}{w(t)}. \tag{2.34}$$

The initial and boundary conditions for saturated flow thickness, h, are:

$$h(x,0) = 0, 0 \leq x \leq L; h(0,t) = 0, t \geq 0. \tag{2.35}$$

The initial condition assumes an initially zero water table height along the hillslope. The boundary condition assumes a no-flow boundary at the topographic divide.

The primary focus of the following analysis is on the solution of (2.34) for arbitrary changes in the input rate, $w(t)$. The solution can be represented parametrically for two flow domains:

1. $x \leq x_e = \alpha_s t/\phi_n$

$$x = \frac{\alpha_s}{\phi_n}(t - t_0), \tag{2.36}$$

$$h(x,t) = \frac{1}{\phi_n}\int_{t_0}^{t} w(\xi)d\xi, \quad q(x,t) = \alpha_s h(x,t); \tag{2.37}$$

2. $x > x_e = \alpha_s t/\phi_n$

$$h(t) = \frac{1}{\phi_n}\int_{0}^{t} w(\xi)d\xi, \quad q(t) = \alpha_s h(t). \tag{2.38}$$

For the first domain, h and q are connected with x through the parameter t_0 (2.36). In the second domain, h and q are independent of x. Parameter t_0 characterizes the time that passes between the celerity wave arrival at the observation point x, $t_x = \phi_n x/\alpha_s$, and the current time, t: $t_0 = t - t_x$.

2.4.2 *Lateral Flow Generated by Recharge in a Soil Profile Under Nonequilibrium Conditions*

In works by Beven and Germann (1982), Sloan and Moore (1984), Duffy (1996), Zhang et al. (2006) and many other later publications, several mathematical models for predicting subsurface flow (stormflow) on impermeable sloping soil-bedrock interface have been developed. From the practical point of view, these models are of greater interest in studying the subsurface flow contribution to storm runoff in steep forested catchments. The governing mass balance equations implement water fluxes out of the unsaturated zone represented by homogeneous or macropore media. However the most relevant equations are formulated in terms of an average saturation and a steady flow velocity through the unsaturated zone below the seepage surface. In this regard, to adjust model setup to the field conditions, we will focus on one particular aspect related to the generation of subsurface flow during infiltration in the form of the preferential flow prior to the wetting front reaching the soil–bedrock interface. After the wetting front reaches the bedrock, a shallow flow parallel to the bedrock is generated. We assume that such flow can be described in the framework of the previously introduced kinematic wave concept.

2.4.2.1 Dynamics of the Saturation Profile (Under an Assumption of Unlimited Matrix Capacity)

It is assumed that the wetting front is formed by a moisture parcel from the ponded surface. Such flow regime involves unsteady moisture transfer between neighboring domains, and the process as a whole is nonequilibrium (Gerke 2006; Jarvis 2007). To be specific, as earlier (Sect. 1.4.1.2), such two domains in structured soils will be associated with low-permeability porous matrix and fractures. As follows from (1.34) and (1.34a), the recharge rate in unsaturated structured soils, $w_f(z,t)$, can be determined through an analytic solution of kinematic wave Eq. (1.35) written for effective saturation, $S_f(z,t)$, and thus describing the transient water movement ("bypass flow") in the fracture domain after the start of surface ponding. We will consider the behavior of the saturation function within time interval $t \geq \Omega(z) \geq z/\bar{k}$, measuring the current time from the time coordinate (1.45), $\Omega(z) = z/\bar{k} + \pi\left(\sigma S_b D_{we}/2\bar{k}z\right)^2$, which is determined by the arrival time of the fracture front at the depth z from the upper inlet boundary (1.45). Now, equation (1.35) becomes:

$$\frac{\partial S_f}{\partial t} + \bar{k}\frac{\partial S_f}{\partial z} = -\sigma S_b\sqrt{\frac{D_{we}}{\pi(t-\Omega(z))}}; \tag{2.39}$$

here S_b is the specific surface of the fracture domain; D_{we} is the matrix diffusivity; $\sigma = \Delta\theta_m/\phi_f$ (Sect. 1.4.1.2). Equation (2.39) is valid for duration of the infiltration process which is small in comparison with the time required for the imbibition front (forming at the contact on the fracture network with matrix) to reach the center of the matrix blocks. This is not a strong limitation, since the rainfall events have usually short durations.

The Eq. (2.39) can be reduced to a system of ordinary differential equations:

$$\frac{dt}{1} = \frac{dz}{\bar{k}} = -\frac{dS_f}{\sigma S_b\sqrt{D_{we}/\pi(t-\Omega(z))}}. \tag{2.40}$$

In particular, the second equation in (2.40) is related to changes in the effective saturation along characteristics defined by initial value problems of the first ordinary differential equations in (2.40):

$$\frac{dS_f}{dz} = -\frac{\sigma S_b \sqrt{D_{we}}}{\bar{k}\sqrt{\pi(t - \Omega(z))}}. \tag{2.41}$$

Introduce a new coordinate system (Castaing 1991): $x' = x$, $t' = t - z/\bar{k} + z'/\bar{k}$. Thus, instead of (2.41), we come to the differential identity:

$$dS_f = -\frac{\sigma S_b}{\bar{k}}\sqrt{\frac{D_{we}}{\pi}}\frac{dz'}{\sqrt{(t - z/\bar{k}) - (\Omega(z') - z'/\bar{k})}}. \tag{2.42}$$

Integrating the left- and right-hand sides of identity (2.42) within intervals $[1,\ S_f]$ and $[0, z]$ yields the solution to the problem in the form

$$S_f = 1 - \frac{\sigma S_b}{\bar{k}}\sqrt{\frac{D_{we}}{\pi}}\int_0^z \frac{dz'}{\sqrt{(t - z/\bar{k}) - (\Omega(z') - z'/\bar{k})}}. \tag{2.43}$$

The integral in (2.43) has an analytical representation; therefore, the solution (2.43) can be given in the closed form:

$$S_f = 1 - \frac{2}{\pi}\arctan\left(\frac{\Omega(z) - t_{0z}}{t - \Omega(z)}\right)^{1/2}, \quad t_{0z} = z/\bar{k}, \quad t \geq \Omega(z) \geq t_{0z}, \tag{2.44}$$

or, after some transformations and passage to dimensionless coordinates

$$\eta = \lambda_m t_{0z}, \quad \tau = \lambda_m t, \quad \tau_\xi = \lambda_m \Omega(z), \quad \lambda_m = S_b^2 D_{we}, \tag{2.45}$$

the solution for a *continuous input* (an infinite period of rainfall application) becomes

$$S_f = 1 - \frac{2}{\pi}\arctan\frac{\sqrt{\pi}\sigma\eta}{2\sqrt{\tau - \Omega(\eta)}}, \quad \tau \geq \Omega(\eta), \tag{2.46}$$

$$\Omega(\eta) \equiv \tau_\xi = \eta + \frac{\pi\sigma^2}{4}\eta^2. \tag{2.47}$$

Using the superposition principle, the solution (2.46) can be extended for a *rectangular input* assuming a finite period, T, of rainfall application:

$$S_f = \begin{cases} S_f(\eta,\ \tau), & \Omega(\eta) \leq \tau \leq \tau_0 + \Omega(\eta), \\ S_f(\eta,\ \tau) - S_f(\eta,\ \tau - \tau_0), & \tau > \tau_0 + \Omega(\eta), \end{cases} \tag{2.48}$$

where $\tau_0 = \lambda_m T$.

2.4.2.2 Subsurface Stormflow Characteristics and an Illustrative Example

The model is based on the concept that the vertical input rate to the soil base or the lower saturated zone of the hillslope, w_f, is determined by the dynamics of water stored in the unsaturated zone (Eqs. 1.34 and 1.34a):

$$w_f(S_f) = k_{fs} S_f, \tag{2.49}$$

where k_{fs} is the hydraulic conductivity of the fracture domain at saturation; S_f is the effective saturation. Substituting the solution for the function S_f from (2.46) into (2.49), yields a relationship

$$w_f = k_{fs}\left(1 - \frac{2}{\pi}\arctan\frac{\sqrt{\pi}\sigma\eta}{2\sqrt{\tau\text{-}\Omega(\eta)}}\right), \tag{2.49a}$$

which can be associated with the recharge rate when the wetting front reaches the bedrock surface or the water table (if it is already present), casing a rapid water table rise under continuous rainfall application. This model differs from that of Germann (1990) by the fact that the reference work is based on the assumption that a steady-state regime prevails behind the advancing wetting front (there is a maximum depth to which a wetting front can proceed under given flow conditions) while the solution (2.49) is fully transient with respect to S_f, w_f and $\Omega(\eta)$.

Now, considering $z = z_0$ (the thickness of the soil mantle) and assuming that the recharge flux to the water surface, $w \equiv w_f$, is independent of h (Eq. 2.33), a general solution of the problem (Eqs. 2.36, 2.37, and 2.38) can be represented as:

(1) in the domain $x \le x_e = \alpha_s(t - \Omega(z_0))/\Delta\theta_m$:

$$x = \frac{\alpha_s}{\Delta\theta_m}(t - t_0), \tag{2.50}$$

$$h(x,t) = \frac{\overline{k}}{\sigma}\int_{t_0}^{t}\left(1 - \frac{2}{\pi}\arctan\frac{\sigma t_{0z}\sqrt{\pi\lambda_m}}{2\sqrt{y - \Omega(z = z_0)}}\right)dy, \quad q(x,t) = \alpha_s h(x,t); \tag{2.51}$$

(2) in the domain $x > x_e = \alpha_s(t - \Omega(z_0))/\Delta\theta_m$:

$$h(t) = \frac{\overline{k}}{\sigma}\int_{\Omega(z_0)}^{t}\left(1 - \frac{2}{\pi}\arctan\frac{\sigma t_{0z}\sqrt{\pi\lambda_m}}{2\sqrt{y - \Omega(z = z_0)}}\right)dy, \quad q(t) = \alpha_s h(t); \tag{2.52}$$

here $\overline{k} = k_{fs}/\phi_f$, $\sigma = \Delta\theta_m/\phi_f$, $\Delta\theta_m = \theta_s - \theta^0$. It is assumed that the effective drainable porosity for the lateral flow, ϕ_n, is equal to the initial moisture deficit, $\Delta\theta_m$.

An example of calculation is given in Fig. 2.8. As seen from the plot, the time lag between the rainfall and water-table rise is controlled by the transfer rate coefficient, λ_m, which is a measure of matrix influence on infiltration process. In this illustrative example, there is a delay of about three to six hours between rainfall and water table response.

The accuracy of calculations by (2.50), (2.51), and (2.52) can be improved, if these formulas are considered as transcendent with the substitution of $z = z_0 - h(t)$. This allows the value of recharge flux at the transient water table to be corrected depending on the aquifer thickness $(z_0 - z)$. The calculations show that, for the chosen parameter values, the two approaches can be seen to differ at the initial stages of the infiltration process.

The position of water table at the recession phase, when $t > T$, can be calculated by analogy with the procedure from Eqs. (2.50), (2.51), and (2.52), where the second formula in the solution (2.48) for a rectangular input is used (Fig. 2.9). The application of formulas (2.50), (2.51), and (2.52) shall take into account whether the vertical flux starts declining before quasi steady-state lateral flow is achieved at x $(t \geq x\Delta\theta_m/\alpha_s + \Omega(z_0))$ or not.

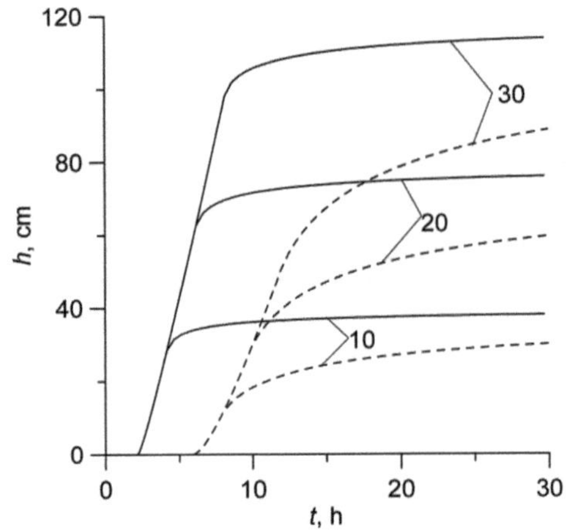

Fig. 2.8 Dynamic response of water table (at different x, numbers at curves, m) to continuous rainfall application ($r = 2$ cmh^{-1}) to sloping surface represented by a fractured-porous media for two values of the transfer coefficient: $\lambda_m = 0.01$ h^{-1} (*solid curves*) and $\lambda_m = 0.05$ h^{-1} (*dashed curves*); other characteristics: $z_0 = 100$ cm, $\bar{k} = 100$ cmh^{-1}, $\sigma = 5$, $\alpha_s/\Delta\theta_m = 500$ cmh^{-1}

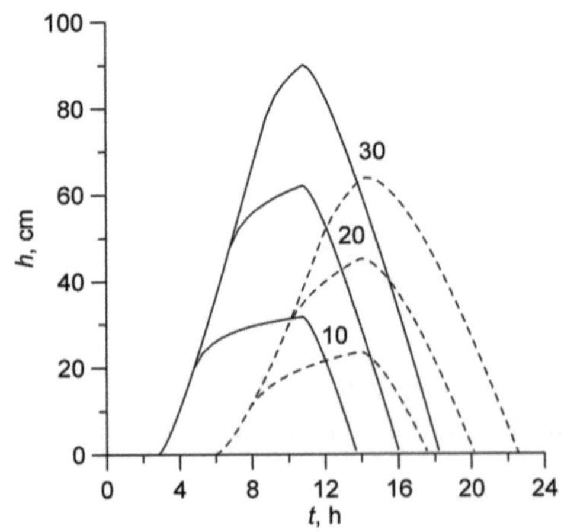

Fig. 2.9 Dynamic response of water table (at different x, numbers at curves, m) to rectangular rainfall input ($r = 2$ cmh^{-1}, $T = 8$ h) to a sloping surface represented by a fractured-porous media for two values of the transfer coefficient : $\lambda_m = 0.01$ h^{-1} (*solid curves*) and $\lambda_m = 0.05$ h^{-1} (*dashed curves*); other characteristics are given in the caption to Fig. 2.8

2.4.3 Lateral Flow Under Triggered Recharge Condition

This section gives the basis for a general expression for finding the discharge of water flowing over a sloped bedrock surface when matrix flow of rainwater through initially unsaturated media is complicated by bypass flow phenomena. Such phenomena are more often observed in cracked soils and unsaturated zone represented by fractured rocks during prolonged rainstorm periods, especially under surface water ponding conditions (Jones and Cooper 1998; Beckers and Alila 2004; Gerke 2014). Subsurface lateral stormflow in the form of preferential flow is most likely to occur as well (Mulholland et al. 1990). Field and modeling studies show that vertical preferential flow contribution to the subsurface runoff becomes greater than matrix flow contribution when certain thresholds in rainfall amount (rainfall depth), soil moisture deficit or matrix potential are exceeded. Antecedent conditions are important in determining flow distribution between matrix and preferential pathway as well.

To determine the triggering thresholds of the phenomena, a lumped-parameter model assuming dual storage capacity of the soil and neglecting the dynamics of moisture front movement is used (Sect. 1.4.1.3). The lumped-parameter model does not contain an infiltration equation and therefore most likely overestimates the degree of soil matrix saturation and the time required for water to flow through macropores or fractures.

In this analysis, we consider outflow components (Eqs. 1.52), activated during rainfall, as recharge in (1.58) for water-saturated domain beneath the surface:

$$w(t) = \begin{cases} w_m, & t < t_m \ (S_m < S_{ms}), \\ w_{ms} + w_f, & t_m \le t < t_p \ (S_m = S_{ms}, \ S_f < S_{fs}), \\ w_{ms} + w_{fs}, & t_p \le t < t(x_d) \ (S_m = S_{ms}, \ S_f = S_{fs}), \\ w_{ms} \exp[-(t-T)/\tau_m] + w_{fs} \exp[-(t-T)/\tau_f], & t \ge t(x_d), \end{cases} \tag{2.53}$$

where w_m, w_f and w_{ms}, w_{fs} are the rates of recharge generated by infiltration in the matrix and fracture domains (all are functions of time, t) under unsaturated and saturated conditions (Sect. 1.4.1.3); $t(x_d)$ is the time when the slope becomes dry in point x_d; this time is determined by formula (2.31) with $i = w_{ms} + w_{fs}$.

Now the solution (2.36), (2.37), and (2.38) can be used to determine the behavior of functions $h(x,t)$ and $q(x,t)$, which characterize the dependence of subsurface stormflow on rainfall intensity. Moreover, since rainfall excess is also determined, [$r_e = r - (w_{ms} + w_{fs}), \ t \ge t_p$], surface runoff hydrograph can also be readily calculated as required (Sect. 2.2).

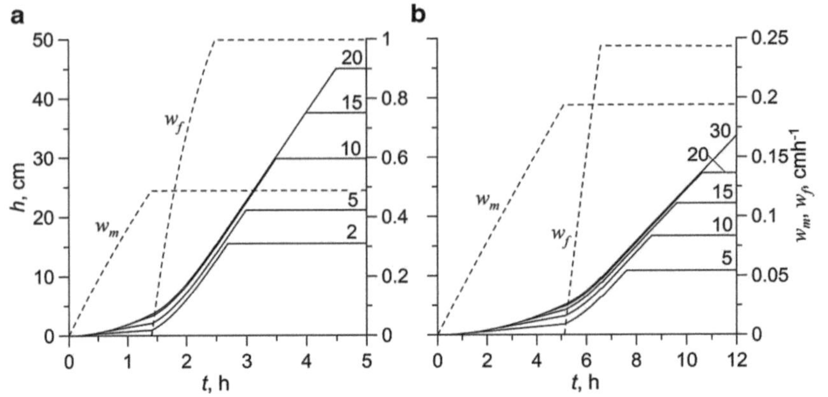

Fig. 2.10 Quick (**a**) and slow (**b**) water table response (h, *solid curves*) and the change in recharge (w_m, w_f, *dashed curves*) due to the triggering effects, generated by continuous rainfall application ($r = 2$ cmh^{-1}) to sloping surface represented by a structured soil. (**a**) $\omega = 0.02$, $z_0 = 50$ cm, $\theta_m^0 = 0.4$, $\theta_{ms} = 0.45$, $\theta_{mc} = 0.4$, $\theta_{fs}^0 = 1$, $\theta_{fc} = 0$, $\tau_m = 5$ h, $\tau_f = 1$ h, $\alpha_s = 100$ cmh^{-1}, $\phi_n = 0.1$; (**b**) $\omega = 0.025$, $z_0 = 100$ cm, $\theta_m^0 = 0.4$, $\theta_{ms} = 0.5$, $\theta_{mc} = 0.4$, $\theta_{fs}^0 = 1$, $\theta_{fc} = 0$, $\tau_m = 50$ h, $\tau_f = 10$ h, $\alpha_s = 50$ cmh^{-1}, $\phi_n = 0.1$. Numbers at the curves are distances, x, m

An example of such calculation (for rising limbs of two subsurface flow hydrographs, $h(t)$) is given in Fig. 2.10. With the parameter values given in the caption, subsurface hydrograph (in terms of water flow depth, Fig. 2.10) has several characteristic segments: (1) $t < \phi_n x/\alpha_s$ (unsteady flow dynamics is due to unsteady kinematics of flow wave and an increase in soil matrix flow recharge rate in time); (2) $\phi_n x/\alpha_s \leq t < t_m$ (the so-called quasi-steady-state phase, when the increase in the lateral flow discharge, $q = \alpha_s h$, is determined by an increase in matrix recharge rate, w_m); (3) $t_m < t < t_p$ (quasi-steady-state phase when the increase in the lateral flow discharge is determined by constant matrix recharge rate and an increase in the fracture (macropore) recharge rate); w_f (4) $t_p < t < t_p + \phi_n x/\alpha_s$ (the linear rise of lateral flow discharge under constant total recharge rate, $w_m + w_f$); (5) $t > t_p + \phi_n x/\alpha_s$ (steady-state subsurface flow phase, which is characterized by unchangeable flow rate and water depth).

Figure 2.10 demonstrates that subsurface flow is a function of a range of soil characteristics. Plots in the figure show the dominance of subsurface stormflow on the hydrological response of the hillslope at the initial phase of the continuous rainfall.

As it is expected, the presence of preferential flow significantly accelerates the rate of water table rise after rainfall application; however, after ponding of the land surface, subsurface runoff may contribute less to the total runoff than overland flow.

References

Agiralioglu N (1981) Water routing on diverging–converging watersheds. J Hydr Div ASCE 107(8):1003–1017
Agiralioglu N (1984) Effect of catchment geometry on time of concentration. Proc Urban Storm Drainage 1:177–184
Agiralioglu N (1988) Estimation of the time of concentration for diverging surfaces. Hydrol Sci 33(2):173–179
Baiamonte G, Agnese A (2010) An analytical solution of kinematic wave equations for overland flow under Green–Ampt infiltration. J Agric Eng Riv Ing Agric 1:41–48
Beckers J, Alila Y (2004) A model of rapid preferential hillslope runoff contributions to peak flow generation in a temperate rain forest watershed. Water Resour Res 40(3)
Beven KJ (1981) Kinematic subsurface stormflow. Water Resour Res 17(5):1419–1424
Beven KJ, Germann PF (1982) Macropores and water flow in soils. Water Resour Res 18(5): 1311–1325
Campbell SY, Parlange JY, Rose CW (1984) Overland flow on converging and diverging surfaces–kinematic model and similarity solutions. J Hydrol 67:367–374
Castaing R (1991) Un modele simple pour la migration de radionucléides par transport colloidal dans un milieu fracture. J Hydrol 125:55–92
Chow VT (1959) Open-channel hydraulics. McGraw-Hill, New York, p 680
De Lima JLMP, van Der Molen (1988) An analytical kinematic model for rasing limb of overland flow on infiltrating parabolic shaped surface. J Hydrol 104:363–370
Downer CW, Ogden FL (2004) GSSHA: a model for simulating diverse streamflow generating processes. J Hydrol Eng 9(3):161–174
Duffy CJ (1996) A two-state integral-balance model for soil moisture and groundwater dynamics in complex terrain. Water Resour Res 32(8):2421–2434
Eagleson PS (1970) Dynamic hydrology. McGraw-Hill, New York
Fan Y, Bras RL (1998) Analytical solutions to hillslope subsurface storm flow and saturation overland flow. Water Resour Res 34(4):921–927
Freeze R (1972a) A role of subsurface flow in generating surface runoff. Base flow contribution to channel flow. Water Resour Res 8(3):609–623
Freeze R (1972b) A role of subsurface flow in generating surface runoff. Upstream source areas. Water Resour Res 8(5):1272–1283
Gerke HH (2006) Review article: preferential flow descriptions for structured soils. J Plant Nutr Soil Sci 169:382–400
Gerke HH (2014) Bypass flow in soil. In: Glinski J, Horabik J, Lipiec J (eds) Encyclopedia of agrophysics (Encyclopedia of Earth Sciences Series). Springer, pp 100–105
Giraldez JV, Woolhiser DA (1996) Analytical integration of the kinematic equation for runoff on a plane under constant rainfall rate and Smith and Parlange infiltration. Water Resour Res 32(11):3385–3389
Govindaraju RS, Jones SE, Kavvas ML (1988) On the diffusion wave modeling for overland flow. Solution for steep slopes. Water Resour Res 24(5):734–744
Govindaraju RS, Kavvas ML (1991) Dynamics of moving boundary overland flows over infiltrating surfaces at hillslopes. Water Resour Res 27(8):885–889
Govindaraju RS, Kavvas ML, Jones SE (1990) Approximate analytical solutions for overland flows. Water Resour Res 26(12):2903–2912
Guo JCY (1998) Overland flow on a pervious surface, IWRA. Int J Water 23(2):1–8
Jarvis NJ (2007) A review of non-equilibrium water flow and solute transport in soil macropores: principles, controlling factors and consequences for water quality. Eur J Soil Sci 58:523–546
Jones HK, Cooper JD (1998) Water transport through the unsaturated zone of the Middle Chalk: a case study from Fleam Dyke lysimeter. In: Robins NS (ed) Groundwater pollution. Aquifer recharge and vulnerability. Geological Society, London, pp 117–128

Julien PY, Simons DB (1985) Sediment transport capacity of overland flow. Am Soc Agric Eng 28(3):755–762
Hjelmfelt AT Jr (1978) Influence of infiltration on overland flow. J Hydrol 36:179–185
Leu JM, Liu CL (1988) Overland flow computation with the characteristics method for a kinematic catchment model. Water Resour Manag 2(4):269–288
Luce CH, Cundy TW (1992) Modification of the kinematic wave–Philip infiltration overland flow model. Water Resour Res 28(4):1179–1186
Mulholland PJ, Wilson GV, Jardine PM (1990) Hydrogeochemical response of a forested watershed to storms: effects of preferential flow along shallow and deep pathways. Water Resour Res 26(12):3021–3036
Parlange JY, Rose CW, Sander G (1981) Kinematic flow approximation of runoff on a plane: an exact analytical solution. J Hydrol 52:171–176
Rivlin J, Wallach R (1995) An analytical solution for the lateral transport of dissolved chemicals in overland flow. Water Resour 31(4):1031–1040
Rose CW, Parlange JY, Sander GC et al (1983) Kinematic flow approximation to runoff on a plane: an approximate analytical solution. J Hydrol 62:363–369
Sabzevari T, Saghafian B, Talebi A (2013) Time of concentration of surface flow in complex hillslopes. J Hydrol Hydromech 61(4):269–277
Sander GC, Parlange J-Y (2000) Comment on "Analytical integration of the kinematic equation for runoff on a plane under constant rainfall rate and Smith and Parlange infiltration" by Giráldez JV and Woolhiser DA. Water Resour Res 36(3):825–826
Sander GC, Parlange JY, Hogarth WL (1990) Kinematic flow approximation to runoff on a plane: solution for infiltration rate exceeding rainfall rate. J Hydrol 113(1–4):193–206
Sander GC, Rose CW, Hogarth WL et al (2009) Mathematical soil erosion modeling. Mathematical models. Encycl Life Support Syst (EOLSS) 2:389–439
Sherman B, Singh VP (1976) A distributed converging overland flow model. Mathematical solutions. Water Resour Res 12(5):889–896
Shokoohi A, Saghafian B (2012) A semi analytical solution for rising limb of hydrograph in 2D overland flow. Int J Civil Eng 10(1):43–50
Singh VP (1996) Kinematic wave modeling in water resources: surface-water hydrology. Wiley-Interscience, New York, p 1400
Singh VP (1997) Kinematic wave modeling in water resources: environmental hydrology. Wiley-Interscience, New York, p 830
Singh VP (2002a) Kinematic wave solutions for pollutant transport by runoff over an impervious plane, with instantaneous or finite-period mixing. Hydrol Process 16:1831–1863
Singh VP (2002b) Kinematic wave solutions for pollutant transport over an infiltrating plane with finite-period mixing and mixing zone. Hydrol Process 16:2441–2477
Singh VP, Woolhiser DA (1976) A nonlinear kinematic wave model for watershed surface runoff. J Hydrol 31:221–243
Sloan PG, Moor ID (1984) Modeling subsurface stormflow on steeply sloping forested watersheds. Water Resour Res 20(12):1815–1822
Smith RE, Hebbert RHB (1983) Mathematical simulation of interdependent surface and subsurface hydrologic processes. Water Resour Res 19(4):987–1001
Stone JJ, Lane LJ, Shirley ED (1992) Infiltration and runoff simulation on a plane. Trans ASAE 35:61–170
Troch P, van Loon E, Hilberts A (2002) Analytical solutions to a hillslope-storage kinematic wave equation for subsurface flow. Adv Water Res 25:637–649
Weiler M, McDonnell JJ (2006) Testing nutrient flushing hypotheses at the hillslope scale: a virtual experiment approach. J Hydrol 319:339–356
Weill S, Mouche E, Patin J (2009) A generalized Richards equation for surface subsurface flow modeling. J Hydrol 366:9–20
Wooding RA (1965) A hydraulic model for the catchment-stream problem: kinematic wave theory. Hydrology 3(3):254–267

Woolhiser DA, Liggett JA (1967) Unsteady, one-dimensional flow over a plane – the rising hydrograph. Water Resour Res 3(3):753–771

Zhang GP, Savenije HHG, Fenicia F et al (2006) Modelling subsurface storm flow with the Representative Elementary Watershed (REW) approach: application to the Alzette River Basin. Hydrol Earth Syst Sci 10:937–955. www.hydrol-earth-syst-sci.net/10/937/2006/

Chapter 3
Models of Dissolved Component Transport at the Hillslope Scale

Rain water on the landscape of a drainage basin can be contaminated by substances that have accumulated on the surface of soil or in its top layer, thus making water flow the major transport and redistribution factor of chemicals (solutes, chemical components, contaminants) at the solid and air interphase. In the rain periods, the contamination hazard is the greatest for surface water streams and bodies. In agricultural regions, fertilizers and pesticides are washed out from fields. In urbanized areas, surface runoff supplies surface waters with dissolved oil products, combustion products of transport fuel, heavy metals, as well as bacteria-polluted waters from emergency sewage spills. A specific class of problems is associated with forecasting radionuclide washout from zones of radioactive pollution, i.e., the areas subject to fallouts of gas-aerosol emissions from facilities of nuclear industry or power engineering, primarily, during emergencies, as well as areas of emergency spills of liquid radioactive wastes.

Chemicals of different origin may occur in soil or on its surface as solutes or in adsorbed state, that is, as components of fixed mineral phase (porous matrix) or on mobile particles (suspensions and colloids) (Fig. 3.1). Thus, the transport of solutes across the landscape involves both dissolved components and those adsorbed by mobile particles (suspension). Dissolved and particulate matter can be transported downslope with the rain water moving over the land surface and/or downslope through the soil layer (troughflow, e.g., in form of stormflow), as well as in the vertical direction in the soil (percolation, infiltration). Stormflow occurs most commonly in forested areas in the wet seasons in a drainage basin where the rocks underlying the soil are impermeable (Sects. 1.2.3 and 2.4.1).

When water begins to build up on the soil surface, the transport of contaminants in the unsaturated zone and the aquifer interacting with it becomes more active. This process activates the paths of rapid transport of solutes in the unsaturated zone, which are due to its heterogeneity, that is, macroporosity, double storage capacity, and permeability. On the other hand, overland flow generated by saturated excess mechanism within variable source areas can also be responsible for solute transport. In this case, after water table intersects the land surface and the flowing water

V.G. Rumynin, *Overland Flow Dynamics and Solute Transport*, Theory and Applications of Transport in Porous Media 26, DOI 10.1007/978-3-319-21801-4_3

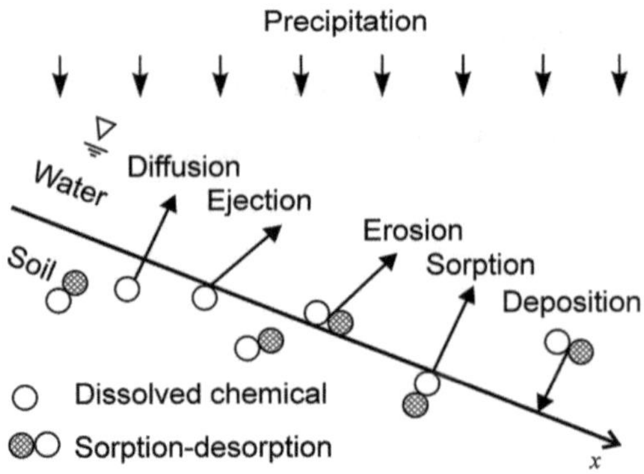

Fig. 3.1 Schematic representation of mass exchange process between soil and hillslope flow (Shi et al. 2011)

becomes surface water, contaminated soil immediately distributes solutes to overland flow region (Govindaraju 1996).

The description of the process as a whole requires interdisciplinary approach. Thus, the model being developed shows all features of *hydrological* models, since they are based on a physicomathematical description of surface water movement over land slopes. However, as far as the liquid/solid interface is represented by permeable porous material (soil), the surface water interacts with subsurface environment where *hydrogeological* processes dominate.

As mentioned above, chemicals migrate in the runoff water either in solution or on dispersed particles, that is, erosion products of soil materials. Therefore, in the former case chemical transport is associated with *liquid runoff*, while in the latter, with *solid/sediment runoff*. From this viewpoint, it is clear that the full formulation of the problem implies the mathematical formalization of three categories of processes:

(a) hydrodynamic flow (movement of rain water over the soil surface);
(b) mass transfer, governed by diffusion, infiltration, and erosion impact of raindrops and water flow onto the soil;
(c) solute transport (water transport of dissolved components and suspended particles – carriers of those components in adsorbed state).

In this chapter, we consider the transport of *conservative components* in dissolved state both in the surface runoff and in the shallow aquifer (during periods of stormflow production). We will focus on different concepts of the mass exchange at the interface of soil and runoff water and discuss mechanisms responsible for such chemical emission. The main emphasis will be on the transient effects that arise from both the unsteady-state surface and subsurface flow conditions and kinetics of solute transfer from soil solution to rainfall-induced runoff water. Traditional approaches for development of analytical solutions of coupled flow and solute transport problems are supported by illustrations of how published or obtained results of concern are consistent with numerical modeling. A quantitative description of the motion of *reactive components* with solid runoff is given in Chap. 4.

3.1 Mathematical Problem Formulation

The variable rainfall conditions (short duration of the surface wetting by rain following a period of soil drying through evaporation) determine the transient effects of rainfall–runoff–infiltration partitioning and chemical response of catchments to excess precipitation. Therefore special attention shall be given to the coupled transient flow and solute transport analysis to make the physical and mathematical problem more precisely formulated.

3.1.1 Basic Representation of the Solute Mass and Fluid Balance Equations

For describing solute transport by runoff water, it is very common to solve the depth-averaged solute transport equation using hydraulic (flow) characteristics, such as water depth, h, and discharge, q, from the basic flow continuity equation. The reason for this is the physical assumption that, for low concentrations, the solute dynamics does not influence the flow behavior. Thus, a general enough representation of the governing system of differential equations includes unsteady mass conservation equation

$$\frac{\partial (AC)}{\partial t} + \nabla(QC) - F_s = 0, \tag{3.1}$$

and unsteady flow continuity equation (Sect. 2.1)

$$\frac{\partial A}{\partial t} + \nabla Q - F_w = 0, \tag{3.2}$$

where C ($\equiv C_d^r$ – see Sect. 4.1) is the concentration of a solute (dissolved chemical) [ML^{-3}]; A is cross section aria of the overland flow [L^2]; Q is the total discharge [L^3T^{-1}]; F_w [L^2T^{-1}] and F_s [$ML^{-1}T^{-1}$] are water and solute source/sink functions, associated with atmospheric precipitation, infiltration, dissolution of matter (minerals and salts), deposited onto land surface or occurring as components of soil solution; all variables are functions of coordinates and time. In this formulation, dispersion effects and hydrodynamic inertia of the flow are neglected.

In the particular case of 1D chemical transport in runoff water flowing over pervious soil surface, Eqs. (3.1) and (3.2) can be rewritten for a conventional flow band with a unit width $B = \text{const} = 1$, where $A = Bh$, $Q = Bq$, in the following manner (Wallach et al. 2001; Singh 2002b; Johnson and Zhang 2007; Walter et al. 2007; Turnbull et al. 2010):

$$\frac{\partial hC}{\partial t} + \frac{\partial qC}{\partial x} = J_s + rC_r, \tag{3.3}$$

$$\frac{\partial h}{\partial t} + \frac{\partial q}{\partial x} = r - i, \tag{3.4}$$

where C_r is the concentration of a chemical in precipitation [ML^{-3}]; $h = h(x, t)$ is the current water depth [L]; $q = q(x, t)$ is the flow discharge in the x direction per unit width [L^2T^{-1}], $q = uh$, u is the flow rate [LT^{-1}]; r is the rainfall rate [LT^{-1}]; i is the infiltration rate [LT^{-1}]; J_s is the total solute flux from the land surface or/and soil water in runoff; in a simplified mathematical setting, J_s is considered as a function of only time, $J_s(t)$ [$ML^{-2}T^{-1}$].

As can be seen, Eqs. 3.1, 3.2, 3.3, and 3.4 contain characteristics of surface flow, A (or h) and Q (or q), which, in general case, are functions, depending on spatial coordinates and time. Those equations are supplemented by appropriate initial and boundary conditions. It is crucial that rainfall rate is highly variable in time; nevertheless, analytical solutions obtained for a stepwise rainfall input function are quite functional for the purposes of evaluating key mechanisms related to the effect of rainfall on runoff contamination. In its turn, it is reasonable to split the analysis of solute transport into several sub-tasks related to different phases of the runoff dynamics, using ideas gained from the kinematic wave theory (Sects. 2.2 and 2.3).

3.1.2 Main Concepts of the Mass Exchange at the Interface of Soil and Runoff Water

After surface ponding takes place, soil surface and soil solution are able to transfer chemicals to surface water. An additional source of the chemicals can be the rainwater itself. Runoff water may carry chemicals, such as fertilizers and pesticides, as well as heavy metals and radionuclides, to wherever it drains. The major interactions in precipitation period are taking place in a thin surface layer not more than a few millimeters in thickness. This accounts for the higher sensitivity of the process to spatial variations of soil and landscape characteristics and makes it difficult to study its physical parameters in experiments.

The exchange at the interface between the media is controlled by a number of natural mechanisms and characteristics, including precipitation rate, natural–climatic and landscape conditions, capillary–diffusion (responsible for infiltration) and sorption properties of soil, its micro- and macro-structural features, as well as the chemical forms of matter attachment to soil particles. The mathematical formalization of the process is based on the two distinct approaches (Fig. 3.2):

- Interfacial diffusion-controlled approach (or boundary layer approach);
- Mixing (exchange) layer approach.

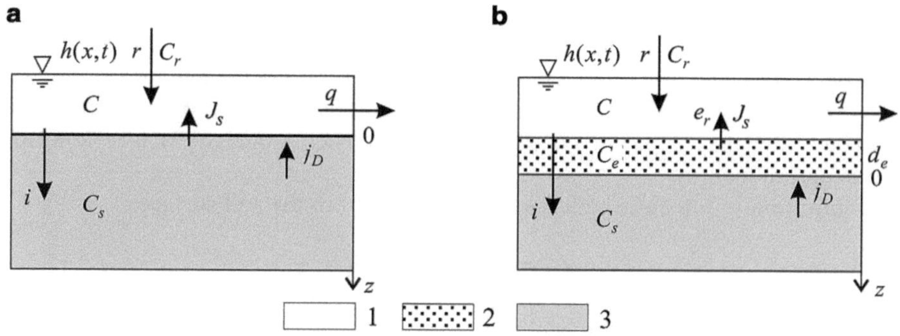

Fig. 3.2 Conceptual models for mass exchange at the interface between runoff and soil: (**a**) diffusion model of boundary layer; (**b**) mixing (exchange) layer model. (*1*) water layer, (*2*) mixing (exchange) layer, (*3*) soil. j_D is the diffusion flux

These models are based on assumptions which are most often working hypotheses, tested in the laboratory but not adequately tested in field studies.

3.1.2.1 Boundary Layer Approach for Transfer of Chemicals from Soil to Surface Runoff

According to the physical concepts underlying boundary-layer diffusion models, the immobile soil solution and the water flow on the surface are separated by some (theoretically, infinitely thin) stagnant water layer, through which solute transfer to runoff by diffusion (Fig. 3.2a).

A model of boundary layer utilizes two scenarios of chemical release from contaminated surface: (a) emission of solutes (chemicals, contaminants) from soil water into surface water (first scenario); (b) dissolution of solid-phase chemicals, which are considered to be laying on the soil surface (second scenario). The first scenario presumes that the solute flux is proportional to the concentration difference between solutes in soil water and surface water (Rivlin and Wallach 1995); the second scenario presumes that solute flux is controlled by the saturation deficiency of the solution that forms the runoff depth (Downer and Ogden 2004). Both scenarios provide the same description for the mass flux, which serves as a mass transfer function in case of absence of infiltration ($i = 0$):

$$J_s = k_e(C_s - C), \tag{3.5}$$

where C_s is the concentration of a chemical in the soil solution at the soil surface interface (first scenario) or the aqueous solubility of the chemical (second scenario) [ML^{-3}]; C is the concentration of a chemical in the surface runoff; k_e is the mass transfer coefficient between the soil surface and the surface runoff [LT^{-1}], the structure of which for the two characteristic hydrodynamic regimes of water flow (laminar or turbulent) can be derived from the boundary layer theory (Wallach et al. 1989). The value of the mass transfer coefficient depends on a number of factors, including the diffusivity of the solute in water, mixing induced by rainfall kinetic energy, and other flow parameters such as runoff depth and mean velocity. The soil surface conditions and the soil properties would also be expected to influence the mass transfer coefficient. The transfer model (3.5) is a lumped-parameter model.

Infiltration ($i > 0$ case) reduces the mass flux from the soil surface:

$$J_s = k_e(C_s - C) - iC. \tag{3.5a}$$

Chemical emission from the soil pore solution Combining the Eqs. 3.3, 3.4 and 3.5a, we obtain a coupled flow and solute transport equation

$$h\frac{\partial C}{\partial t} + q\frac{\partial C}{\partial x} = k_e(C_s - C) - rC + rC_r, \tag{3.6}$$

which is consistent with Walton et al. 2000. The same mathematical formulation for the hydrological process of rainfall-infiltration on a contaminated slope was suggested by Singh (2002b) for a modified Akan (1987) washout model, $J_s = Kh - iC$ (K is a mass transfer constant [$ML^{-3}T^{-1}$]).

Formally, the concentration C_s can be assumed constant meaning unchangeable content of a solute in the soil at the phase boundary ($z = 0$). In such formulation, with C_s assumed equal to the soil surface concentration in the absence of infiltration, the equation (3.6) is "self-sufficient". Moreover, to avoid some uncertainty in the determination of the initial concentration condition and adjust the mathematical model to the experimental setup and data analysis, Eq. (3.6) can be recasted to one dimension "transport-rate-based equation" (Deng et al. 2005).

A more strict formulation implies $C_s = C_s(0, t)$, i.e., the concentration at the phase interface is a function, depending on the solution of transport equations for moisture and solutes in the soil (Shi et al. 2011; Dong and Wang 2013). The mathematical formulation of the problem involves the incorporation into the system (3.3), and (3.4) of the following equation, which approximates the boundary condition at the soil surface, $z = 0$ (Wallach and van Genuchten 1990; Wallach et al. 2001):

$$-J_s(0,t) = J_s(t) = k_e[C_s(0,t) - C] = -\left[iC_s(0,t) - D_s\frac{\partial C_s(0,t)}{\partial z}\right], \tag{3.7}$$

where D_s is the effective diffusion coefficient [L^2T^{-1}], i is the infiltration rate [LT^{-1}]; z is the vertical coordinate (positive downwards, Fig. 3.2a) [L].

Equations (3.6) and (3.7) are written for infiltration under ponding condition, $t > t_p$ (t_p is the time of surface ponding [T]). During this period, a boundary condition for the subsurface model is specified as a zero pressure head (expressed as saturation). As long as there is no ponding (when $t < t_p$), $i = r$ and therefore $q = 0$

(there is no runoff): rain water flux determines a boundary condition for the subsurface model.

Dissolution (or desorption) of chemicals from the contaminated surface The kinetic model (3.5), describing the rate of change in the mass of a nonadsorbable dissoluble chemical, N [ML^{-2}], laying at the soil surface, is represented by the equation

$$-\frac{\partial N}{\partial t} = k_e(C_s - C), \tag{3.8}$$

where C_s is the solubility. Thus, the flux of the dissolved chemical constituent which is transferred in runoff determined by the following equation

$$J_s = -\frac{\partial N}{\partial t} - iC. \tag{3.8a}$$

If the dissolving substance is on the soil surface, at large k_e and $C_s >> N_0/h_{\max}$ ($h_{\max}$ is the maximum flow depth), the transfer model (3.8) degenerates into the "large pulse impulse" model, in which the specific (per unit area) mass flux of the solute from the solid surface can be described by the equality:

$$J_s = \delta(t)N_0, \quad i = 0, \tag{3.9}$$

where N_0 is the initial mass of the chemical per unit surface area [ML^{-2}]; $\delta(t)$ is the delta-function [T^{-1}]. Such formulation of the problem implies that the contaminant, which is uniformly distributed over soil surface, instantaneously passes into the aqueous phase once a water layer of any depth, h, appears on the surface ("nonlimited" dissolution), resulting in an increase in solute concentration to $C = N_0/h$ (Sect. 3.2.1).

The main ideas of the diffusion boundary layer theory can also be used for mathematical formalization of mass transfer of reactive components, which occur on the land surface at some (initial) moment. If the transport of components into the soil layer during the time of runoff generation can be neglected ($i = 0$), the solute flux from the surface, as a first approximation, can be described by the expression

$$J_s = -k_e\left(C - K^*N\right) = -\frac{\partial N}{\partial t}, \tag{3.10}$$

which is similar to the equation of kinetic surface sorption; here, C is the current solute concentration in the flowing water [ML^{-3}], N is the current density of surface contamination [ML^{-2}], k_e is the mass transfer coefficient [LT^{-1}], K^* is an entrainment (distribution) coefficient [L^{-1}], whose physical meaning will be elucidated later (Sect. 6.2.1).

Such formalization of the exchange process is best suited to describe the washout from a surface of highly absorbable components, such as radionuclides, which fall onto soil as a result of accidents at nuclear facilities (Sect. 7.1.2).

3.1.2.2 Mixing Layer Approach for Transfer of Chemicals from Soil to Surface Runoff

Basic Assumption The literature on hydrological processes within drainage basins shows the relative popularity of a concept based on the assumption that an active narrow layer exists in the upper part of soil column (Fig. 3.2b), in which rainwater interacts (mixes) with soil solution (Ahuja et al. 1981; Ahuja 1982; Singh 2002b; Gao et al. 2004; Tong et al. 2010). The depth of the mixing (exchange) layer, d_e, is assumed constant, and, in the most general formulation of the problem, chemicals are allowed to enter runoff water from underlying soil layers. The value of d_e is commonly not higher than 10 mm, and, according to estimates made by different authors, most often it is a few mm (Zhang et al. 1999; Shi et al. 2011).

In the simplest model of this type, the mixing of rainwater, which forms water layer on the surface, with soil solution in the active layer, θd_e, is assumed instantaneous and complete, allowing us to write the following homogeneous equation:

$$(h + \theta d_e)\frac{\partial C}{\partial t} + q\frac{\partial C}{\partial x} = r(C_r - C), \quad (3.11)$$

where C is the concentration of a chemical in the runoff water, equal to the concentration of the chemical in the exchange layer (C_e); h is the water depth; θ is the saturated volumetric water content ($\equiv \theta_s$). In the case of equilibrium sorption, parameter θ is replaced by the sum $\theta + K_d\rho_b$, where K_d is the equilibrium sorption distribution coefficient, ρ_b is the dry bulk density of the soil. The possibility to take sorption into account explicitly is an important advantage of this model as compared with the boundary layer model (Sect. 3.1.2.1).

Experimental testing of model (3.11) has shown that, in fact, the concentration in the surface soil layer, C_e, is higher than that in water flow on the surface, C (Snyder and Woolhiser 1985). In this context, it was suggested that the mixing of pore and rain water in the model element $h + d_e$ (Fig. 3.2b) is not complete. Therefore, the further improvement of the model consisted in the incorporation of a mechanism aimed to account for the kinetic character of soil solution and runoff water exchange in flow profile. Such kinetic model can be represented by the following system of solute mass balance equations (Havis et al. 1992; Singh 1997, 2002b):

$$\frac{\partial hC}{\partial t} + \frac{\partial qC}{\partial x} = k_e(C_e - C) - iC + rC_r, \quad (3.12)$$

$$\frac{\partial \theta d_e C_e}{\partial t} = -k_e(C_e - C) + i(C - C_e). \quad (3.13)$$

This model, including elements of the diffusion boundary layer model, discussed above, takes into account the difference between the chemical concentrations in the two media in contact, C and C_e, which determines, in turn, the exchange rate of the solute between the active layer and overland flow (Deng et al. 2005). Equations

(3.12) and (3.13) also reflect the advective exchange of chemicals between the media (flux component iC) and chemical export from the mixing layer into the underlying soil layers due to infiltration (flux component iC_e).

Taking into account the continuity equation for overland flow (3.4), equation (3.12) can be transformed into the form similar to (3.6), and the system of equations (3.12), and (3.13) becomes

$$h\frac{\partial C}{\partial t}+q\frac{\partial C}{\partial x}=k_e(C_e-C)-rC+rC_r, \tag{3.14}$$

$$\frac{\partial \theta d_e C_e}{\partial t}=k_e(C-C_e)+i(C-C_e). \tag{3.15}$$

A Particular Solution A number of solutions for a system of the governing equations (3.14), (3.15) added by the flow and transport equations for the soil column has been derived for studying the chemical transport at the hillslope scale. Some of them are discussed in the second part of this chapter. Here, we focus, for illustrative purpose, on developing a particular analytical solution that may be favorable to an expert analysis of some experimental data.

The considered limiting case, which is often mentioned in the literature, is associated with the assumption that the concentration in the surface runoff is much lower than those in the near-surface zone of the soil (Wallach and van Genuchten 1990; Wallach 1991; Walton et al. 2000). In this case, Eq. (3.15) becomes

$$\theta d_e\frac{dC_e}{dt}=-(k_e+i)C_e, \tag{3.16}$$

which allows it to be solved independently of Eq. (3.14):

$$\overline{C}_e=\frac{C_e}{C_0}=\exp\left(-\frac{k_e+i}{\theta d_e}t\right), \tag{3.17}$$

where C_0 is the initial concentration in the mixing layer. Now the solution of Eq. (3.14) becomes simpler, because it now contains a source term, which is a damped function of time, t, not depending on the space coordinate, x.

As an example, we consider a particular solution of equation (3.14), written in dimensionless form

$$(1-\gamma)\tau\frac{d\overline{C}}{d\tau}+\overline{C}-\beta e^{-\mu\tau}=0, \tag{3.18}$$

and valid at the initial stage of the rainfall event in a domain remote from the water divide line ($x \gg 0$), where water depth linearly depends on time $h=r(1-\gamma)t$ and $\partial C/\partial x=0$; here $\overline{C}=C/C_0$, $\tau=rt$, $\mu=(\beta+\gamma)/\theta d_e$, $\beta=k_e/r$, $\gamma=i/r$. The solution of the ordinary differential equation (3.18) has an integral representation:

$$\overline{C}=\frac{\beta}{(1-\gamma)\tau^{1/(1-\gamma)}}\int_0^{\tau}u^{\gamma/(1-\gamma)}\exp(-\mu u)du, \tag{3.19}$$

which is valid under the condition $\tau < \tau_e = (1-\gamma)^{1/n-1}(rx/\alpha)^{1/n}$. If infiltration can be neglected ($\gamma = 0$), the solution (3.19) can be simplified

$$\overline{C} = \frac{\beta}{\mu\tau}[1 - \exp(-\mu\tau)], \quad \mu = \beta/\theta d_e. \tag{3.20}$$

At $\tau \to 0$ ($t \to 0$) the relative concentration $\overline{C}$, determined by (3.19) and (3.20), tends to the ratio $\beta = k_e/r$, i.e., $C = \beta C_0$. The solutions (3.19) and (3.20) can be of use for the diagnostics and interpretation of laboratory and field data from experiments with artificial tracers applied to the soil surface of plots followed by the application of simulated rainfall.

We performed such expert appraisal for the experimental data given in works by Walton, Walter, Dong and Wang (Walton et al. 2000; Walter et al. 2007; Dong and Wang 2013). Despite the different conditions under which the experiments have been carried out, the obtained breakthrough curves, $C(t)$, which characterize the variations of concentrations of ion-tracers (Br^{-1}, Cl^{-}) at the outlets of the experimental slopes (flumes, plots), are similar: at the very beginning of runoff, the concentration is usually low but then increases quickly to a maximum; next, the concentration starts dropping exponentially. The measured peak concentrations were used to calculate $\overline{C}_{\max}(t) = C_{\max}/C_s$, where C_s is the initial (experimentally determined) concentration of a tracer in the soil solution. We assumed that $\beta = k_e/r \approx \overline{C}_{\max}(t)$. Since the rainfall rate was known in all experiments, approximate variation ranges of the transfer rate coefficient were calculated (cm/h): $k_e = 0.05 - 0.08$ (Dong and Wang 2013); $k_e = 0.02 - 0.07$ (Walton et al. 2000); $k_e = 1.3 - 3.4$ (Walter et al. 2007); the variation range of r was from 0.9 to 6.8 cm/h. Thus, the coefficient k_e may vary within a wide range of values $k_e = N(0.01 - 1)$ cm/h.

Influence of Raindrop Chemical Transfer The models considered above are applicable to the description of exchange processes taking place at interphases, when soil surface is protected against the direct impact of raindrops, for example, by grass cover or a water layer, when the latter forms rapidly enough on the land surface.

In fact, the solute exchange between the soil and runoff water on the surface can be initiated by the mechanical impact of drops on the land surface. The collision of a moving drop with the surface is accompanied by the penetration of rainwater into soil pores and mechanical displacement of mineral grains with respect to one another, altogether contributing to the mixing of rainwater and soil solution. At the same time, part of this mixture is ejected onto the surface, introducing chemicals into runoff water. The manifestation of this interaction mechanism is especially significant at the beginning of rainfall, or in the case of weak inundation of the surface.

Obviously, such process causes soil erosion: the release of the energy of falling raindrops contributes to the detachment of soil particles and loosening the soil (Chap. 4). The collision of drops with soil surface facilitates washing out fine (dust or silt) particles, represented by clay minerals, humic substances, oxides of iron, aluminum, manganese, phosphates, etc. Those particles enter the water flow that forms on the slope and become active transporters of some *absorbable* components

(Herbert and Przepiora 1995). The erosion effect is largest during showers (when precipitation rate exceeds 0.3 in/h). As mentioned above, the mechanical impact of falling drops onto the soil can be largely smoothed in the presence of a dense vegetation cover and mulch.

The ascending solute flux from the mixing layer through the interface, induced by the release of kinetic energy of falling drops can be defined as

$$J_s = k_{er}(C_e - \lambda C), \tag{3.21}$$

where k_{er} is a transfer coefficient characterizing the rate of soil water ejection (release) into runoff due to raindrop impact (raindrop chemical transfer rate) [LT^{-1}]; λC is the concentration of the water entering the mixing layer, $0 < \lambda \leq 1$ (Ahuja and Lehman 1983; Gao et al. 2004): $\lambda = 0$ means that only rainwater containing no solute enters the mixing zone, $\lambda = 1$ means that the water entering the mixing layer shows the same composition as the surface flow water.

The transfer coefficient, k_{er}, depends on the rainfall rate, r, and the physical properties of the soil. Gao et al. (2004) and Walter et al. (2007) suggest using a simple linear relationship between k_{er} and r proposed before in a form:

$$k_{er} = e_r\theta/\rho_b, \quad e_r = a_0 r, \tag{3.22}$$

where e_r is the net rate of soil erosion caused by raindrop impact [$ML^{-2}T^{-1}$] (Sect. 4.1.1); a_0 is the soil erodibility (detachability) coefficient, which determines to what extent the impact of drops can weaken the bonds between mineral–organic complexes [ML^{-3}].

The literary sources give the values of the coefficient a_0, varying within the range 0.3–2 g/cm^3 obtained from the laboratory tests conducted with a simulated rainfall of rather high intensity (Ahuja and Lehman 1983; Gao et al. 2004; Walter et al. 2007). Experiments also show (Gao et al. 2003), that $a_0 = a_m$ is a constant value until some critical flow depth h_0 is exceeded. With an increase in h at $h > h_0$, the values of a_0 show a decrease in the form $a_0 = a_m(h_0/h)^\delta$, $\delta \approx 0.6 - 0.7$, and h_{cr} is about three times the diameter of raindrops (Proffitt et al. 1991). Thus, the equality (3.22) is valid for "shallow" flow. In the case of "deep" flow, when the water layer on soil surface contributes to the protection of the soil and weakens the rain-induced erosion, the model may overestimate the exchange ability of the soil.

Thus, the solute transport with the overland flow is governed by soil erosion characteristics. Moreover, the surface migration of erosion products, that is, suspended particles (suspension solutions), can be associated with transport of contaminants adsorbed on those particles. Therefore, strictly speaking, to describe soil solute migration, one needs more complete physico-mathematical formalization of the process as a whole, taking into account the combined effect of several physical mechanisms (Sects. 4.2 and 4.3). Meanwhile, in the following, we assume the exchange coefficient k_{er} to be constant (given the rainfall rate), and neglect the sorption of solutes by suspended particles that enter the flow due to soil erosion.

The mass balance equation for hillslope flow (3.3), which takes the form

$$\frac{\partial hC}{\partial t} + \frac{\partial qC}{\partial x} = k_{er}(C_e - \lambda C) - iC + rC_r, \tag{3.23}$$

is considered along with the kinematic wave equation (3.4). Combining these two equations, (3.23) and (3.4), yields

$$h\frac{\partial C}{\partial t} + q\frac{\partial C}{\partial x} = k_{er}(C_e - \lambda C) - rC + rC_r. \tag{3.24}$$

The above model does not take into account the diffusion exchange between the runoff water and the mixing layer, since diffusion is considered a process with secondary significance against the background of rain (erosion) reinjection of pore solution. However, this may lead to underestimation of mass exchange at the stage of surface ponding, when both processes become significant. On the other hand, with the drop-erosion effect neglected at the initial runoff at the beginning of rainfall, the rate of the exchange process (controlling pollution intensity) would be underestimated as well.

Also, the mixing layer contributes to the exchange of dissolved chemicals not only with surface water, but also with underlying soil layers ($z \geq 0$, Fig. 3.2b). Such exchange is based on advection induced by water infiltration, as well as diffusion. The mass balance equation in the mixing layer can be presented in a form (Gao et al. 2004; Walter et al. 2007):

$$\frac{\partial \theta d_e C_e}{\partial t} = -k_{er}(C_e - \lambda C) + i(C - C_e) + j_D, \tag{3.25}$$

where $j_D = D_s \partial C_s / \partial z$ is the diffusion solute flux from the underlying soil [$ML^{-2}T^{-1}$]; C_s is the concentration of a chemical in soil water; D_s is the effective diffusion coefficient [LT^{-2}]; i is the infiltration rate [LT^{-1}].

Equations (3.24) and (3.25) are solved along with the advection–diffusion equation written for the concentration function C_s characterizing the solute transport in the underlying soil layer (Walter et al. 2007). For the solute being sorbed, $\theta d_e \rightarrow (\theta + K_d \rho_b) d_e$ (K_d is the equilibrium sorption distribution coefficient for partitioning solute between soil particles and water; ρ_b is the dry bulk density of the soil). Relatively similar, through the incorporation into the transport model of the retardation factor, it can be taken into account chemical sorption within the whole soil profile as well (Freeze and Cherry 1979; Van der Perk 2006; Rumynin 2011). As said above, the incorporation into the model of sorption increases the model applicability and robustness.

One can readily see that the terms that account for the kinetics of solute exchange between runoff water and underlying soil in the mixing layer model (3.23) and in the interfacial diffusion-controlled model (3.12) are similar. They differ only in the physical interpretation of transfer coefficients k_{er} and k_e (Shi et al. 2011). Therefore, a physically sound idea is to combine the two mechanisms in a single (effective) transfer coefficient, k_e^* (Gao et al. 2005),

$$k_e^* = k_e + k_{er}; \tag{3.26}$$

in that study, it is proposed to assume the solute flux through the interphase from the soil side to be proportional to the current solute concentration in soil solution, C_s. The incorporation of a coefficient responsible for the raindrop effect into the effective mass exchange parameter, k_e^*, makes the description of the early behavior of the system more reliable (Gao et al. 2005).

Overall, we can note that the considered solute transfer models, which incorporate a series of kinetic, storage, and sorption parameters (such as, k_e, k_{er}, d_e, K_d) are of a conceptual character. In most cases, the coefficients have been obtained under idealized laboratory conditions. The passage to a field scale requires the development of new effective approaches to studying the process, which is a difficult experimental problem. In this case, an alternative to physically based (distributed) parameters can be an empirical model with effective coefficients incorporated in especially chosen functions or combinations of such functions (Chap. 6).

3.2 Solute Transport in the Rain Water Flowing Over Impermeable Soil Surface

The large number of assumptions and simplifications underlying the development of analytical models restricts their predictive ability under actual field conditions characterized by a wide variability of soil parameters and micro- and macro-scale landscape particularities. However, analytical solutions appear to be the only possible means of obtaining insight into the fundamental nature of the studied process and trends in chemical hydrograph generation at the slope scale. In this section, we closely follow the analytical approaches developed by V.P. Singh (1997, 2002a), R. Wallach with co-authors (1988, 1989, 1990, 1991, 2001) and some other researchers in their pioneer works.

3.2.1 Instantaneous Dissolution of a Chemical From the Soil Surface in Runoff Water

In this formulation of the problem, it is assumed that

- at the initial stage of the rain, a chemical is uniformly distributed (with a known density N_0) over the sloped surface;
- the kinetics of the chemical substance passage from soil surface into water can be neglected, that is, the substance instantaneously dissolves in water;

- the substance is not absorbed by soil;
- the infiltration losses and the diffusion exchange with the underlying rocks (soil) are negligible.

For zero values of the infiltration rate, i, and a chemical's concentration in precipitation, C_r, and with a delta-type source term J_s determined by (3.9), the original system of equations (3.3) and (3.4) becomes:

$$\frac{\partial hC}{\partial t}+\frac{\partial qC}{\partial x}=N_0\delta(t), \tag{3.27a}$$

$$\frac{\partial h}{\partial t}+\frac{\partial q}{\partial x}=r. \tag{3.27b}$$

Combining these two equations, (3.27a) and (3.27b), yields

$$h\frac{\partial C}{\partial t}+q\frac{\partial C}{\partial x}+rC=N_0\delta(t). \tag{3.27c}$$

Analytical solution for Eq. 3.27c can be developed using the method of characteristics (Singh 1997; Singh 2002a). The method involves rewriting Eq. 3.27c as a system of ordinary differential equations for solute transport in terms of the concentration at a distance on the sloped plane and time. Partial solutions of these equations can be adopted for several regions (domains) identified in the t–x diagram (Fig. 3.3), in other words, the solution for (3.27c) is associated with distinct regimes (phases) of water flow in different zones on the plane ($0 \le x \le L$), where the regimes are controlled by ratios of the current process time, t, and the characteristic times, T, t_e (Eq. 2.12) and t_{se} (definition of t_{se} see below).

3.2.1.1 Solute Transport Description When Flow Wave Reaches Equilibrium

In a domain of a *linear dependence* of water depth on time,

$$h=h(t)=rt, \quad q=q(t)=\alpha(rt)^n, \tag{3.28}$$

the partial differential equation (3.27c) becomes

$$\frac{\partial C}{\partial t}+\alpha(rt)^{n-1}\frac{\partial C}{\partial x}+\frac{C}{t}=N_0\frac{\delta(t)}{rt}. \tag{3.29}$$

This equation corresponds to two characteristic equations: the first,

$$\frac{dx}{dt}=\alpha(rt)^{n-1}, \tag{3.30}$$

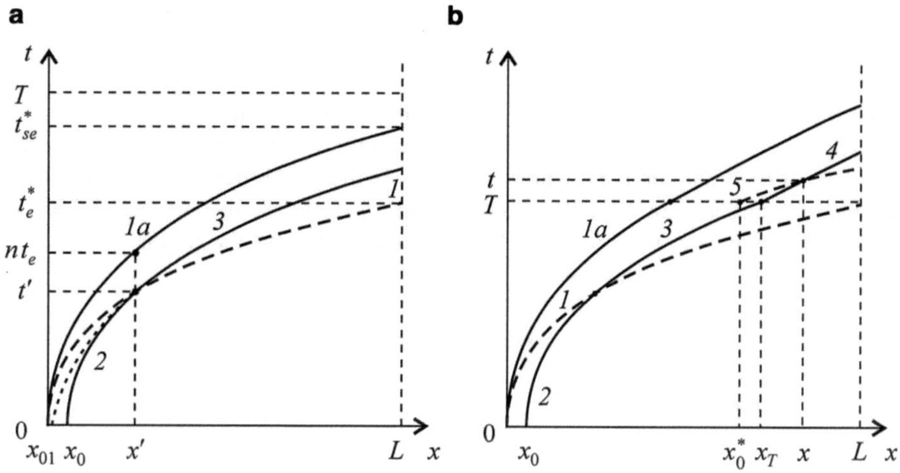

Fig. 3.3 Characteristics: (*1*) the characteristic (2.12), (*1a*) (3.40), (*2*) (3.32), (*3*) (3.43), (*4*) (3.50a), (*5*) (3.50b). (**a**), (**b**) equilibrium ($T > t^*_{se}$) and nonequilibrium ($t^*_{se} > T > t^*_e$) profiles of the transport wave, respectively

yields a family of characteristic curves, along which the concentration is described by the solution of the second characteristic equation (Singh 2002a):

$$\frac{dC}{dt} + \frac{C}{t} = N_0 \frac{\delta(t)}{rt}. \tag{3.31}$$

Integrating (3.30), we can evaluate the solute transport time, $t_s(x, x_0)$ within the interval between points x_0 and x (Fig. 3.3a, curve *2*) from the equation:

$$x - x_0 = \frac{\alpha}{n} r^{n-1} t_s^n(x, x_0). \tag{3.32}$$

The comparison of (3.32) and (2.10) yields

$$t_s(x, x_0) = n^{1/n} t(x, x_0), \tag{3.33}$$

i.e., the time of solute migration between two arbitrary points, x_0 and x, at this (unsteady) stage of the process is $n^{1/n}$ times greater than the time of motion of a hydrodynamic wave between the same points.

The solution of the mass transport equation (3.31) has the form

$$C = \frac{N_0}{rt} = \frac{N_0}{h(t)}, \quad t \le t_e, \tag{3.34}$$

i.e., the concentration, which remains constant in each section $x > \alpha t^n r^{n-1}$, drops with time. The solution (3.34) can be represented in a dimensionless form:

$$\overline{C} = \frac{Ch_x}{N_0} = \frac{\tau_e}{\tau}, \quad \tau \leq \tau_e, \tag{3.35}$$

where $\tau = t/T$, $\tau_e = t_e/T = h_x/rT$; h_x is the maximal flow depth attained at $t = t_e$.

The process in the period when the hydrodynamic wave has a *steady-state profile* at $t > t_e$ (Fig. 3.3a, curve *3*), with $t_{se} < T$ (t_{se} is the solute transport time between point $x = 0$ and a point, x, further downstream), is also described by Eq. (3.27c), in which, however, the right-hand part is set equal to 0, and the initial condition is derived from (3.34). Assuming

$$h = h(x) = (rx/\alpha)^{1/n}, \quad q = q(x) = rx, \tag{3.36}$$

we come to the equation

$$\frac{\partial C}{\partial x} + \frac{h}{rx}\frac{\partial C}{\partial t} + \frac{C}{x} = 0. \tag{3.37}$$

The partial differential equation (3.37) corresponds to a characteristic equation

$$\frac{dt}{dx} = \frac{h}{rx}, \tag{3.38}$$

which yields a family of characteristic curves. Variations of concentration along a characteristic is described by another ordinary differential equation:

$$\frac{dC}{dx} + \frac{C}{x} = 0. \tag{3.39}$$

Integrating (3.38) within intervals $[0, t_s]$ and $[0, x]$ at flow depth, $h = h(x)$, derived from (2.15a), we obtain

$$t_s(x, 0) = \frac{n}{r}\left(\frac{rx}{\alpha}\right)^{1/n}, \tag{3.40}$$

that is, the maximal migration time of solute particles from the boundary $x = 0$ to a point x.

Comparing the obtained result with solution (2.12), one can see that the hydrodynamic wave moves ahead of the transport front and the migration of the substance is taking place against the background of a stationary distribution $h(x)$ all over the flow domain. The time

$$t_s(x, 0) = nt(x, 0) = \frac{n}{r}h(x) \tag{3.41}$$

can be identified with the period of export of the entire mass of the substance that have precipitated within the segment (0, x) of the watershed through a section $h(x)$, provided that the rate of precipitation remains unchanged, i.e., $T \geq t_{se}^*$.

The integration of (3.39) yields the general solution of the problem of concentration distribution in the flow

$$C = A/x, \tag{3.42}$$

where A is an integration constant.

This constant can be determined from the condition of the equality of concentrations, determined by solutions of (3.34) (in which $rt = (rx'/\alpha)^{1/n}$) and (3.42), at the moment of passage of the hydrodynamic wave through the point $(x', t' = t_e)$ of intersection of two characteristics (2.12) and (3.32) (Singh 2002a)

$$\frac{A}{x'} = \frac{N_0}{(rx'/\alpha)^{1/n}} \tag{3.42a}$$

(see Fig. 3.3a), whence the constant A can be obtained. Now the solution (3.42) can be rewritten as

$$C = \frac{N_0}{(rx/\alpha)}\left(\frac{rx'}{\alpha}\right)^{1-\frac{1}{n}}. \tag{3.42b}$$

The solution (3.42b) is valid along the characteristic (3.38), $t = f(x > x') > t_e$. The remaining problem is to relate the dimensionless group rx'/α in the solution (3.42b) with the process time t. The integration of (3.38) yields

$$t - t' = \frac{n}{r}\left(\frac{rx}{\alpha}\right)^{1/n} - \frac{n}{r}\left(\frac{rx'}{\alpha}\right)^{1/n}, \tag{3.43}$$

or, with denotation $(1/r)(rx'/\alpha)^{1/n} = t'$,

$$t = t'(1-n) + \frac{n}{r}\left(\frac{rx}{\alpha}\right)^{1/n}. \tag{3.43a}$$

At $t = 0$, we obtain an expression containing the coordinate of intersection of characteristic (3.43a) with the x-axis (Fig. 3.3a):

$$rt' = \left(\frac{rx'}{\alpha}\right)^{1/n} = \frac{n}{(n-1)}\left(\frac{rx_{01}}{\alpha}\right)^{1/n}. \tag{3.43b}$$

The coordinate x_{01} is auxiliary: it helps us find the equation of characteristic (3.38) and can be further eliminated. Indeed, integrating (3.38), we obtain

$$t - 0 = \frac{n}{r}\left[\left(\frac{rx}{\alpha}\right)^{1/n} - \left(\frac{rx_{01}}{\alpha}\right)^{1/n}\right], \tag{3.43c}$$

whence $(rx_{01}/\alpha)^{1/n}$ is determined. Substituting this relationship into Eq. 3.43b, we obtain

$$\left(\frac{rx'}{\alpha}\right)^{1/n} = \frac{n}{(n-1)}\left[\left(\frac{rx}{\alpha}\right)^{1/n} - \frac{rt}{n}\right]. \tag{3.43d}$$

Thus, the solution of the problem (3.42b) becomes

$$C = \frac{N_0}{(rx/\alpha)}\left\{\frac{n}{(n-1)}\left[\left(\frac{rx}{\alpha}\right)^{1/n} - \frac{rt}{n}\right]\right\}^{n-1}. \tag{3.44}$$

The solution (3.44) can be represented not only as a functional dependence $C = C(x,t)$, but also in other forms, e.g., $C = C(h_x, t)$:

$$C = \begin{cases} \dfrac{N_0}{h_x}\left[\dfrac{1}{h_x}\dfrac{nh_x - rt}{n-1}\right]^{n-1}, & t_e \le t \le nt_e, \\ 0, & t > nt_e, \end{cases} \tag{3.45}$$

where water depth h_x ($h_x \le rt \le nh_x$) is associated with the equilibrium condition and corresponds to the maximal (steady-state) thickness of water layer on the surface at point x. Since the depth h_x is a univalent function of characteristic time t_e, formula (3.45) can be rewritten in another, dimensionless form:

$$\overline{C} = \frac{C}{N_0/h_x} = \begin{cases} \left[\dfrac{n - \tau/\tau_e}{n-1}\right]^{n-1}, & \tau_e \le \tau < n\tau_e, \\ 0, & \tau \ge n\tau_e. \end{cases} \tag{3.45a}$$

As can be seen, at $\tau = \tau_e$, the solution (3.45a) yields an initial concentration distribution in the section $h(x) = h_x$, i.e., solution (3.35). At $\tau \ge n\tau_e$ $(t > t_{se})$ we obtain zero concentration, that is, all the solute that have fallen onto slope surface is being exported through the model section, and the front of clear water, which starts forming in section $x = 0$ at moment $t = 0$, reaches this section.

Now let us consider the case when *precipitation ceases* before the entire mass of the solute is exported from the domain. This is the case when the movement of the rear part of the concentration wave is taking place during the recession of hydrodynamic wave, i.e., under transient hydrodynamic conditions: $T < t^*_{se}$ (Fig. 3.3b).

It has been shown, that under this condition the dynamics of the flow is described by the function $h(x, x_0^*)$ (2.18) and a characteristic equation for time $t(x, x_0^*)$ (2.19). Solute transport equation (3.27c) becomes:

$$\frac{1}{\alpha h^{n-1}}\frac{\partial C}{\partial t} + \frac{\partial C}{\partial x} = 0. \tag{3.46}$$

This equation corresponds to characteristic curves:

$$\frac{dt}{dx} = \frac{1}{\alpha h^{n-1}}, \tag{3.47}$$

along which the concentration follows the law:

$$dC/dx = 0. \tag{3.48}$$

Consider first the solution of the equation of characteristics (3.47), following the transformations suggested by Singh (2002a). Here $h = h(x, t)$ is a function varying over both time and space and defined by the transcendent formula (2.20). To obtain a closed analytical solution, it is more convenient to use equations (2.18) and (2.19), where x_0^* is a parameter.

Suppose that $y = rx_0^*/\alpha$. Now equations for the kinematic wave (2.18) and (2.19) become

$$h = y^{1/n}, \tag{3.49a}$$

$$x = \frac{\alpha y}{r} + n\alpha y^{(n-1)/n}(t - T). \tag{3.49b}$$

Equation (3.47) for the solute transport characteristic becomes

$$\frac{dt}{dx} = \frac{1}{\alpha} y^{-(n-1)/n}. \tag{3.49c}$$

To solve it, we represent the differential as

$$\frac{dx}{dt} = \frac{\partial x}{\partial t} + \frac{\partial x}{\partial y}\frac{dy}{dt}. \tag{3.49d}$$

Next, differentiating (3.49b), we obtain

$$\frac{\partial x}{\partial y} = \frac{\alpha}{r} + (n-1)\alpha y^{-1/n}(t - T), \tag{3.49e}$$

$$\frac{\partial x}{\partial t} = n\alpha y^{(n-1)/n}. \tag{3.49f}$$

Substituting (3.49c), (3.49e) and (3.49f) into (3.49d), we come to an ordinary differential equation

$$\frac{dt}{dy} = -\frac{t - T}{y} - \frac{1}{(n-1)r} y^{-(n-1)/n}, \tag{3.49g}$$

which has the general solution

$$t - T = \frac{A}{y} - \frac{n}{(n^2-1)r} y^{1/n}, \tag{3.49h}$$

where A is an integration constant, or

$$t - T = \frac{A\alpha}{rx_0^*} - \frac{n}{(n^2-1)r}\left(\frac{rx_0^*}{\alpha}\right)^{1/n}. \tag{3.49i}$$

The constant A can be evaluated from the condition: at $t = T$, the equality $x_0^* = x_T$ is to hold (Fig. 3.3b). Now

$$A = \frac{nx_T}{\alpha(n^2-1)}\left(\frac{rx_T}{\alpha}\right)^{1/n}, \tag{3.49j}$$

and expression (3.49i) becomes:

$$t - T = \frac{n}{(n^2-1)r}\left[\frac{x_T}{x_0^*}\left(\frac{rx_T}{\alpha}\right)^{1/n} - \left(\frac{rx_0^*}{\alpha}\right)^{1/n}\right]. \tag{3.50a}$$

In accordance with (3.49b),

$$x = x_0^* + n\alpha(t-T)\left(\frac{rx_0^*}{\alpha}\right)^{(n-1)/n}. \tag{3.50b}$$

Thus, the equalities (3.50a) and (3.50b), can be regarded as a system of equations whose solution relates the coordinates (x, t) with parametric points (x, T), x_T, and x_0^*. We rewrite (3.50a) and (3.50b) in dimensionless form:

$$\frac{(n^2-1)(\tau-1)}{n\overline{h}_0\tau_e} = \left[\frac{\overline{h}_T}{\overline{h}_0}\right]^{n+1} - 1, \tag{3.51a}$$

$$\frac{n(\tau-1)}{\overline{h}_0\tau_e} = \left[\frac{1}{\overline{h}_0}\right]^{n} - 1; \tag{3.51b}$$

here $\tau = t/T$, $\tau_e = t_e/T = h_x/rT$; $\overline{h}_0$ and $\overline{h}_T$ are water depth characteristics normalized by the parameter h_x associated with the equilibrium condition (where $t = t_e$ is the time when equilibrium is attained at point x), $\overline{h}_0 = h_0/h_x$, $\overline{h}_T = h_T/h_x$; $h_0 = h(x_0^*)$ is the water depth at $x = x_0^*$ at moment $t = T$; $h_T = h(x_T)$ is the same at point x_T, characterizing the displacement $(x - x_T)$ of fluid particles over time $t - T$. Thus, $h_x \geq h_T \geq h_0$. The relative decrease of the flow depth can be derived from the approximate equality

$$\overline{h}_0 = \frac{h_0}{h_x} \approx \exp\left(-\frac{t-T}{t_e}\right) = \exp\left(-\frac{\tau-1}{\tau_e}\right), \tag{3.51c}$$

which is valid at $n = 5/3$.

The solution of equation (3.48) can be represented as (Singh 2002a):

$$C = \frac{N_0}{h_T}\left[\frac{1}{h_T}\frac{nh_T - rT}{n-1}\right]^{n-1}. \tag{3.52}$$

Passing to dimensionless groups, we obtain

$$\overline{C} = \frac{Ch_x}{N_0} = \frac{1}{\overline{h}_T}\left[\frac{n\overline{h}_T - 1/\tau_e}{\overline{h}_T(n-1)}\right]^{n-1}, \tag{3.52a}$$

$\overline{h}_T = h_T/h_x$. The system of equations (3.51) is used to find $\overline{h}_T$. Solution (3.52a) is valid within the domain $\tau > 1$ $(t > T)$; solution (3.45a) is used to calculate the concentration function within the dimensionless time interval $\tau_e < \tau \leq 1$.

Thus, the dimensionless concentration function $\overline{C}(\tau)$ depends on dimensionless parameter τ_e, equal to the ratio of the characteristic time t_e, required for equilibrium to be reached in the section h_x, to the duration of precipitation event T. Characteristic plots are given in Fig. 3.4a. Figure 3.4b gives a plot of dimensionless mass flux:

$$\overline{q}_s = \overline{q}\overline{C}, \quad \overline{q} = \frac{\alpha h^n(t)}{\alpha h_x^n} = \left(\frac{h(t)}{h_x}\right)^n, \quad \overline{C} = \frac{Ch_x}{N_0}. \tag{3.53}$$

To determine $\overline{q}$, one can use the transcendent formula (2.20), rewritten as:

$$1 = \overline{q}^{1-1/n}\left(\overline{q}^{1/n} + \frac{n}{\tau_e}(\tau - 1)\right). \tag{3.53a}$$

Interestingly, the inflexion point in the plot of $\overline{C}(\tau)$ has a coordinate τ_e, at which $\overline{C}_{\max} = 1$ (Fig. 3.4a). This point also shows the maximal value of the relative mass flux $\overline{q}_{s,\max} = 1$ (Fig. 3.4b).

3.2.1.2 Solute Transport Description Under Partial Equilibrium of Water Profile

As mentioned above, short-time rainfalls on long enough slopes may have too short duration for an equilibrium water depth profile to form (Sect. 2.2). Suppose that water depth characteristics $h(\overline{x})$ (related to the front of water wave at $t = T$), where $\overline{x}$ is determined by Eq. (2.12) at $t_e = T, \overline{x} = \alpha T^n r^{n-1}$, separates the upper zone of the slope flow, $0 < x \leq \overline{x}$ (here, function $h(x)$ is steady-state, see Eq. (2.15a)), from the time-dependent lower zone, $\overline{x} < x \leq L$, where function $h(t)$ grows linearly (2.11) with time, until moment $t = T$ (Fig. 2.3c, d). In points $0 < x \leq \overline{x}$, function $C(t)$ is determined similarly to the case considered above: formula (3.34) is used for $t \leq t_e$ and formula (3.45) or (3.52), for $t > t_e$.

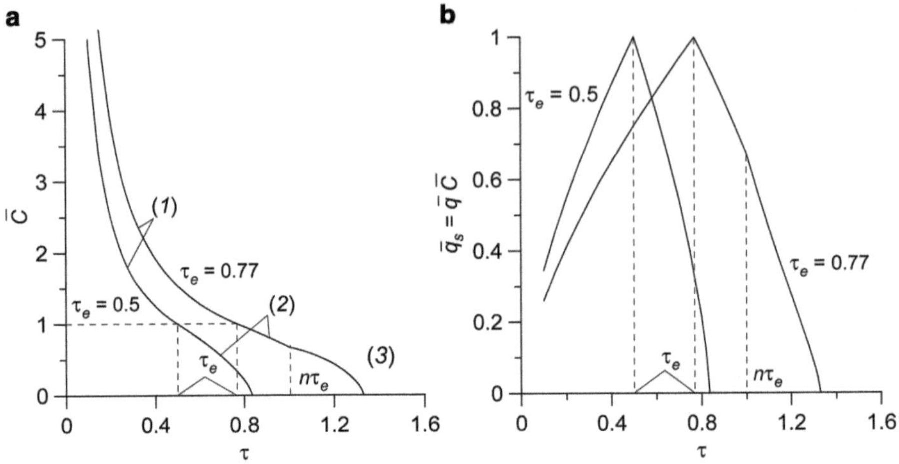

Fig. 3.4 Functions (**a**) $\overline{C}(\tau)$, parenthetical numbers at the curves are the numbers of analytical solutions: (*1*) (3.35), (*2*) (3.45a), (*3*) (3.52a); (**b**) $\overline{q}(\tau)$ at different τ_e

The concentration function in segment $\overline{x} < x \leq L$ (the lower zone of the slope flow) in period $0 \leq t \leq T$ is also constructed based on solution (3.34), which at $t = T$ yields $C = N_0/rT$. A specific feature of flow regime at a later stage is the existence of a time interval $T \leq t \leq t_p(x)$, during which the flow depth, $h(x)$, remains constant until the recession front of water wave reaches this point (Fig. 2.3c, d). Since at $t = T$, the entire segment $\overline{x} < x \leq L$ shows the same concentration

$$C = \frac{N_0}{h(T)} = \frac{N_0}{rT}, \tag{3.54}$$

we will have the same constant concentration for some time in the section under consideration. The chemicals that pass through this section at moment t, have been captured in the section $h(x_T)$, where x_T is determined by

$$t - T = \frac{x - x_T}{\alpha(rT)^{n-1}}. \tag{3.55a}$$

The displacement of point x_T, which characterizes the capture zone, can be regarded as the motion of a solute transport wave. Since the motion of flow wave (coordinate x') is determined by the equality

$$t - T = \frac{x' - \overline{x}}{\alpha n(rT)^{n-1}}, \tag{3.55b}$$

we can calculate the time and coordinate when and where the flow and solute waves will meet,

$$t_{p1} = T\frac{x/\bar{x} + n}{1+n}. \tag{3.56}$$

Further changes in the concentration are determined by the spatial distribution $C(x)$ within the segment $0 < x < \bar{x}$ at $t = T$. The required calculations can be based on formulas (3.45) and (3.52).

3.2.1.3 Examples and Their Comparison with Numerical Modeling Data

An illustration of the tendencies considered above can be the calculation of concentration variations in an outlet section of a flow that forms on a gently sloping plane ($S_0 = 0.01$ m/m) 2000 m long (Sect. 2.2.2). The exponent in the equation (2.3) for flow rate q, is $n = 5/3$. Manning parameter, characterizing surface roughness, $m = 0.05$ s/m$^{1/3}$, i.e., $\alpha = 2$ m$^{1/3}$/s. The rainfall rate $r = 5{\cdot}10^{-6}$ m/s (18 mm/h). The surface pollution density $N_0 = 0.01$ kg/m^2. Two variants with different duration of precipitation event are considered: (a) $T = T_1 = 300$ min, i.e., rainfall duration is greater than the formation time of an equilibrium profile $T > t^*_{se} = nt^*_e = (5/3) \cdot 138.8$ min $= 230.4$ min (see Eq. 3.40), and (b) $T = T_2 = 170$ min, i.e., $T < t^*_{se}$. As before, taking into account the test nature of the considered task, to make it more impressive, values r and T_i were increased in comparison with the mean statistical background values; the selected r and T_i are closer to the values corresponding to heavy-storm periods. The same reason underlay the selection of a rather large hillslope length, L.

In the former case, the initial ($t \le t^*_e$) points of the concentration curve, $C(t)$, are calculated by (3.34), while $C(t)$ for $t^*_e < t \le nt^*_e$ is evaluated by (3.45), see. Fig. 3.5a. The time $t_e(L) \equiv t^*_e$ is calculated by (2.13). In the latter case, (Fig. 3.5b) the calculation of concentration curve segment within the range $t > T = T_2$ is based on solution (3.52). Figure 3.5 also gives numerical solutions of the problem (software package GSSAH, Downer and Ogden 2004). Figure 3.5 highlights a good agreement between the analytical calculations and the data of numerical modeling.

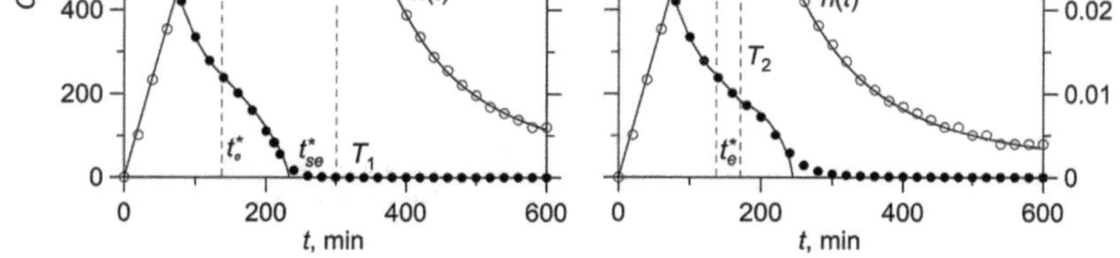

Fig. 3.5 Functions $C(t)$ and $h(t)$ for the outlet of the hillslope ($x = L$): (**a**) $t^*_{se} < T$; (**b**) $t^*_{se} > T$. The *full lines* are analytical calculation, *dots* in plot (**a**) are the result of numerical modeling (code GSSHA, $\Delta x = 10$ m, $\Delta t = 2$ s)

Based on the obtained data, the solute export through the outlet flow section within the identified unit-length flow band can be readily calculated by the formula

$$q_s = Cq = C\alpha h^n. \tag{3.57}$$

The results of calculations are given in Fig. 3.6. The analysis of the plot shows that at an impulse fallout of the solute onto soil surface, the curve $q_s(t)$ has an extreme character with peak concentrations coinciding with the formation time of equilibrium kinematic wave. The integration of the curve that corresponds to long-time fallout (the case of $T = T_1 = 500$ min), yields the solute mass which exactly corresponds to the initial distribution, $N(t = 0) = N_0$. On the other hand, it can be clearly seen that at short-time precipitation (the case of $T = T_2 = 170$ min), a part of solute remains in the watershed.

Finally, the plot in Fig. 3.7, based on numerical modeling data, illustrates the behavior of functions $h(t)$ and $C(t)$ in the case of a partial equilibrium of water depth profile ($T < t_e^*$).

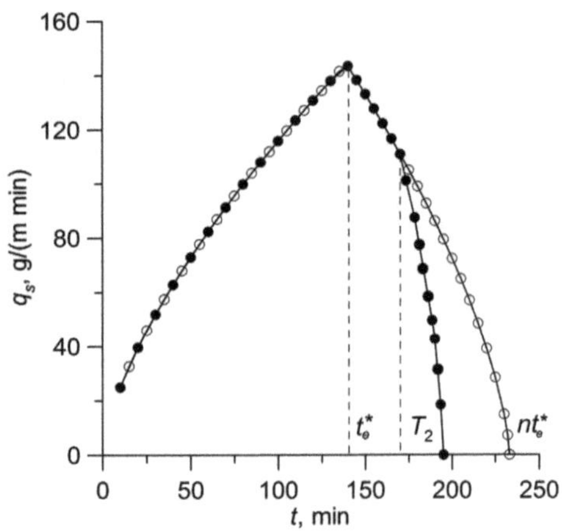

Fig. 3.6 Time variations of solute flux in the outlet section of the hillslope. The *hollow circles* show the first model situation $T > nt_e^*$, the *black circles* show the second model situation $T < nt_e^*$

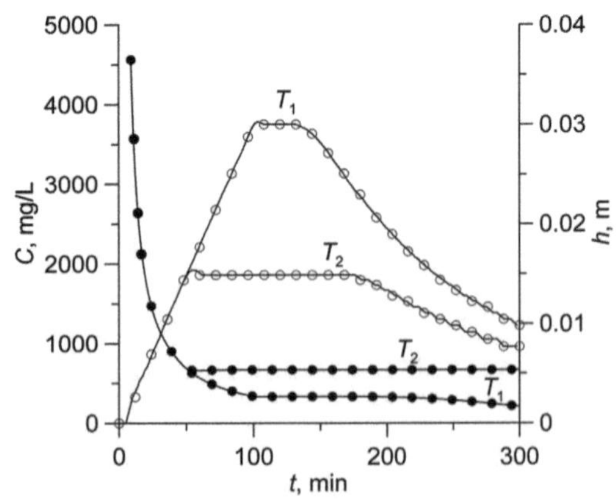

Fig. 3.7 The behavior of functions $h(t)$ (*hollow dots*) and $C(t)$ (*black dots*) in the case of partial equilibrium water depth profile ($T_i < t_e = 138.8$ min). $T_1 = 50$ min, $T_2 = 100$ min

The plots $h(t)$ show plateau-type segments in the period $t > T$, as it was mentioned in Sect. 2.2. Such inertial feature of the system accounts also for the behavior of function $C(t)$ (Sect. 3.2.2).

3.2.2 Kinetically-Controlled Exchange Between Contaminated Soil and Runoff Water

With the kinetics of the transfer of chemicals from the soil pore water (or from the soil surface) into surface water described by equation (3.5) ($i = 0$ case), the governing system of equations consisting of partial differential equations (3.3) and (3.4) is reduced to a single (coupled) partial differential equation:

$$h\frac{\partial C}{\partial t} + \alpha h^n \frac{\partial C}{\partial x} - k_e(C_s - C) + rC = 0, \tag{3.58}$$

where C_s is the concentration of a chemical in the soil solution at the soil surface interface or aqueous solubility of the chemical in the case of dissolution of solid contaminants laying on the soil surface. In this analysis, the concentration C_s is assumed constant over time, and, without a loss of generality, we can assume $C_r = 0$.

The rising stage of the overland flow ($0 \leq t < T$) Consider the behavior of the concentration function under the conditions of equilibrium and partial-equilibrium water depth profiles ($t_e^* < t_{se}^* < T$) (Fig. 3.8).

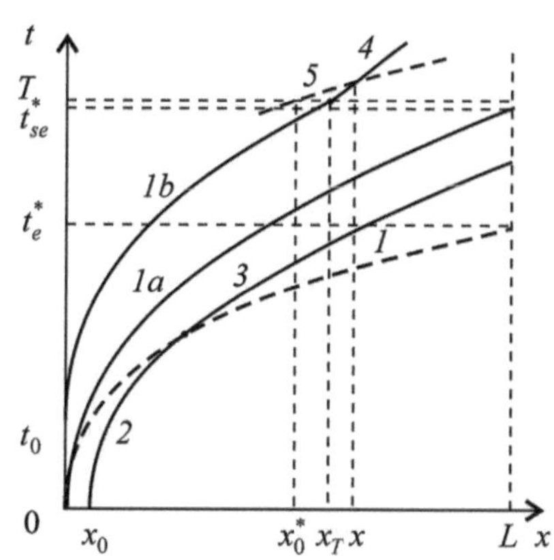

Fig. 3.8 Plots of characteristics: curve *1* is the characteristic (2.12), *1a* – (3.40), *1b* – (3.67), *2* – (3.32), *3* – (3.43), *4* – (3.50a), *5* – (3.50b)

Equation (3.58) corresponds to a characteristic system of equations:

$$\frac{dx}{\alpha h^n} = \frac{dt}{h} = \frac{dC}{k_e(C_s - C) - rC}. \tag{3.59}$$

Attempts to obtain the solution of equation (3.59) with zero initial condition, $h(x, t = 0) = 0$, by the method of characteristics face a problem of singularity. To overcome this problem, we, following Rivlin and Wallach (1995), introduce an initial water depth, $h(x, 0) = h^0$, which can be chosen arbitrarily, in the initial formulation of the problem.

Now, in the domain of *linear growth* of function h with time

$$x > \frac{\alpha}{r}\left[\left(rt + h^0\right)^n - h^{0n}\right], \tag{3.60}$$

the equation of solute characteristic curves becomes

$$x - x_0 = \frac{\alpha}{nr}\left[\left(h^0 + rt\right)^n - h^{0n}\right]. \tag{3.61}$$

Variations of concentration along those characteristics can be described by the second equation of the system (3.59), which can be conveniently rewritten in the dimensionless form

$$\left(\tau + h^0\right)\frac{d\overline{C}}{d\tau} + (1 + \beta)\overline{C} - \beta = 0, \tag{3.62}$$

where $\overline{C} = C/C_s$, $\tau = rt$, $\beta = k_e/r$. Solving this equation with the initial condition $\overline{C}(\tau = 0) = \overline{C}^0$, we obtain:

$$\overline{C} = \frac{\beta}{1 + \beta} + \left(\overline{C}^0 - \frac{\beta}{1 + \beta}\right)\left(\frac{h^0}{\tau + h^0}\right)^{1+\beta}, \quad \beta = k_e/r. \tag{3.63}$$

The limit $h^0 \to 0$ can be seen to lead to the steady-state solution

$$\overline{C} = \frac{\beta}{1 + \beta}, \tag{3.64}$$

i.e., in the domain $x \geq (\alpha/r)(rt)^n$, where the flow is unsteady-state (Fig. 3.8, curve 2), the concentration remains constant over both time and space.

In the domain of the *steady-state distribution* of water depth function, $h(x) = \left[rx/\alpha + \left(h^0\right)^n\right]^{1/n}$, i.e., when the condition

$$x \le \frac{\alpha}{r}\left[\left(rt + h^0\right)^n - h^{0n}\right] \tag{3.65}$$

is satisfied, the equation for solute characteristic curves takes the form

$$\frac{rx}{\alpha} = \left[\left(\frac{rx_0}{\alpha} + h^{0n}\right)^{1/n} + \frac{rt}{n}\right]^n - h^{0n}. \tag{3.66}$$

Because of the kinetics of dissolution process, the concentration distribution exists in both the domain lying below the limiting characteristic curve for the dissolved solute ((3.66), $x_0 = 0$) (Fig. 3.8, кривая *1a*) and in the domain above this characteristic,

$$t(x, t_0) = t_0 + \frac{n}{r}\left[\left(\frac{rx}{\alpha} + h^{0n}\right)^{1/n} - h^{0n}\right]. \tag{3.67}$$

This distribution is described by the ordinary differential equation

$$\left(h_x^n + h^{0n}\right)\frac{d\overline{C}}{dh_x^n} + (1+\beta)\overline{C} - \beta = 0, \quad h_x^n = \frac{rx}{\alpha}, \tag{3.68}$$

as follows from the basic system of equations (3.59), written with the initial water depth, $h = \left[rx/\alpha + \left(h^0\right)^n\right]^{1/n}$ taken into account. Solving this equation, we obtain:

$$\overline{C} = \frac{\beta}{1+\beta}\left[1 - \left(1 - \frac{1+\beta}{\beta}\overline{C}^0\right)\left(\frac{h^{0n}}{h_x^n + h^{0n}}\right)^{1+\beta}\right], \quad \overline{C}\left(h_x^n = 0\right) = \overline{C}^0. \tag{3.69}$$

From this it can be seen that at $h^0 \to 0$, Eq. (3.68) has a single steady-state solution, coinciding with (3.64) for any characteristic (3.67) specified by parameter t_0 (Fig. 3.8, curve *1b*), i.e., the concentration in the domain $t_e^* < t < T$ is also a steady-state function, constant over the space. The solution (3.64) is also valid for the characteristic lying below the line (3.40) (Fig. 3.8, curve *3*). To prove the invariance of solution (3.64) in three ranges $t \le t_e$, $t_e < t \le t_{se}$, and $t > t_{se}$, we can use the solution of the problem developed for a general constant initial water depth case (Rivlin and Wallach 1995) with the subsequent passage to $h^0 = 0$.

An Application of the Steady-State Solution The obtained result can be expertly used to evaluate the natural attenuation of contaminated areas as applied to the conditions of uniform distribution on the land surface (at some initial time, $t = t_0$) of a final mass of a chemical (initially present in solid form) with a density of N_0. It is considered that the chemical change to a dissolved form can be transferred in runoff from the surface through which the vertical infiltration is negligible (Sect. 3.1.2.1).

Since under idealized conditions of continuous precipitation with intensity r, the concentration in runoff water, C, is constant (steady-state solution, Eq. (3.64)), the dissolution kinetic equation (3.8) becomes

$$\frac{dN}{dt} = -\frac{\beta}{1+\beta} C_s r, \tag{3.70}$$

where C_s is the chemical solubility (saturation concentration).

The integration of Eq. (3.70) allows us to obtain a solution for quantifying the effect of reduction of the surface contamination due to runoff:

$$N(t) = N_0 - \frac{k_e r C_s}{k_e + r}(t - t_0). \tag{3.71}$$

Let us assume further that such washout effect is negligible during the short recession phase following the long-duration fallout (or irrigation) phase. Under this assumption, we can rewrite Eq. (3.71) to obtain a solution for prediction of the washout effect caused by a series of storms, approximated by a step-wise periodical function r:

$$N_m = N_0 - k_e C_s \sum_{i=1}^{m} \frac{r_i}{k_e + r_i}(t_i - t_{i-1}). \tag{3.71a}$$

where m is the total number of the fallout periods with known rate r_i that is fixed within the time-interval $t_i - t_{i-1}$.

If we assume that precipitation for a particular region can be characterized by a mean value, r, the "total rain time", t^*, required for the entire mass of the chemical to be washed out of the hillslope surface, can be evaluated:

$$t^* = \frac{k_e + r}{k_e r} \frac{N_0}{C_s}, \tag{3.71b}$$

After this time, the soil contaminant is reduced to negligible concentration.

The obtained results may support and contribute to the proposals related to environment protection issues.

The Falling Stage of the Overland Flow ($t > T$) The migration during the recession of hydrodynamic wave ($t > T$) is described by the equation

$$h\frac{\partial C}{\partial t} + \alpha h^n \frac{\partial C}{\partial x} - k_e(C_s - C) = 0, \tag{3.72}$$

which follows from (3.58) at $r = 0$.

Equation for solute transport along characteristics

$$\frac{dt}{dx} = \frac{1}{\alpha h^{n-1}} \tag{3.73}$$

(corresponding to Eq. (3.72)) has the form

$$\frac{dC}{dx} = \frac{k_e}{\alpha h^n}(C_s - C). \tag{3.74}$$

Here $h = h(x, t)$ is an unsteady-state, spatially varying function, determined by the transcendent formula (2.20) or Eqs. (2.18) and 2.19, where x_0^* is a parameter (see Sect. 3.2.1).

The solution of equation (3.73) is given by formulas (3.50a) and (3.50b). The substitution $y = rx_0^*/\alpha$ (Singh 2002a) transforms Eqs. (2.18) and (2.19) to the form (3.49a)–(3.49b), and Eq. 3.74 becomes

$$\frac{dC}{dy} = \frac{k_e}{\alpha y}\frac{dx}{dy}(C_s - C) = \frac{k_e}{\alpha y}\left(\frac{\partial x}{\partial y} + \frac{\partial x}{\partial t}\frac{dt}{dy}\right)(C_s - C). \tag{3.75}$$

Determining the derivatives in the right-hand part of Eq. (3.75) in accordance with Eqs. (3.49e), (3.49f) and (3.49g), we come to the equation:

$$\frac{dC}{dy} = -\frac{k_e}{y}(C_s - C)\left[\frac{t - T}{y^{1/n}} + \frac{1}{(n-1)r}\right]. \tag{3.76}$$

Substituting $t - T$ from Eq. (3.49i) into Eq. (3.76), we obtain

$$\frac{dC'}{dy} = -\frac{k_e}{y}C'\left[\frac{1}{(n^2 - 1)r} + Ay^{-(1/n+1)}\right], \quad C' = C_s - C \tag{3.77a}$$

(A is a constant determined by Eq. 3.49j). Integrating Eq. (3.77a) yields

$$\ln C' = \frac{k_e}{(n^2-1)r}\ln y - A\frac{k_e n}{n+1}y^{-(n+1)/n} + B, \tag{3.77b}$$

where B is an integration constant.

We rewrite the solution (3.77b) in the explicit form:

$$\ln C' = \frac{k_e}{r}\left[\frac{1}{n^2-1}\ln(rx_0^*/\alpha) - \frac{n^2}{(n^2-1)(n+1)}\frac{(rx_T/\alpha)}{(rx_0^*/\alpha)}\frac{(rx_T/\alpha)^{1/n}}{(rx_0^*/\alpha)^{1/n}}\right] + B. \tag{3.77c}$$

The concentration in point $x_0^* = x_T$ is determined by solution (3.64), i.e.,

$$\ln\frac{C_s}{1+\beta} = \beta\left[\frac{1}{n^2-1}\ln\frac{rx_T}{\alpha} - \frac{n^2}{(n^2-1)(n+1)}\right] + B, \tag{3.77d}$$

therefore

$$B = \ln\frac{C_s}{1+\beta} - \frac{\beta}{n^2-1}\left(\ln\frac{rx_T}{\alpha} - \frac{n^2}{(n+1)}\right). \tag{3.77e}$$

Further algebraic transformations yield a solution of the problem, which can be given in a dimensionless form

$$\overline{C} = \frac{C}{C_s} = 1 - \frac{1}{1+\beta}\left(\frac{\overline{h}_0}{\overline{h}_T}\right)^{\beta n/(n^2-1)} \exp\left\{-\frac{\beta n^2}{(n^2-1)(n+1)}\left[\left(\frac{\overline{h}_T}{\overline{h}_0}\right)^{n+1} - 1\right]\right\}, \tag{3.78}$$

where $\overline{h}_T/\overline{h}_0$ is a function of dimensionless time τ, and, with fixed parameter τ_e, it can be determined from the solution of the system of Eqs. (3.51a), (3.51b).

At $\tau \to 1$, which corresponds to the limit $\overline{h}_T/\overline{h}_0 \to 1$, the solution leads us to the formula (3.64), i.e., it is in agreement with the steady-state phase of the process. At large τ ($\tau >> 1$), solution (3.78) tends to an asymptote $\overline{C} = 1$ ($C = C_s$). Physically, such behavior of the function is due to a decrease in the flow depth ($h \to 0$), which, in the absence of precipitation, supplying fresh water, causes the concentration of soluble salts in solution up to the saturation level C_s.

An illustrative example and finding an approximate solution The above features in the behavior of the concentration function are illustrated by the plot in Fig. 3.9a. The accompanying plot in Fig. 3.9b gives the behavior of the function of dimensionless mass flux, $\overline{q}_s(\tau)$ (3.53).

The numerical solution of the problem confirms the described trends in the process and is in agreement with analytical calculations (Fig. 3.9a). Function $C(\tau)$ shows nonsteady behavior at the first moments $\tau < 0.01$, thus confirming the acceptability of the steady-state solution (3.64) for describing the migration process.

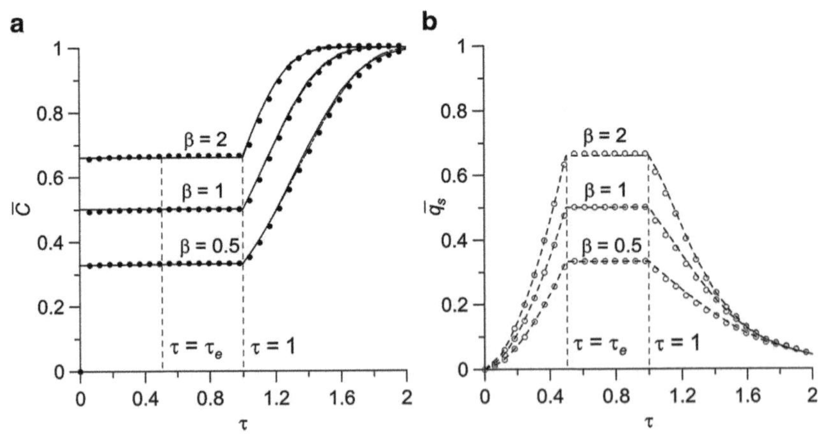

Fig. 3.9 (**a**) Functions $\overline{C}(\tau)$ (*full lines*) calculated by formulas (3.64, $\tau = t/T \leq 1$) and (3.78, $\tau > 1$) (the *dashed line* shows the approximate solution (3.79)), and (**b**) specific mass flux, $\overline{q}_s$ (3.53) (*dashed curves*). Dots are results of numerical modeling (code GSSHA; the parameters of the grid model are $\Delta\overline{x} = 5 \cdot 10^{-4}$, $\Delta\overline{t} = 1.8 \cdot 10^{-3}$). $\tau_e = 0.5$

The approximate equality (3.51c) for parameter $\overline{h}_0$ allows the formula (3.78) to be rewritten as

$$\overline{C}(\tau) = 1 - \frac{1}{1+\beta}\left(1 + \frac{n^2-1}{n}\frac{\alpha(\tau)}{\exp[-\alpha(\tau)]}\right)^{-\beta n/((n+1)(n^2-1))} \exp\left[-\frac{\beta n}{(n+1)}\frac{\alpha(\tau)}{\exp[-\alpha(\tau)]}\right], \quad (3.79)$$

where $\alpha(\tau) = (\tau - 1)/\tau_e = \bar{t} - \overline{T}$ $(\bar{t} = rt/h_m, \overline{T} = rT/h_m)$. Calculations (Fig. 3.9) suggest the high accuracy of the proposed approximation.

3.3 Solute Transport in Surface Runoff Coupled with Infiltration into Underlying Soil

In the previous section, we considered the problem of chemical transport in the overland flow derived under an assumption of very low permeability of the underlying soil, preventing infiltration water losses. In fact, the rate of surface water flow during rainfall decreases due to water imbibition by soil cover, followed by its infiltration in the aeration zone (Sect. 2.3); this should be reflected in the character of concentration functions. In the general case, the infiltration is known to be a nonsteady-state function (Sect. 1.3), and only when rainfall events are long enough for the soil surface to be ponded, the values of function i can be assumed to be independent of overland flow characteristics and moisture content distribution in the unsaturated zone. Nevertheless, some useful trends in overland solute transport process can be established by solving a simplified problem assuming a fixed infiltration rate, $i = \text{const}$.

3.3.1 *Solute Transport with Surface Exchange Kinetics Under a Constant Infiltration Rate*

The solution of the problem can be conveniently related with the two main stages of overland flow, i.e., the rising ($r > 0$) and falling ($r = 0$) stages.

The Rising Stage of the Overland Flow Suppose that the duration of a storm rainfall event is long enough for the formation of a steady-state (equilibrium) hydrodynamic water depth profile, $T > t_e^*$. The partial differential equation of kinematic wave (2.4) corresponds to a system of ordinary differential equations

$$\frac{dt}{1} = \frac{dx}{\alpha n h^{n-1}} = \frac{dh}{r - i}. \quad (3.80)$$

The parametric form of representation of solution (3.80), similar to (2.10) and (2.9), is as follows:

$$t(x, x_0) = (r - i)^{-1}\left[\frac{(r - i)}{\alpha}(x - x_0)\right]^{1/n}, \tag{3.81}$$

$$h(x, x_0) = \left[\frac{r - i}{\alpha}(x - x_0)\right]^{1/n}, \tag{3.82}$$

where x_0 is a parameter, representing the coordinate of intersection of the characteristics curve and the x-axis. The equation of the main characteristic of the water flow, passing through the origin ($x = 0, t = 0$) and dividing the plot of characteristics into two zones (an unsteady-state zone with uniform flow depth distribution and a steady-state zone with non-uniform flow depth distribution), has the form:

$$t(x, 0) = \frac{1}{r - i}\left(\frac{r - i}{\alpha}\right)^{1/n} x^{1/n}. \tag{3.83}$$

Combining mass balance equation for solute transport (3.3) and continuity equation for overland flow over an infiltrating soil surface (3.4) with a source-term equation (3.5a) we come to a coupled equation:

$$h\frac{\partial C}{\partial t} + \alpha h^n \frac{\partial C}{\partial x} = k_e(C_s - C) - rC + rC_r, \tag{3.84}$$

which formally coincides with the equation of solute transport over impervious soil surface (3.58).

Equation (3.84) can be rewritten as a system of ordinary differential equations for solute transport in terms of the concentration as a function of distance and time. Partial solutions of these equations can be adapted to three domains identified in the t–x plane, separated by the main flow characteristic curve (3.83), and the following main characteristic curve for the dissolved chemical

$$t_s(x, 0) = \frac{n}{r - i}\left(\frac{r - i}{\alpha}\right)^{1/n} x^{1/n}. \tag{3.85}$$

Analysis of all particular solutions obtained by the method of characteristics (Sect. 3.2.2, $h^0 \to 0$) and describing the process in the domains

1. $x > \frac{\alpha}{r - i}[(r - i)t]^n$,
2. $x < \frac{\alpha}{r - i}\left[\frac{(r - i)t}{n}\right]^n$,
3. $\frac{\alpha}{r - i}\left[\frac{(r - i)t}{n}\right]^n \le x < \frac{\alpha}{r - i}[(r - i)t]^n$,

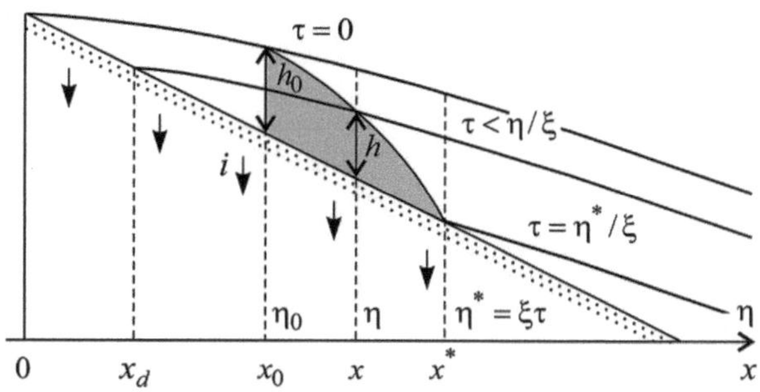

Fig. 3.10 A conceptual sketch showing the main features of the falling stage of the overland flow. Shown in *gray* is the domain of variations of water depth for the case where the front of chemicals started to migrate from point $x_0(\eta_0)$

shows that all solutions are described by a single stationary concentration function (3.64) as in the previous scenario ($i = 0$, Sect. 3.2.2). Notwithstanding the equality of concentrations of chemicals in water, runoff solute flux per unit width, $q_s = Cq$ can be easily shown to be always less than the solute flux, calculated at $i = 0$.

Note that obtained result differs from that of in the work by Rivlin and Wallach (1995) due to lack of coincidence in the solute source-term J_s representation in the governing system of equations (3.3), and (3.4).

The Falling Stage of the Overland Flow After the cessation of precipitation event ($t > T$) the flow condition is unsteady and nonuniform. Moreover, a drying free surface (Fig. 3.10) starts forming: the process shows the presence of a moving boundary, x_d, between the liquid and solid phases, i.e., a contact between the zone where the surface is underflooded and the zone where soil is fully drained. Equations for determining x_d (2.31) and $h(x, t)$ (2.32) are transcendent, which makes it difficult to find an exact solution for equation (3.84) in the concentration function. Therefore, we consider an approximate analytical solution of the problem, which can be derived from the following reasoning.

Suppose that function h in any point, x, of flow decreases linearly from its maximum equilibrium value $h = h_x$, attained at $t = t_e < T$, to the value $h = 0$, and the time of such decrease is determined by the travel time, $\Delta t = t(x) - T$, to the point x of the interface, determined by (2.31), i.e.

$$h = h_x\left(1 - \frac{t - T}{\Delta t(x)}\right). \tag{3.86}$$

After some algebraic transformations, we come to an equation for flow depth:

$$h = \eta - \tau, \tag{3.87}$$

where $\tau = \xi r(t - T)$, $\eta \equiv h_x = [(r - i)x/\alpha]^{1/n}$, $\xi = \gamma/\gamma^{1/n}$, $\gamma = i/r$.

Then the solute transport Eq. (3.84) is transformed as follows:

$$\frac{d\eta}{a(1-\tau/\eta)^n} = \frac{d\tau}{(1-\tau/\eta)} = \frac{\xi}{\beta}\frac{\eta d\overline{C}}{(1-\overline{C})}, \tag{3.88}$$

$\overline{C} = C/C_s, \quad \beta = k_e/r, \quad a = (1-\gamma)/\xi n.$

The first equality in (3.88) determines an equation for solute characteristic curves:

$$\frac{d\eta}{d\tau} = a(1-\tau/\eta)^{n-1}. \tag{3.89}$$

Substitution $\tau = z\eta$ yields an ordinary differential equation:

$$\eta\frac{dz}{d\eta} + z = \frac{1}{a}(1-z)^{1-n}, \tag{3.90}$$

with the general solution in the form

$$\ln\eta - \int\limits_0^{z(\eta)} \frac{adu}{(1-a)^{1-n} - au} + A = 0, \tag{3.91}$$

where A is a constant of integration, which can be determined from the condition: $\tau = 0,\ \eta = \eta_0$. Thus, the solute characteristics can be determined from the transcendent integral equation:

$$\eta = \eta_0 \exp\left(\int\limits_0^{\tau/\eta} \frac{adu}{(1-a)^{1-n} - au}\right). \tag{3.92}$$

It determines the coordinate (position) of the front of dissolved chemicals, $\eta(\tau > 0)$, given that at the initial moment, its coordinate was $\eta(\tau > 0)$. Thereafter, a solution describing the decrease of flow depth along the characteristic curves after when rainfall ceases is as follows:

$$h(\eta,\tau) = h_0(1-\tau/\eta), \quad h_0 = [(r-i)x_0/\alpha]^{1/n}. \tag{3.93}$$

Equation (3.87) describes the decrease in flow depth in any flow section, while equation (3.93) describes flow depth variations at the point where the front of dissolved chemicals is situated at $t - T$. At $\tau = 0$ $(t = T)$ the value of h corresponds to the flow depth h_0 at the moment when rainfall ceases. At $\tau/\eta = 1$, we have $h = 0$ (in the general case, $\tau = \eta \geq 1$).

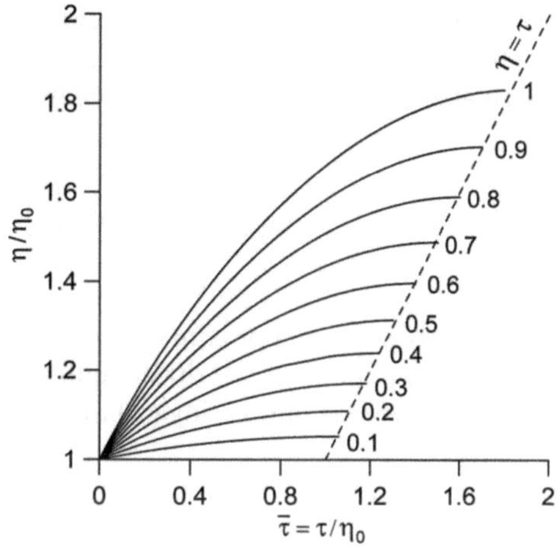

Fig. 3.11 Curves characterizing the dependence of the dimensionless coordinate $\overline{\eta}$ of the front on time $\overline{\tau}$. The figures at *curves* are the values of parameter $a. n = 2$

Formula (3.92) is transcendent. The value of η at any moment τ at known η_0 is found by iterations (Fig. 3.11). The plot in Fig. 3.11 characterizes the motion of the front of dissolved chemicals along characteristic curves. We can see that at small values of the dimensionless time $\overline{\tau} = \tau/\eta_0$, the function $\overline{\eta}(\overline{\tau})$ is linear, and $d\overline{\eta}/d\overline{\tau} \approx a$; at large $\overline{\tau}$, the curves flatten, in such a way that at $\overline{\tau} \geq 1$, they tend to some asymptotic values ($d\overline{\eta}/d\overline{\tau} \to 0$, and at $d\overline{\eta}/d\overline{\tau} = 0$, $\overline{\eta} = \overline{\tau}$). A good approximation for function $\overline{\eta}$ is the formula:

$$\overline{\eta}(\overline{\tau}) = 1 + (0.889a + 0.0223)\overline{\tau} - \left(0.0047 + 0.464 - 0.223a^2\right)\overline{\tau}^2, \quad a = (1-\gamma)/\xi n. \tag{3.94}$$

Variations of the concentration along solute transport characteristic curves is described by the second equation of the system (3.88), which can be rewritten as:

$$\frac{\xi}{\beta}\frac{d\overline{C}}{1-\overline{C}} = \frac{d\tau}{\eta - \tau} = \frac{d\tau}{h(\eta, \tau)}. \tag{3.95}$$

The solution of equation (3.95) that describes the relative concentration along the characteristic (3.92), emanating from (η_0, $\tau = 0$), can be represented in the following dimensionless form:

$$\overline{C}(\eta, \tau) = 1 - (1 - \overline{C}_T)\exp\left(-\frac{\beta}{\xi}\int_0^{\overline{\tau}} \frac{d\overline{s}}{1 - \overline{s}/\overline{\eta}(\overline{s})}\right), \tag{3.96}$$

where $\overline{C}_T = \overline{C}(\tau = 0)$ is the concentration at the moment when the rain ceases, determined by Eq. (3.64), $\overline{\tau} = \tau/\eta_0$, $\overline{\eta}(\overline{s}) = \eta(\overline{s})/\eta_0$, and the integration function is calculated from the transcendent equation

$$\overline{\eta}(\overline{s}) = \exp \int\limits_0^{\overline{s}/\overline{\eta}(\overline{s})} \frac{a du}{(1-a)^{1-n} - au}. \tag{3.97}$$

We also give an approximate solution, which is an asymptotic form of solution (3.95), when $\overline{\eta}(\overline{s})$ does not differ significantly from 1. In that case, the integrand in formula (3.96) becomes a logarithmic function, and we obtain

$$\overline{C} \approx 1 - \left(1 - \overline{C}_T\right)\left(1 - \frac{\tau}{\eta}\right)^{\beta/\xi} = 1 - \frac{1}{1+\beta}\left(1 - \frac{\tau}{\eta}\right)^{\beta/\xi}. \tag{3.98}$$

As seen, the process is controlled by a limited set of parameters: $\xi(\gamma, n)$ and $\beta = k_e/r$.

Based on solution (3.64), one can readily calculate the specific solute flux, $q_s = Cq$, i.e., a characteristic, which is of main interest in terms of the impact of surface runoff on water bodies. Thus, maximal values of q_s that can be attained at the formation of equilibrium flow profile ($t > t_e$), are:

$$q_s = C_s \frac{\beta}{1+\beta}(r - i)x. \tag{3.99}$$

Some Calculation Examples They illustrate the behavior of the dimensionless concentration function under above conditions.

The plots in Fig. 3.12 are constructed for two fixed values of γ, $\gamma = 0.2$ and 0.7, using different values of coefficient β, which determines the kinetics of the transfer of chemicals from solid to liquid phase. In particular, the plot shows that the concentration $\overline{C}(\overline{\tau})$ starting from the initial value (3.64) (reached during the rising stage of overland flow) increases gradually to a value less than or equal to 1.

When the kinetic coefficient of dissolution is greater than the runoff rate ($\beta > 1$), the approximate model (3.98) yields results quite acceptable for practical estimates within all range of the dimensionless time, τ. This suggests that variations of concentration are largely due to the dynamics of a decrease in water depth and dissolution of solid chemicals, while the effect of the lateral motion of water over the surface is far less significant. A power approximation (3.98) is also appropriate for $\beta < 1$ if $\tau \leq 0.5 - 0.7$.

Finally, the approximate solution (3.98) was compared with a solution obtained numerically (numerical code GSSHA), Fig. 3.13. The input data are slope length, $L = 1600$ m; slope gradient, $S_0 = 0.01$ m/m; the exponent in the relationship for water discharge (2.3), $n = 5/3$; Manning's roughness coefficient, $m = 0.05$ c/м$^{1/3}$, i.e., $\alpha = 2$ м$^{1/3}$/c; the rainfall rate, $r = 18$ mm/h ($5{\cdot}10^{-6}$ m/s); the concentration of the chemical in the soil, $C_s = 1$ mg/L; the mass transfer coefficient, $k_e = 35.83$ mm/h ($\beta = 2$); fallout period, $T = 4.617$ h (277 min). Three scenarios with different infiltration rate, i, are considered:

Scenario	i, mm/h	$\gamma = i/r$	$\xi = \gamma/\gamma^{1/n}$	h_x, mm
1	15	0.83	0.93	13.7
2	10	0.56	0.79	24.7
3	5	0.28	0.59	33

The agreement between the numerical and analytical solutions can be seen to be good at $\gamma > 0.5$ and satisfactory at $\gamma < 0.5$, i.e., the accuracy of analytical calculations increases with increasing infiltration capacity of soils.

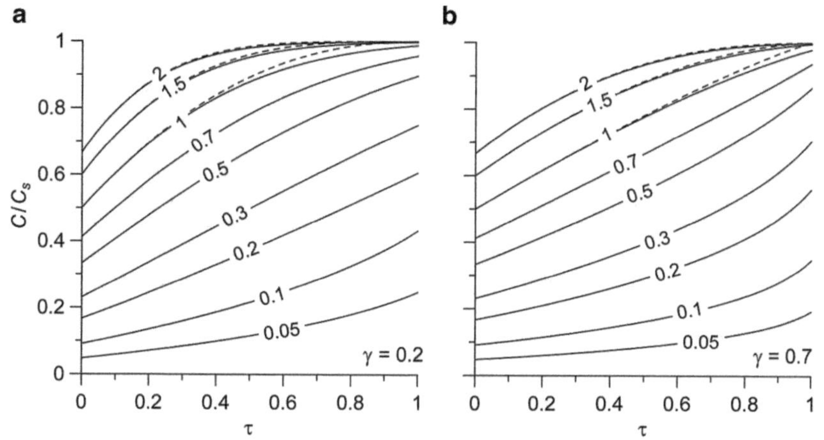

Fig. 3.12 Variations of the relative concentration along characteristic curves. *Solid lines* give solution (3.96) with approximation (3.94) for function $\bar{\eta}$, *dashed lines* give approximate solution (3.98, $\tau \leq 0.7$). The figures at curves are the values of $\beta = k_e/r$. (**a**) $\gamma = 0.2$, (**b**) 0.7. $n = 2$

Fig. 3.13 Comparison of numerical (*circles*) and analytical, Eqs. (3.64, 3.98) (*solid curves*), solutions. Numbers at the curves are the values of k_s, mm/h (see a Table above)

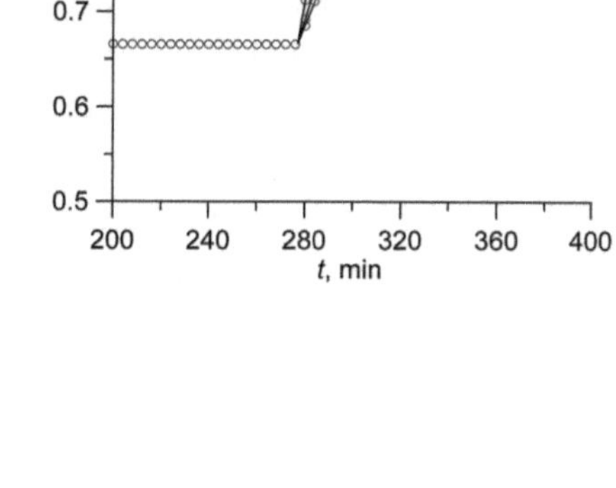

3.3.2 A Numerical Solution for Solute Transport in Runoff Coupled with Transient Infiltration

The author does not know any analytical solution of the problem of solute transport in overland flow under the conditions of unsteady-state infiltration. Therefore, to analyze general trends at the qualitative level, we will use particular solutions of the problem obtained by numerical methods (code GSSHA).

It is assumed that the input for the overland flow/solute transport model is determined by the condition (2.24), where the infiltrability is approximated by Green–Ampt solution (1.18). An application will be carried out for the same parameters of the rainfall–runoff condition used in the previous Sects. 2.3.1 and 3.3.1, namely: $n = 5/3$, $r = 18$ mm/h, $L = 1600$ m, $\alpha = 2\,\mathrm{m}^{1/3}/\mathrm{c}$, $T = 277$ min. Contrary to an example in Sect. 3.3.1 where fixed values of infiltration rate were assumed, in this test the soil infiltrability is calculated as a time-dependent function using soil characteristics k_s and h_c for a fixed value of the saturation deficit, $\Delta\theta = 0.05$ ($\theta_s = 0.1$, $\theta^0 = 0.05$). Two scenarios are considered. Under the first scenario, at a fixed value of the suction pressure, $h_c = 10$ cm, the effect of changes in the hydraulic conductivity, k_s on solute concentration measured at the hillslope outlet is studied (Fig. 3.14). Under the second scenario, the coefficient k_s is assumed constant, and the concentration response is assessed at different values of h_c (Fig. 3.15).

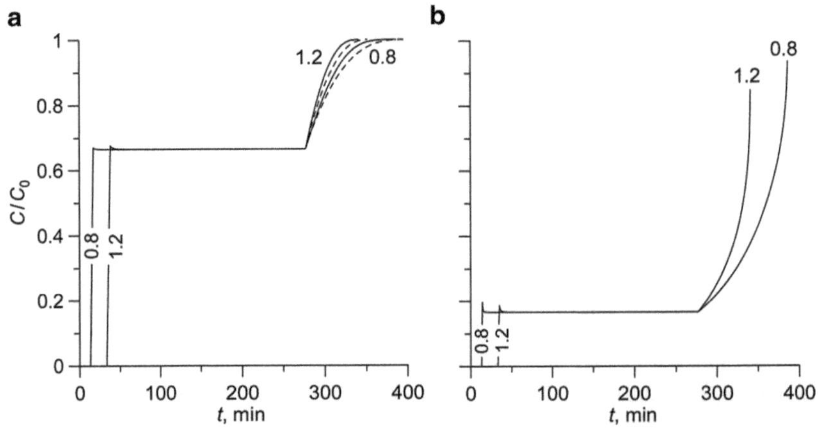

Fig. 3.14 The effect of k_s at $h_c = 10$ cm on the plots of function $C(t)$. (**a**) $\beta = 2$, (**b**) $\beta = 0.2$. Calculation were performed using GSSHA numerical code. Dashed curve corresponds to approximate solution described by Eq. (3.98)

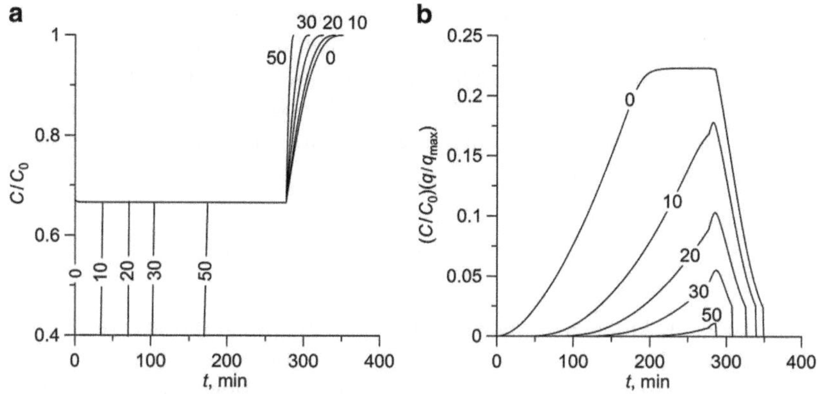

Fig. 3.15 The effect of suction pressure at the wetting front on dimensionless concentration curves (**a**) and normalized solute flux discharge (**b**). $k_s = 1.2$ cm/h, $\beta = 2$, $x = L = 1600$ m. Numbers at the *curves* are the values of h_c, cm. Calculations were performed using GSSHA numerical code

Figures 3.14 and 3.15 illustrate the following tendencies in the behavior of the breakthrough curves.

1. The time of ponding, t_p, and, accordingly, the beginning of concentration changes in the system, is determined with a high accuracy by formula (1.19a).
2. At moment t_p, the concentration instantaneously reaches the value $\overline{C}$, which remains unchanged until the moment $t = T$, when the precipitation event ceases (Figs. 3.14 and 3.15a). This value of $\overline{C}$ does not depend on the parameters of Green-Ampt model; it is determined by formula (3.64), i.e., the ratio $\beta = k_e/r$ of the exchange rate coefficient to rainfall rate, i.e., we have come to the same result of concentration independency of the soil infiltration capability as in the previous Sect. 3.2.
3. The moment of rain cessation and, accordingly, a decrease in the depth of water layer on the surface show an increase in the concentration of the chemical in this water up to the value of $\overline{C} = 1$ $(C = C_0)$. The rate of attainment of the steady-state value $\overline{C} = 1$ (Figs. 3.14, 3.15a) is the higher, the higher is the drying of the surface, i.e., the higher are k_s and h_c.
4. Such character of the plots $\overline{C} = f(t)$ determines the similarity of normalized solute flux discharge, $\overline{C}(q/q_{\max})$ (Fig. 3.15b), to hydrographs in Fig. 2.6b. Here $q_{\max} = rL$ is the outlet discharge maximum in the absence of infiltration.

3.3.3 *An Asymptotic Mixing-Layer Model for Solute Transport under a Constant Infiltration Rate*

The models with active (mixing) layer (Sect. 3.1.2.2) should be preferred in the description of surface runoff pollution in watersheds with intense farming, accompanied by the use of fertilizers and pesticides. Those substances generally show a uniform distribution in the surface soil layer and, involved in the exchange process, they determine the character of pollution of surface runoff.

Adding the left and right parts of Eqs. (3.14) and (3.15) (or 3.24 and 3.25 at $j_D = 0$), we come to the equilibrium model of solute transport in the overland flow (Emmerich et al. 1989, Eq. [7]):

$$(h + \theta d_e)\frac{\partial C}{\partial t} + q\frac{\partial C}{\partial x} = r(C_r - C), \tag{3.100}$$

which neglects the kinetics of matter exchange between the surface water layer and the mixing layer. In an other words, the Eq. (3.100) corresponds to the limiting case of no "film" resistance, $k_e \to \infty$, when the two concentration functions, C and C_e, in the system of equations (3.14) and (3.15) become closer ($C \to C_e$). A term θd_e in the Eq. (3.100) can be transformed to account for the sorption capacity of the soil (Sect. 3.1.2.2).

The specific feature of the present illustrative task is that the solute transport process is analyzed under the condition of flow continuity

$$\frac{\partial h}{\partial t} + \frac{\partial q}{\partial x} = r_e, \quad r_e = (r - i), \tag{3.101}$$

accounting for infiltration with a constant rate, i.

Equation (3.100) corresponds to the system of ordinary differential equations:

$$\frac{dt}{h + \theta d_e} = \frac{dx}{q} = \frac{dC}{r(C_r - C)}. \tag{3.102}$$

The main (limiting) flow characteristics for the hillslope runoff model (3.101)

$$t(x, 0) = \frac{1}{r}\left(\frac{r_e x}{\alpha}\right)^{1/n}, \tag{3.103}$$

given in (x–t) diagram (Fig. 3.3, curve *1*), divides the problem domain into two domains with an unsteady-state, $h = h(t)$, and steady-state, $h = h(x)$, behavior of the function of water layer depth. Let us consider the behavior of the concentration function in those domains, assuming $C_r = 0$.

In the *unsteady* domain, where $h = r_e t$, the process is described by the differential equation

$$\frac{dC}{dt} = -\frac{rC}{r_e t + \theta d_e}. \tag{3.104}$$

Integrating (3.105) yields

$$\overline{C}(r_e t) = \left(\frac{\theta d_e}{r_e t + \theta d_e}\right)^{\omega}, \quad \overline{C} = \frac{C}{C_0}, \quad \omega = \frac{r}{r_e} = \frac{1}{1 - \gamma}, \quad (\gamma = i/r). \tag{3.105}$$

The solution (3.105) is valid along solute characteristic curves:

$$\frac{dx}{dt} = \frac{\alpha(r_e t)^n}{r_e t + \theta d_e}. \tag{3.106}$$

To describe the solute transport in the domain of *steady-state* behavior of the function $h = h(x)$, when $q = r_e x$, another differential equation from the system (3.102) is to be solved, namely:

$$\frac{dC}{dx} = -\frac{rC}{r_e x}. \tag{3.107}$$

Integrating (3.107), we obtain a solution for the relative concentration ($\overline{C} = C/C_0$) in a general form:

$$\overline{C} = A/x^{\omega}, \tag{3.108}$$

where A is integration constant. The latter can be determined from the condition of the equality of concentrations determined by solution (3.105) (in which $r_e t = (r_e x'/\alpha)^{1/n}$) and (3.108) at the moment of flow wave passage through the point x', t', where the characteristics (2.12) and (3.106) intersect:

$$\frac{A}{x'^{\omega}} = \left(\frac{\theta d_e}{(r_e x'/\alpha)^{1/n} + \theta d_e}\right)^{\omega}, \tag{3.109}$$

whence we have constant A. Now the solution (3.108) can be written as

$$\overline{C} = \left(\frac{\theta d_e x'}{(r_e x'/\alpha)^{1/n} + \theta d_e}\right)^{\omega} \frac{1}{x^{\omega}}. \tag{3.110}$$

Since the solution (3.110) is valid along solute characteristic curves, represented by a solution of the following ordinary differential equation

$$\frac{dx}{dt} = \frac{r_e x}{(r_e x/\alpha)^{1/n} + \theta d_e}, \tag{3.111}$$

we have to relate the coordinate points (x', t') to a solution of Eq. (3.111). Integrating (3.111) within intervals $[0, t]$ and $[x_{01}, x]$ yields:

$$t = \frac{n\theta d_e}{r_e} \ln\left(\frac{x}{x_{01}}\right) + \frac{n}{r_e}\left[\left(\frac{r_e x}{\alpha}\right)^{1/n} - \left(\frac{r_e x_{01}}{\alpha}\right)^{1/n}\right]. \tag{3.112}$$

Since (3.112) passes through the intersection point of characteristics (x', t'), we have:

$$t' = \frac{n\theta d_e}{(1-n)r_e} \ln\left(\frac{r_e t'}{(r_e x_{01}/\alpha)^{1/n}}\right) - \frac{n}{(1-n)r_e}\left(\frac{r_e x_{01}}{\alpha}\right)^{1/n}, \tag{3.112a}$$

$$x' = \frac{\alpha}{r_e}(r_e t')^n. \tag{3.112b}$$

Coupling (3.112a) and (3.112b), we obtain:

$$\left(\frac{r_e x'}{\alpha}\right)^{1/n} = \frac{n\theta d_e}{(1-n)} \ln\left(\frac{(r_e x'/\alpha)^{1/n}}{(r_e x_{01}/\alpha)^{1/n}}\right) - \frac{n}{(1-n)}\left(\frac{r_e x_{01}}{\alpha}\right)^{1/n}, \qquad (3.113)$$

i.e., equation for x' is transcendental.

For the current value of x, relating to the characteristic (3.111) and being an argument of solution (3.110), the time t is determined by equality (3.112).

The final solution can be represented in the dimensionless form

$$\overline{C}(h_x, r_e t) = \left(\frac{\theta d_e (h'/h_x)^n}{h' + \theta d_e}\right)^{\omega}, \qquad (3.114)$$

where h' depends on h_{01} and can be determined from the relationship

$$h' = \frac{n\theta d_e}{1-n} \ln\left(\frac{h}{'h_{01}}\right) - \frac{n}{1-n} h_{01}, \qquad (3.115a)$$

while h_{01} is determined by the process time

$$r_e t = n\theta d_e \ln(h_x/h_{01}) + n(h_x - h_{01}); \qquad (3.115b)$$

here $h_x = (r_e x/\alpha)^{1/n}$, $h_{01} = (r_e x_{01}/\alpha)^{1/n}$, $h' = (r_e x'/\alpha)^{1/n}$. At $h_x = h'$, we come to Eq. 3.105 with $h_x = r_e t$.

It can be easily shown that at $\theta d_e \ll h_x$ and $i = 0$, the solutions (3.105) and (3.114) transform into the previously obtained formulas (3.35) and (3.45a) for the conditions of an "impulse" pollution of surface (3.9), where the fallout density $N_0 = \theta d_e C_0$.

The obtained solutions can be used to evaluate the breakthrough concentration curves (Fig. 3.16a), and the reduced mass flux of the chemical in the section h_x (Fig. 3.16b), see

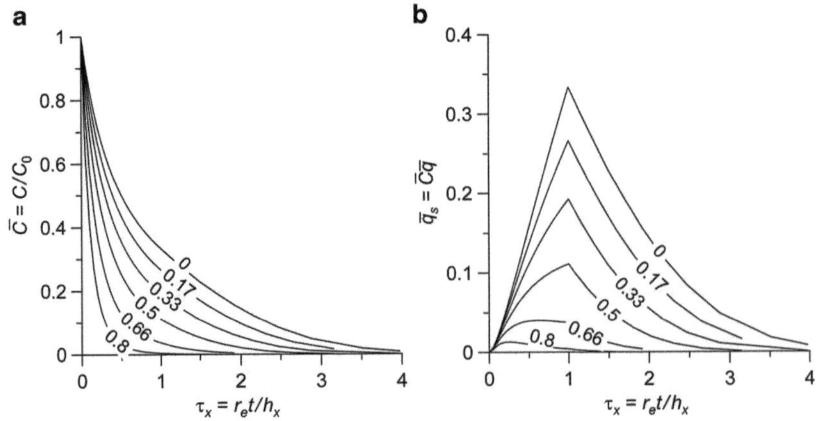

Fig. 3.16 Variations of the relative concentration of chemical (**a**) (solution (3.105) for $\tau_x \leq 1$ and solution (3.114) for $\tau_x > 1$) and the reduced mass flux (**b**) (solution (3.53) in the hillslope flow in the mass exchange with soil layer. The numbers at *curves* are the values of $\gamma = i/r$. A particular case $\overline{d}_e = \theta d_e/h_x = 0.5$. $n = 5/3$

the first relationship in (3.53) at $\overline{q} = q/\alpha h_x = \tau^n$ (when $0 \leq \tau \leq 1$) or $\overline{q} = 1$ (when $\tau \geq 1$). The curves in the plots show a relatively high sensitivity to parameter γ, which characterizes infiltration water losses. With increasing γ, the rate of drop in the concentration function increases, and the mass flux accordingly decreases.

3.4 Solute Transport along Lateral Subsurface Flowpaths

As it has been already thoroughly discussed (Sects. 1.2.3 and 2.4.1), subsurface stormflow is an important component of the rainfall–runoff response, especially in steep-sloped and wet areas (mountain terrain, valleys, landscape depressions) with forest cover. Being also an important contributor of dissolved species, including contaminants, to the surface water bodies, subsurface stormflow is also responsible for their chemistry.

In particular, the issues that have been actively discussed in the two or three recent decades include the age, origin, and pathways of subsurface stormflow (Sklash and Farvolden 1979; McDonnell 1990; Peters et al. 1995; Brutsaert 2005; Liggett et al. 2014), and, consequently, the contribution of new (event) and old (pre-event) water to event runoff. The majority of published studies of hydrograph separation (based on flow, isotope and geochemical tracer data) indicate that, under wet antecedent conditions, pre-event water dominates in hillslope runoff despite event-based chemical dilution (Sklash and Farvolden 1979; Buttle 1994; Kirchner 2003; Weiler et al. 2003). As pointed out by Kirchner (2003), in many small catchments, streamflow responds promptly to rainfall inputs, but fluctuations in passive tracers are often strongly damped. This indicates that storm flow in these catchments is mostly old water. Under dry antecedent conditions, the role of event water commonly increases (Brown et al. 1999).

Currently, the relative contributions of surface and subsurface water to streamflow are often determined with the use of sophisticated numerical models, which are in turn used as an effective tool for storm hydrograph separation. Many attempts were made to study numerically mechanisms and effects related to the rapid groundwater response to precipitation and domination of pre-event water in stream hydrographs. For example, using fully integrated surface and subsurface flow and solute transport models, it was shown that hydrodynamic mixing can dramatically influence the estimates of pre-event water contribution by a tracer-based separation (Jones et al. 2006; Liggett et al. 2014). Unfortunately, the progress in numerical simulation shaded the potentialities of analytical methods of analysis, maybe, except for the simplest two- or three-component balance (mixing) models used as a main analytical framework for the interpretation of streamwater chemistry.

In our model, pre-event water storage is associated with the retention of a certain amount of water in soil above a sloped interface between soil and bedrock (Fig. 3.17). There can also be a thin weathered zone near the soil–bedrock interface. The saturation zone (pre-event groundwater), shown in the diagram (Fig. 3.17), is

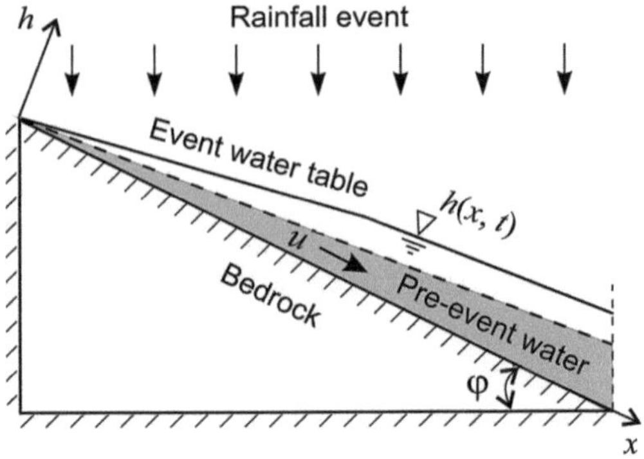

Fig. 3.17 A conceptual model of mixing process in subsurface stormflow (SSF) above the bedrock surface as a base of the steep slope

generated mostly through annual net recharge, which is characterized by the average rate i. Such steady-state profile of saturation in terms of kinematic-wave approximation (Sect. 1.2.3)

$$\alpha_s \frac{dh(x)}{dx} = i \tag{3.117}$$

is simply a linear function of x (Beven 1981):

$$h(x) \equiv h^0(x) = ix/\alpha_s, \tag{3.117a}$$

where $\alpha_s = k_s \sin \varphi$ (a coefficient which is associated with the specific discharge). Beven (1981) found that the kinematic-wave approximation is a good representation of the Dupuit–Forchhmier flow theory in terms of predicting subsurface stormflow hydrographs, if some criterion for a dimensionless parameter depending on the slope conditions and recharge rate is fulfilled (Sect. 1.2.3). Thus, the amount of water held within the soil profile is determined by a combination of soil and hillslope parameters, such as k_s, i and φ.

There are three major assumptions about subsurface flows. The first is that the event water falling onto a hillslope infiltrates rapidly to the saturated zone via vertical preferential flow paths. Hydrological system behavior is consistent with many observations which show how subsurface flow through macropores can play a major role in storm runoff generation (Beven and Germann 1982; Germann 1990; Peters et al. 1995; Brutsaert 2005). The second assumption is that the lateral flow induced by infiltration at any point in the slope occurs in accordance with Darcy's law where hydraulic gradient is equal to the slope of the impermeable base. The third assumption concerns the intensity of the recharge: in model setup, the event recharge, w, is equal to or exceeds the average value i.

As for formalization of solute transport process, it is assumed that a solute (chemical component) in concentration C_r, exceeding the background concentration C^0, enters the saturated zone only in the rainfall event period. The solute can be dissolved in the precipitation water or scavenged from the soil surface during infiltration. Recharge water between rainfalls is free of this component or shows solute concentration $C = C^0$. Thus, annual recharge dilutes the contaminated groundwater on the hillslope between storms. The hydrodynamic dispersion and diffusion are neglected.

The hydrodynamic response to the precipitation event is described by an equation as follows

$$\phi_n \frac{\partial h}{\partial t} + \frac{\partial q}{\partial x} - w = 0, \tag{3.118}$$

where $q = \alpha_s h$ is the discharge; $h = h(x, t)$ is water head height measured in the vertical direction from the impermeable sloping bed (Fig. 3.17); ϕ_n is the drainable porosity. Equation describing mass balance of the solute in the lateral flow is

$$\phi_n \frac{\partial hC}{\partial t} + \frac{\partial qC}{\partial x} - wC_r = 0, \tag{3.119}$$

where C_r is solute concentration in the recharge water. Combining (3.118) and (3.119) leads to an equation

$$h\frac{\partial C}{\partial t} + uh\frac{\partial C}{\partial x} = -\frac{w}{\phi_n}(C - C_r), \tag{3.120}$$

where $u = \alpha_s/\phi_n$ is the stormwater flow velocity.

A system of two ordinary differential equations written in a chain form,

$$\frac{dt}{h} = \frac{dx}{uh} = -\frac{\phi_n}{w}\frac{dC}{C - C_r}, \tag{3.121}$$

is mathematically consistent with the solute-balance Eq. (3.120). For a period before the rainfall event secession, $t < T$, two space–time domains with distinctive behavior of the flow depth, h (and, consequently, flow discharge, q), in each can be distinguished:

1. the domain where h behaves transiently in response to the flow input, w,

$$h = h^0 + (w - i)t/\phi_n, \quad t \le t_e = \phi_n x/\alpha_s; \tag{3.122a}$$

2. the domain where h behaves similarly to a steady-state function

$$h = wx/\alpha_s, \quad t > t_e. \tag{3.122b}$$

Here, t_e is the time required to reach the steady-state equilibrium at the point x.

Integrating Eq. (3.121) yields the same equations of the problem with respect to changes in solute concentrations in the mixture of event and pre-event water in both distinguished domains (Eqs. (3.122a) and (3.122b)). This solution can be conveniently represented in a dimensionless form:

$$\overline{C} = 1 - \left(\frac{1}{1 + (\overline{w} - 1)\tau}\right)^{\overline{w}/(\overline{w}-1)}, \quad 0 \le \tau \le \tau_T, \tag{3.123a}$$

where $C = (C - C^0)/(C_r - C^0)$, $\tau = it/\phi_n h^0 = \alpha_s t/\phi_n x$, $\tau_T = \alpha_s T/\phi_n x$, $\overline{w} = w/i$. The identity of the solutions is due to the linear character of the process (the first equation in the system (3.121)).

If during the input of the solute into a shallow aquifer, the event recharge rate, w, only slightly differs from its average annual value, i ($\overline{w} \approx 1$), the solution of the problem can be described by an exponential function

$$\overline{C} = 1 - \exp(-\tau), \quad 0 \le \tau \le \tau_T. \tag{3.123b}$$

The recession hydrograph and the behavior of solute in a mixture of event and pre-event waters after the rain stops are described by a system of equations:

$$\phi_n \frac{\partial h}{\partial t} + \frac{\partial q}{\partial x} - i = 0, \tag{3.124}$$

$$\phi_n \frac{\partial hC}{\partial t} + \frac{\partial qC}{\partial x} - iC^0 = 0, \tag{3.125}$$

which shows that the water depth of the saturation zone generated during the event is controlled by "reference" recharge rate, i, and that the infiltrating water contains a tracer component in background concentration, C^0. Combining the two above equations gives an equation

$$h\frac{\partial C}{\partial t} + uh\frac{\partial C}{\partial x} = -\frac{i}{\phi_n}\left(C - C^0\right), \tag{3.126}$$

for which a system of two ordinary equations,

$$\frac{dt}{h} = \frac{dx}{uh} = -\frac{\phi_n}{i}\frac{dC}{\left(C - C^0\right)}, \tag{3.127}$$

is mathematically equivalent. The thickness of the saturation zone is described by the equation depending on both x and t variables:

$$h(x, t) = \frac{wx}{\alpha_s} - \frac{w - i}{\phi_n}(t - T), \quad t > T. \tag{3.128}$$

Solving the first equation of the system (3.127) gives

$$x = x_0 + \frac{\alpha_s}{\phi_n}(t - T). \tag{3.129}$$

Now, the characteristic equation for water depth, h, becomes:

$$h(t) = \frac{wx_0}{\alpha_s} + \frac{i}{\phi_n}(t - T). \tag{3.130}$$

The concentration is found from the solution of an ordinary differential equation

$$\frac{dt}{h} = -\frac{\phi_n}{i}\frac{dC}{C - C^0}, \tag{3.131}$$

under an assumption that at $t = T$, the distribution of concentration is known (Eq. 3.123a):

$$\overline{C} = \left[1 - \left(\frac{1}{1 + (\overline{w} - 1)\tau_T}\right)^{\overline{w}/(\overline{w}-1)}\right]\frac{\overline{w}}{\overline{w} + (\tau - \tau_T)}, \quad \overline{w} = w/i, \quad \tau \geq \tau_T. \tag{3.132}$$

Problem solution in the form of equations (3.123a), (3.132) corresponds to the conditions when storm duration is equal to or greater than the time of concentration ($\tau_T \geq 1$, Fig. 3.18a). This, for example, is typical of longer and larger autumn storms. Contrarily to this situation, for the analysis of short-term processes, for example, summer storm events, of greater use could be solutions obtained for the partially equilibrium hydrograph ($\tau_T < 1$, Fig. 3.18b).

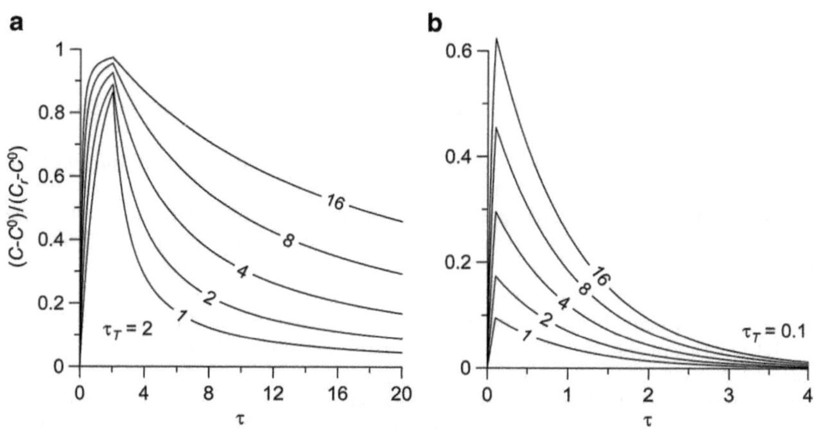

Fig. 3.18 Dimensionless concentration: (**a**) $\tau_T = 2$ and (**b**) $\tau_T = 0.1$. The numbers at the *curves* show the ratio *w*/*i*

This model gives remarkable skew concentration distributions with respect to the peak values (Fig. 3.18). Such behavior is due to the nonpoint solute source, which exhibits a marked variation in the residence times in the shallow aquifer.

The discussed mathematical problem setup corresponds to one-component hydrograph separation, and with this viewpoint, the dimensionless concentration, $\overline{C}$, can be considered as the fraction of the event (new) water in the total discharge:

$$\overline{C} = Q_{\text{event}} / \left(Q_{\text{event}} + Q_{\text{pre-event}} \right), \tag{3.133}$$

where Q_{event} and $Q_{\text{pre-event}}$ are the contributions of event water (rainfall) and pre-event water (groundwater).

According to this formula, the pre-event water always shall dominate in the initial storm hydrograph (Fig. 3.18). Such chemical signature of hillslope water is a quite expectable fact, since it stems from the very mathematical formulation of the problem, which implied that the pre-event water is replaced by the event water throughout the saturated zone and the mixture of the both fluxes discharges through the slope outlet.

Clearly, the idealized model considered above has a number of drawbacks and limitations regarding its practical applicability. In particular, it neglects the role of pre-event water contained in macropores above the water table, the effects of hydrodynamic dispersion, soil heterogeneity and many others. Nevertheless, the relationships considered above give an idea about the effect of (a) hillslope and soil parameters on the subsurface flow and tracer response times, (b) transient phenomena on hydrograph separation, and hence enable the assessment of the applicability of steady-state mass balance models to evaluating the fate of a solute in the subsurface environment.

References

Ahuja LR (1982) Release of soluble chemical from soil to runoff. Trans Am Soc Agric Eng 25:948–953

Ahuja LR, Sharpley AN, Yamamoto M (1981) The depth of rainfall-runoff-soil interaction as determined by 32P. Water Resour Res 17(4):969–974

Ahuja LR, Lehman OR (1983) The extent and nature of rainfall–soil interaction in the release of soluble chemicals to runoff. J Environ Qual 12:34–40. doi:10.2134/jeq1983.00472425001200010005x

Akan AO (1987) Pollutant washoff by overland flow. J Environ Eng 113(4):811–823

Beven KJ (1981) Kinematic subsurface stormflow. Water Resour Res 17(5):1419–1424

Beven KJ, Germann PF (1982) Macropores and water flow in soils. Water Resour Res 18(5):1311–1325

Brown VA, McDonnell JJ, Burns DA et al (1999) The role of event water, a rapid shallow flow component, and catchment size in summer stormflow. J Hydrol 217:171–190

Brutsaert W (2005) Hydrology: an Introduction. Cambridge University Press, Cambridge, UK, p 605

Buttle JM (1994) Isotope hydrograph separations and rapid delivery of pre-event water from drainage basins. Progr Phys Hydrol 18:6–41

Deng ZQ, de Lima JLMP, Singh VP (2005) Transport rate-based model for overland flow and solute transport: parameter estimation and process simulation. J Hydrol 315:220–235

Dong W, Wang Q (2013) Modeling soil solute release into runoff and transport with runoff on a loess slope. J Hydrol Eng 18:527–535

Downer CW, Ogden FL (2004) GSSHA: a model for simulating diverse streamflow generating processes. J Hydrol Eng 9(3):161–174

Emmerich WE, Woolhiser DA, Shirley ED (1989) Comparison of lumped and distributed models for chemical transport by surface runoff. J Environ Qual 18(1):120–126

Freeze AR, Cherry JA (1979) Groundwater. Prentice Hall, Englewood Cliffs, p 609

Gao B, Walter MT, Steenhuis TS et al (2003) Investigating ponding depth and soil detachability for a mechanistic erosion model using a simple experiment. J Hydrol 277:116–124

Gao B, Walter MT, Steenhuis TS et al (2004) Rainfall induced chemical transport from soil to runoff: theory and experiments. J Hydrol 295:291–304

Gao B, Walter MT, Steenhuis TS et al (2005) Investigating raindrop effects on transport of sediment and non-sorbed chemicals from soil to surface runoff. J Hydrol 308:313–320

Germann PF (1990) Preferential flow and generation of runoff. Boundary layer flow theory. Water Resour Res 26(12):3055–3063

Govindaraju RS (1996) Modeling overland flow contamination by chemicals mixed in shallow soil horizons under variable source area hydrology. Water Resour Res 32(3):753–758

Havis RN, Smith RE, Adrian DD (1992) Partitioning solute transport between infiltration and overland flow under rainfall. Water Resour Res 28:2569–2580

Herbert BE, Przepiora (1995) Particle-mediated transport of nonpoint-source pollutants. In: Jordan W, Jensen R (eds) Proceedings of the 24th Water for Texas Conference. Texas Water Research Institute, Austin, pp 367–376

Johnson B, Zhang Z (2007) Development of a distributed source contaminant rransport, transformation, and fate (CTT&F) Sub-model for military installations. Environmental Laboratory US. Army Engineering Research and Development Center. Final report. ERDC/EL TR-07-10, p 65

Jones JP, Sudicky EA, Brookfield AE et al (2006) An assessment of the tracer-based approach to quantifying groundwater contributions to streamflow. Water Resour Res 42. doi: 10.1029/2005WR004130

Kirchner JW (2003) A double paradox in catchment hydrology and geochemistry. Hydrol Process 17:871–874

Liggett JE, Werner AD, Smerdon B et al (2014) Fully integrated modeling of surface-subsurface solute transport and the effect of dispersion in tracer hydrograph separation. Water Resour Res 50:7750–7765. doi:10.1002/2013WR015040

McDonnell JJ (1990) A rationale for old water discharge through macropores in a steep, humid catchment. Water Resour Res 26(11):2821–2832

Peters DL, Buttle JM, Taylor CH et al (1995) Runoff production in a forested, shallow soil, Canadian shield basin. Water Resour Res 31(5):1291–1304

Proffitt A, Rose C, Hairsine P (1991) Rainfall detachment and deposition: experiments with low slopes and significant water depths. Soil Sci Soc Am 55:325–332

Rivlin J, Wallach R (1995) An analytical solution for the lateral transport of dissolved chemicals in overland flow. Water Resour 31(4):1031–1040

Rumynin VG (2011) Subsurface solute transport models and case histories (with applications to radionuclide migration), vol 25, Theory and applications of transport in porous media. Springer Science + Business Media BV, Dordrecht, p 815

Shi X, Wu L, Chen W et al (2011) Solute transfer from the soil surface to overland flow: a review. J Soil Sci Soc Am 75(4):1214–1225

Singh VP (1997) Kinematic wave modeling in water resources: environmental hydrology. Wiley–Interscience, New York, p 830

Singh VP (2002a) Kinematic wave solutions for pollutant transport by runoff over an impervious plane, with instantaneous or finite-period mixing. Hydrol Process 16:1831–1863

Singh VP (2002b) Kinematic wave solutions for pollutant transport over an infiltrating plane with finite-period mixing and mixing zone. Hydrol Process 16:2441–2477

Sklash MG, Farvolden RN (1979) The role of groundwater in storm runoff. J Hydrol 43:45–65

Snyder IK, Woolhiser DA (1985) Effects of infiltration on chemical transport onto overland flow. Trans Am Soc Agric Eng 28:1450–1457

Tong J-X, Yang J-Z, Hu BX et al (2010) Experimental study and mathematical modelling of soluble chemical transfer from unsaturated/saturated soil to surface runoff. Hydrol Process 24:3065–3073

Turnbull L, Wainwright J, Brazier RE (2010) Hydrology, erosion and nutrient transfers over a transition from semi-arid grassland to shrubland in the South-Western USA: a modelling assessment. J Hydrol 388:258–272

Van der Perk M (2006) Soil and water contamination, 2nd ed. Taylor & Francis, London/New York, p 389

Wallach R (1991) Runoff contamination by soil chemicals-time scales approach. Water Resour Res 27:215–223

Wallach R, Jury WA, Spencer WF (1989) The concept of convective mass transfer for prediction of surface-runoff pollution by soil surface applied chemicals. J Trans ASAE 32:906–912

Wallach R, van Genuchten MT (1990) A physically based model for predicting solute transfer from soil to rainfall-induced runoff. Water Resour Res 26(9):2119–2126

Wallach R, William AJ, William FS (1988) Transfer of chemical from soil solution to surface runoff: a diffusion-based soil model. J Soil Sci Soc Am 52:612–617

Wallach R, Grigorin G, Rivlin J (2001) A comprehensive mathematical model for transport of soil-dissolved chemicals by overland flow. J Hydrol 247:85–99

Walter MT, Gao B, Parlange J-Y (2007) Modeling soil solute release into runoff with infiltration. J Hydrol 347:430–437

Walton RS, Volker RE, Bristow KL et al (2000) Solute transport by surface runoff from low-angle slopes: theory and application. Hydrol Process 14:1139–1158

Weiler M, Uchida T, McDonnell J (2003) Connectivity due to preferential flow controls water flow and solute transport at the hillslope scale. Proceedings of MODSIM, Townsville

Zhang XC, Norton LD, Lei T et al (1999) Coupling mixing zone concept with convection-diffusion equation to predict chemical transfer to surface runoff. Trans ASAE 42(4):987–994

Chapter 4
Contaminant Sorption and Transport by Suspended Particles with Runoff

The description of near-surface migration of absorbable chemicals requires more rigorous problem formulation, taking into account erosion phenomena, which always accompany runoff formation. Mobile fine material, which is a product of soil erosion, becomes an active transporter of components adsorbed on the surface of suspended particles. An analogy with the subsurface colloid-facilitated contaminant transport (Rumynin 2011) is appropriate here.

The first part of this chapter is dedicated to the development of an analytical framework, which is needed for the description of soil erosion and particle transport across a hillslope during a rain event independently of chemical transport. First, we will focus on idealized rainfall-runoff soil erosion resulting in the generation and overland transport of a single particle-size suspension class, accompanied by irreversible particle deposition. We rely here on the *sediment transport capacity concept*, which is used in most erosion models. We consider several limiting scenarios related to a combination of several processes, namely, rainsplash and flow-driven (hydraulic) erosion and particle deposition, taking into account some threshold criteria for particle entrainment and gravitational settling. Then, a more representative model of erosion and suspended particle transport considering soil as multi-size class sediment will be introduced. It is based on a theory of a cohesionless deposited layer forming on the water/soil interface, from which particles can be removed again by some erosion process. The governing equations for sediment continuity, detachment, deposition, and transport capacity are presented. The relevant models are important for prediction of the chemical-carrying capacity of the sediment, which is the subject of the second part of the chapter, which examines to what extent the erosion and deposition process can affect the *adsorbed chemical transport*. The derivation of analytic solutions describing sediment and absorbable solutes transport with overland flow is based on the coupled kinematic-wave erosion and solute transport equations. To simplify the analytical considerations, only the case of constant rainfall excess and the rising stage of runoff are considered.

V.G. Rumynin, *Overland Flow Dynamics and Solute Transport*, Theory and Applications of Transport in Porous Media 26, DOI 10.1007/978-3-319-21801-4_4

4.1 Physical and Mathematical Formulation of the Rainfall- and Runoff-Driven Soil Erosion Problem

Most physically based erosion models separate the erosion process descriptions into splash and sheet-flow erosion (summarized in the term "*interrill*") and concentrated flow ("*rill*") erosion (Hairsine and Rose 1992a, b; Al-Hamdan et al. 2012). In this work we focus exclusively on interrill processes which include a complex mixture of rainfall-driven (rainsplash) and runoff-driven (flow entrainment) soil erosion (Bryan 2000).

According to the classic definition (Trimble 2007), interrill erosion is the removal of a relatively uniform and thin layer of soil from the land surface by unchanneled runoff, or sheet flow. Splash erosion is driven by rainsplash kinetic energy, which depends on raindrop characteristics; the effective availability of this energy depends on soil characteristics and conditions. Shallow interrill flow has little entrainment capacity without raindrop impact, but runoff energy is critical for soil particle transport in overland water (Bryan 2000). Interactions between raindrop impact and shallow flow cause the involvement of solid particles, which form the mineral and organic base of the soil, into runoff. Changes in the hydraulic conditions and energetics of overland flow can cause the inverse process – settling (deposition) of dispersed particles (suspension) onto soil surface from the solid phases, resulting in a *newly generated sediment*, whose properties generally differ from those of the *original soil material*. Thus, soil surface erosion and particle deposition are inseparable processes (Fig. 4.1).

The physical-mathematical description of soil erosion and allied processes is the focus of a huge scientific literature, since this phenomenon has a global effect on the preservation of soil productivity in vast regions, thus determining their economic well-being. Here, we will consider only a few of the wide diversity of experimentally validated mathematical models, which can be effectively used to better describe pollutant transport in sheet flows.

Fig. 4.1 A flow chart, representing three interrill components of the soil erosion

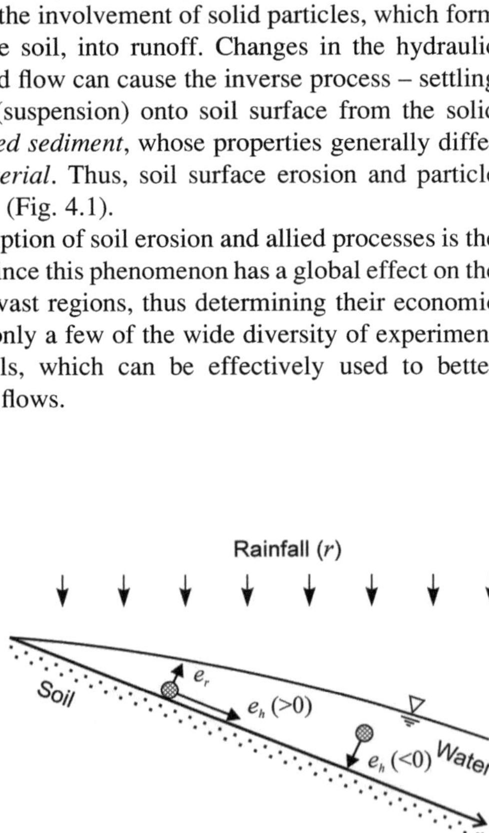

4.1.1 Basic Equations for Interactions in an Idealized Rainfall, Fluid Flow, and Sediment System

Equation of mass conservation for erosion of idealized soil material composed of a single size class with following suspension transport during rainfall, as well as for deposition of sediment at the decline of rainfall (Fig. 4.1) can be represented in the generalized form (Hjelmfelt et al. 1975; Johnson and Zhang 2005; Deng et al. 2008):

$$\frac{\partial hS}{\partial t} + \frac{\partial qS}{\partial x} = e_r + e_h, \tag{4.1}$$

where x is distance in the direction of flow [L]; t is time [T]; h is the flow depth [L]; q is the discharge per unit width of slope [L^2T^{-1}]; S is the concentration of dispersed particles in water [ML^{-3}]; e_r is the net rate of soil erosion caused by raindrop impact (rainsplash erosion) [$ML^{-2}T^{-1}$]; e_h is the rate of hydraulic erosion caused by flowing water or deposition from the water [$ML^{-2}T^{-1}$] (positive for detachment and negative for deposition, thus implying that the two processes, flow erosion and gravity settling, cannot proceed simultaneously).

Formula (4.1) has a steady-state representation:

$$\frac{\partial q_s}{\partial x} = e_r + e_h, \tag{4.1a}$$

where $q_s = qS$ is the suspended sediment load [$ML^{-1}T^{-1}$].

The dependence of the rates of soil erosion, e_r and e_h, on rainfall intensity, r, actual flow velocity, u (or overland discharge, q), the properties of surface deposits, slope geometry, and landscape conditions can be described by a number of empiric relationships.

Rainsplash detachment of soil particles dominates at the first stage of the erosion processes on a hillslope. The falling raindrops have high kinetic energy, and the pressure they exert on the surface when colliding with it reaches several hundred kPa. The accompanying shear stress can be many times greater than the soil shear strength. The result is that the structural bonds in the soil are broken and the newly formed solid particles, as well as pore moisture drops and drops from water layer on the surface, splash as shown schematically in Fig. 4.2. The scatter radius of particles and drops can be as large as several tens of cm. The soil particles drop onto soil surface or the surface of sheet flow, which transports them further downslope. Overall, the erosion process of this type depends on rainfall characteristics (the intensity and rain drop size), water flow depth, and the mechanical and physical properties of the soil (consolidation, cohesion

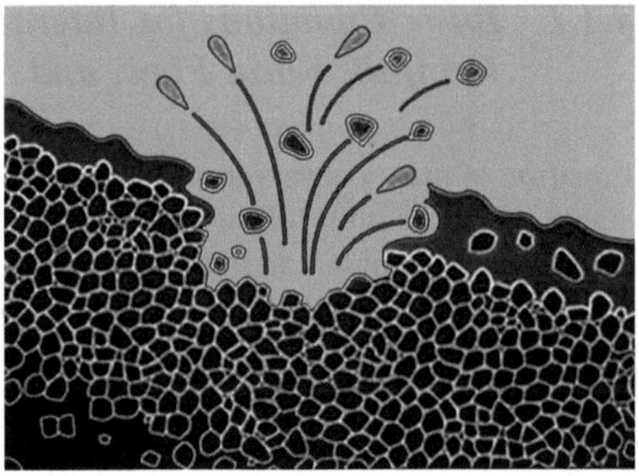

Fig. 4.2 Raindrop impact causing splash (rainfall) erosion (Adopted from: www.partnershipsforchange.cc/planningeduc0042.asp)

and soil particle-size distribution) (Kinnell 2005; Planchon and Mouche 2010). Of particular importance for erosion rate is vegetation, which protects the soil against a direct impact of raindrops.

In an early study (Meyer 1981), the rate of rain splash erosion, e_r, was evaluated by a two-parameter relationship

$$e_r = a_n r^{\beta}, \tag{4.2}$$

where r is the rainfall rate, a_n and β are empirical coefficients, and $\beta \approx 2$. The Eq. (4.2) was modified later by including various correction factors in coefficient a_n, e.g. (Downer and Ogden 2004),

$$a_n = K_r C_w C_G C_i, \tag{4.3}$$

where K_r is the soil erodibility factor for detachment by raindrop impact (J^{-1}); C_w, C_G and C_i are coefficients, accounting for the effects of water layer and vegetation on soil surface and land use.

However, some researchers take $\beta = 1$ in Eq. 4.2, thus the rain splash detachment rate is assumed to be proportional to the rainfall rate (Hjelmfelt et al. 1975; Lisle et al. 1998; Shaw et al. 2006; Barry et al. 2010):

$$e_r = a_0 r, \tag{4.4}$$

where parameter a_0 [ML^{-3}] characterizes soil detachability (in this representation, this is the concentration of soil aggregates in water, produced by rainsplash erosion of the surface).

Obviously, the rate of sediment addition to the surface water by rainfall detachment is reduced as flow depth, h, increases (Hairsine and Rose 1992a; Morgan et al. 1998; Gabet and Dunne 2003). Studying the appropriate physical problems has shown the erosion rate to drop exponentially with h (Morgan et al. 1998):

$$e_r = a_R KE \exp(-\xi h), \tag{4.5}$$

where a_R is an index of the detachability of the soil (kgJ^{-1} s^{-1}); KE is the total rainfall kinetic energy at the ground surface (Jm^{-2}); h is depth of the surface water layer (m); ξ is soil texture characteristic (m^{-1}).

In some software packages, formula (4.5) is replaced by another expression with the same exponential factor (Woolhiser et al. 1990; Deng et al. 2008)

$$e_r = \frac{a_0 r^2}{v} \exp(-\xi h), \tag{4.6}$$

where a_0 is the maximum sediment concentration produced by the raindrop impact in the overlying water at the end of the ponding time (kgm^{-3}); v is the sediment fall velocity (ms^{-1}).

Historically, definition of the second rate-term in mass conservation equation for overland flow (4.1), namely the rate of hydraulic erosion, e_h, has relied on the hydraulic theory developed for stream flows. The basic notions used in this case include sediment transport capacity, shear stress, shear velocity and stream power, determining physical and hydraulic properties of the flow and sediment. Thus, the source/sink term e_h can be represented in terms of the so-called *sediment transport capacity equation*, which has the form of an equation of first-order exchange reaction (Foster and Meyer 1975; Morgan et al. 1998; Sander et al. 2007),

$$e_h = \sigma(T_c - q_s), \tag{4.7}$$

implying that the rate of soil erosion (*particle detachment* from a surface by a hydrodynamic moment) or the deposition of particles from water flow is proportional to the difference between sediment transport capacity of the overland flow, T_c [ML^{-1}T^{-1}], which is the maximum equilibrium sediment load that a flow can transport, and the value of sediment discharge or sediment load, $q_s = qS$ [ML^{-1}T^{-1}]; here σ is an empirical coefficient, which characterizes the erodibility or the settling ability of particles [L^{-1}].

For *flow soil erosion,* this coefficient can be represented as the ratio

$$\sigma = D_c / T_c, \tag{4.8}$$

where characteristic D_c can be referred to as detachment capacity, which characterizes the maximal erosion impact of water flow when the concentration of suspended particles in water is minimal, i.e., $S = 0$. In this case, the relationship between sediment load and sediment transport capacity, can be rewritten as (Foster and Meyer 1975; Foster et al. 1995):

$$\frac{e_h}{D_c} + \frac{q_s}{T_c} = 1, \tag{4.9}$$

or

$$e_h = \frac{D_c}{T_c}(T_c - q_s) > 0. \tag{4.10}$$

In a simplified description of the process, it is proposed to assume the ratio to be constant $D_c/T_c = \text{const} = \sigma$, i.e., not depending on the velocity or discharge of flow.

This concept can be applied to mathematical formalization of the process of *deposition* of suspended particles from water. It was proposed to identify the coefficient σ in Eq. (4.7) with the ratio of the particle settling velocity, v, to the flow discharge, q, leading to the following kinetic equation, describing the *gravitational deposition* of suspended particles:

$$e_h = \frac{v}{q}(T_c - q_s) < 0; \tag{4.11}$$

with measurement units $\text{kgm}^{-1}\,\text{s}^{-1}$ for T_c and $q_s = qS$, ms^{-1} for v, and $\text{m}^2\,\text{s}^{-1}$ for q.

The transport capacity concept cannot be unique for a soil composed of a range of size classes and that uniqueness only occurs for the exceptional case of single size class soil (Sander et al. 2007). As one can see, the absolute values of sediment transport capacity, T_c, and the current values of sediment flux, q_s, control the direction of the process of interaction between water flow and soil in any point of the slope: the process taking place in this case is either the erosion of slope surface or gravitational deposition of suspended sediment.

Sediment transport capacity is strongly related to the shear stress, τ_s, which is defined as the force applied by flowing water on the soil surface per unit area (Fig. 4.3), and acting to detach soil particles from soil mass (Finkner et al. 1989; Foster et al. 1995):

$$T_c \approx K_t \tau_s^{3/2}, \tag{4.12}$$

where K_t is a transport coefficient which gives a measure for flow sediment transport ability. The greater the shear stress, τ_s, the more significant the sediment transport capacity of the surface flow, T_c.

Formula (4.12), which follows from Yalin's model (Yalin 1963), is valid at $\tau_s >> \tau_c$ (τ_s being far in excess of the critical surface shear stress, τ_c, which is a threshold for soil particle entrainment), i.e., (4.12) can be used to evaluate the potential of erosion process. Shear stress (Pa) is calculated as

$$\tau_s = \rho_w g h_e S_0, \tag{4.13}$$

where ρ_w is water density (kgm^{-3}), g is gravitation constant (ms^{-2}), h_e is effective flow depth (the fraction of the full depth, h, that determines the pressure on soil surface) (m), S_0 is the surface slope along the flow direction (m/m).

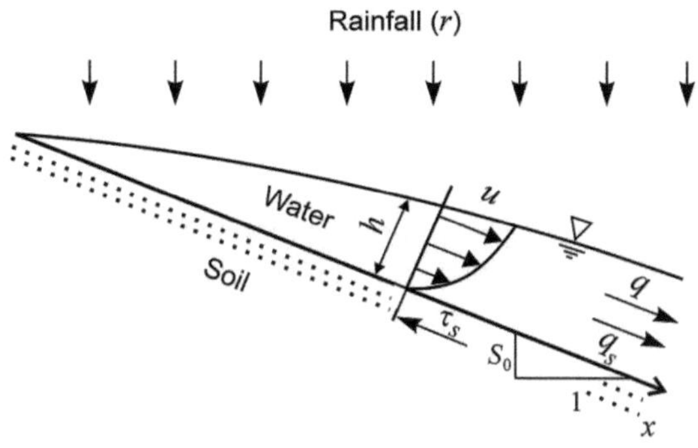

Fig. 4.3 A flowchart, illustrating flow shear stress on a slope under rainfall condition

Experiments show that T_c increases as a power function with discharge and slope gradient (Julien and Simons 1985; Finkner et al. 1989; Zhang et al. 2009):

$$T_c \sim \varepsilon\, S_0^{\gamma} q^{\beta}, \tag{4.14}$$

where ε, β, γ are empirical coefficients, characterizing the erosion process, in particular, ε is sometimes referred to as the erodibility of the sediment (soil); S_0 is the surface slope (m/m); q is the unit discharge rate (m^2/s). Therefore, T_c depends on surface slope, flow discharge, and some transport coefficients (Flanagan and Nearing 1995). The absolute values of T_c may vary within the wide range from 0.0001 to 0.1 kgm^{-1}s^{-1}; on steep slopes, T_c may reach a few or a few tens of kgm^{-1}s^{-1}.

Equations (4.10) and (4.11) suggest that at any moment and in any point, the rate of erosion or gravity settling is proportional to the difference between sediment transport capacity and sediment load. In terms of the net sediment exchange, the hydraulic erosion and the hydraulic deposition cannot occur simultaneously. However, both processes can be coincident in time but separated in space. For example, erosion can take place in higher elevated parts of the slope, where flow velocity is high, while settling can occur in gentler parts of the slope near its bottom, where flow decelerates. The effect of slope flow on the surface is maximal when the concentration of suspended particles in it is maximal.

The rate of hydraulic erosion e_h can be represented in another form (Deng et al. 2008), taking into account both the shear stress, τ_s, affecting the erosion of cohesive sediment, and the deficit of water flow saturation with suspended particles, ΔS (i.e., the difference between sediment concentration under equilibrium conditions, S^*, and the current sediment concentration, S). Introducing shear velocity as

$$u^* = (\tau_s/\rho_w)^{1/2} = (gh_e S_0)^{1/2}, \tag{4.15}$$

we can replace (4.10) and (4.11) by (Deng et al. 2008):

$$e_h = \varsigma\left(u^* - u_c^*\right)\left(S^* - S\right), \tag{4.16a}$$

$$e_h = -\xi\, v\left(S^* - S\right). \tag{4.16b}$$

where u_c^* is the critical shear velocity; in the first equation, $\varsigma > 0$ (dimensionless constant), if $S^* > S$ and $u^* > u_c^*$ (the hydraulic erosion occurs only when the shear velocity is greater than the critical shear velocity), otherwise, $\varsigma = 0$; in the second equation, $\xi > 0$ (dimensionless constant), if $S^* < S$, otherwise $\xi = 0$. Equations (4.10) and (4.11) will be similar to (4.16a) and (4.16b), if we set $S^* = T_c/q = \text{const}$.

Finally, it has been shown experimentally that the combined influence of runoff, rainfall intensity, and slope on the interrill erosion within individual hillslopes can be adequately described by the following empirical equation (Zhang et al. 1998):

$$e_{rh} = K_1 S_0^{\gamma} r q^{\beta}, \tag{4.17}$$

where K_1 is an interrill erodibility coefficient, depending on soil characteristic and the conditions on the surface; α and β are empirical coefficients. The linear term r represents the detachment of soil by raindrop impact and the enhancement of the transport capacity of sheet flow, while the product $S_0^{\gamma} q^{\beta}$ describes sediment transport by sheet flow (Zhang et al. 1998). In the cited work, $\beta = 1/2$ and $\gamma = 2/3$.

4.1.2 Basic Equations for Multi-size Erosion and Deposition (Re-detachment and Re-entrainment Effects)

The models discussed above treated soil as single-size class sediment. However the loss of chemicals bound to eroded soil is strongly dependent on both the size distribution of sediment and its overall concentration in runoff. Therefore, the development of more representative models of erosion and suspended particle transport considering soil as multi-size class sediment seemed to be an important and urgent task (Hairsine and Rose 1991; 1992a). Hairsine and Rose (1991, 1992a) developed a physically based theory treating erosion of soil (represented by multi-size class sediment) and sedimentation from runoff as two simultaneous processes, described by undependable source/sink functions. This model, referred to as HR model, not only takes into account rainfall detachment and/or flow entrainment erosion mechanisms, but also reflects the fact that a significant proportion of the eroded particles returns to the soil surface, forming a cohesionless deposited layer, from which it can be removed again by some erosion process. The process when the freshly deposited material is detached again by the rain and entrained by the flow is called re-detachment and re-entrainment, respectively. A particle is detached only once, but is then subject to repeated deposition/re-detachment cycle (Lisle et al. 1998). In the HR model it is assumed that rainfall detachment is non-selective with respect to the original soil and that rainfall re-detachment is non-selective with respect to the deposited layer (Hairsine et al. 1999) (Fig. 4.4).

HR model is based on a system of two mass conservation equations for individual particle size classes describing soil erosion, suspension transport, and

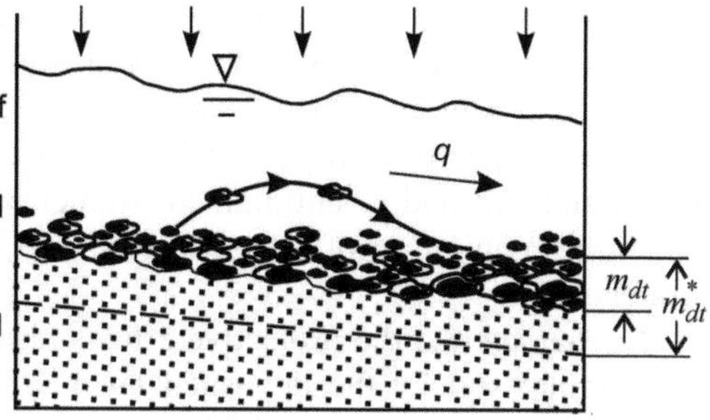

Fig. 4.4 A schematic diagram, illustrating a dynamically forming deposited layer (m_{dt}), and the maximum deposited layer (m^*_{dt}) that would render complete protection to the soil from further erosion (Rose et al. 2007)

accumulation on the soil surface (Hairsine and Rose 1992a; Hairsine et al. 2002; Sander et al. 2007):

$$\frac{\partial hS_i}{\partial t} + \frac{\partial qS_i}{\partial x} = e_i + e_{ri} + r_i + r_{ri} - d_i, \tag{4.18}$$

$$\frac{\partial m_{di}}{\partial t} = -(e_{ri} + r_{ri} - d_i), \tag{4.19}$$

where S_i is the sediment concentration in the mass of the i-th class (kgm^{-3}); m_{di} is the mass of sediment of the i-th size class in the deposited layer per unit area (kgm^{-2}). The total suspended sediment concentration, S, and the total mass in the deposited layer, m_t, are found by summing across all size classes as $S = \sum_1^I S_i$ and $m_{dt} = \sum_1^I m_{di}$ ($I = \sum i$). The right side of the equations is represented by source/sink terms (kgm^{-2}s^{-1}), describing the rates of rain splash detachment (e_i) and flow entrainment of soil particles from the original (uneroded) soil (r_i), the rates of rainsplash re-detachment (e_{ri}) and re-entrainment (r_{ri}) from the deposited layer formed during the current erosion, and the rate of deposition (d_i). These source/sink terms are given by equations for size class i (Sander et al. 2007):

$$e_i = (1 - H)a_0 p_i r, e_{ri} = Ha_d \frac{m_{di}}{m_{dt}} r, \tag{4.20}$$

$$r_i = (1 - H)p_i \frac{F}{J}(\Omega - \Omega_c), r_{ri} = H\frac{F}{gh}(\Omega - \Omega_c)\left(\frac{\rho_s}{\rho_s - \rho_w}\right)\frac{m_{di}}{m_{dt}}, \tag{4.21}$$

$$d_i = v_i S_i, \tag{4.22}$$

where p_i ($0 < p_i \leq 1$) is the proportion of sediment in size class i of the original soil; a_0 and a_d are the detachability coefficients of the original soil and the deposited layer, respectively (kgm^{-3});

$$\Omega = \rho_w g S_0 q = \tau_s u \left(\text{Wm}^{-2}\right) \tag{4.23}$$

($q = uh, h \approx h_e$ – see Eq. 4.13) is the stream power; Ω_c is the critical threshold stream power (as soil erosion due to flow is a threshold process, the stream power Ω must exceed the critical value to entrain soil particles Ω_c); S_0 is the slope; q is the discharge per unit width of slope (m^2s^{-1}); ρ_w and ρ_s are water and sediment densities, respectively (kgm^{-3}); F is the effective fraction of excess stream power in entrainment and re-entrainment (dimensionless); J is the energy expended in entraining a unit mass of cohesive sediment (Jkg^{-1}); v_i is the settling velocity of the i-th class (ms^{-1}).

To characterize the protective effects resulting from the fractional shielding of the original soil by the deposited layer, a protection factor, H $(0 < H \leq 1)$, is introduced (Hairsine et al. 1999):

$$H = m_{dt}/m^*_{dt}, \tag{4.24}$$

or more precisely (Le et al. 2013) $H = \min\{1, m_{dt}/m^*_{dt}\}$, where m^*_{dt} is the critical mass per unit area of the deposited layer to shield completely the original soil (kgm^{-2}).

The model (4.18)–(4.19) which incorporates a multi-size class description, enables one to predict particle size fractionation in the process of particle deposition on hillslope profile from overland flow: the coarser particles settle at a greater rate than the finer particles, and therefore the overland flow is enriched in these finer particles compared with the soil being eroded, and the size distribution of sediment forming the deposited layer becomes coarser than the original soil, with the eroded sediment being proportionately finer (Marshall et al. 1999).

Some specific analytical, semi-analytical and numerical solutions of HR model describing this and other phenomena under a variety of erosion scenarios are available in the literature (Hairsine and Rose 1991, 1992a; Hairsine et al. 2002; Sander et al. 2002, 2007; Hogarth et al. 2004b; Rose et al. 2007; Le et al. 2013). Several illustrations of concern will be given in Sect. 4.3.2.

4.2 Mathematical Formulation of Sediment-Bound Chemical Transport

We present a general model of soil solute (chemical) transport with runoff, incorporating several types of physicochemical and mechanical interactions, namely

- ejection of dissolved soil solutes into runoff due to raindrop-driven and diffusion exchange;
- ejection of solutes in the particulate state from soil surface in the case of a single-sediment-size class because of its raindrop and flow induced erosion;
- transport of solutes in the soluble and in the particulate state with runoff;
- gravitational deposition of suspended particles, carried by water flow and containing adsorbed solutes, onto soil surface;

- losses of dissolved solutes due to water infiltration from the surface;
- adsorption-based interaction between soil solutes and mineral soil matrix.

We will consider a model representing a combination of mass transfer models, including mixing (exchange) layer with an upper boundary represented by boundary diffusion layer (Sect. 3.1.2.2). We introduce the following denotations for concentration functions (Johnson and Zhang 2005): C_d^r is the concentration of dissolved solute in runoff water [ML^{-3}]; C_d^s is dissolved solute in pore water in the mixing layer [ML^{-3}]; C_a^r is the concentration of solute, adsorbed onto disperse particles in surface water [−]; C_a^s is the concentration of solute adsorbed in soil matrix [−]; S is the concentration of suspended particles in runoff [ML^{-3}].

The equation of mass conservation for dissolved solute transport is written as

$$\frac{\partial hC_d^r}{\partial t}+\frac{\partial qC_d^r}{\partial x}=k_e\left(C_d^s-C_d^r\right)-iC_d^r+\frac{e_r\theta}{\rho_b}\left(C_d^s-\lambda C_d^r\right)-k_shS\left(K_d^rC_d^r-C_a^r\right). \tag{4.25}$$

In the right side of Eq. (4.25), the first term describes the diffusion exchange between soil solution and water flowing over the surface, the second term describes matter losses from surface water due to its infiltration into the soil, the third term accounts for soil water ejection from the soil to runoff due to raindrop impact, and the fourth term describes sorption kinetics onto suspended particles in the surface water. Notation for coefficients in Eq. 4.25 will be given later, when the whole system of governing equations is presented.

Equation of mass conservation for transport of solutes adsorbed onto dispersed particles is

$$\frac{\partial hSC_a^r}{\partial t}+\frac{\partial qSC_a^r}{\partial x}=k_shS\left(K_d^rC_d^r-C_a^r\right)+\begin{cases}C_a^s(e_r+e_h), & e_h>0,\\ C_a^se_r+C_a^re_h, & e_h<0;\end{cases} \tag{4.26}$$

the last term in the right side of equation accounts for the flux of a particulate solute formed due to (a) raindrop impact and flow erosion of the soil surface ($e_h > 0$), (b) raindrop impact and gravitational deposition of suspended particles from surface water ($e_h < 0$); e_r, e_h [ML^{-2}T^{-1}]. This form of record shows that the two processes, flow erosion and gravitational deposition of suspension, cannot occur simultaneously.

The balance equation for solute dissolved in the mixing layer is

$$d_e\frac{\partial\theta C_d^s}{\partial t}=iC_d^r-iC_d^s-k_e\left(C_d^s-C_d^r\right)-\frac{e_r\theta}{\rho_b}\left(C_d^s-\lambda C_d^r\right)-d_ek_s\left(C_d^s-\frac{C_a^s}{K_d^s}\right). \tag{4.27}$$

The same for a solute in adsorbed state is

$$d_e \rho_b \frac{\partial C_a^s}{\partial t} = d_e k_s \left(C_d^s - \frac{C_a^s}{K_d^s} \right) - \begin{cases} C_a^s (e_r + e_h), & e_h > 0, \\ C_a^s e_r + C_a^r e_h, & e_h < 0. \end{cases} \tag{4.28}$$

Adding (4.27) and (4.28), we come to a full balance equation for the mixing layer of soil:

$$d_e \frac{\partial \left(\theta C_d^s + \rho_b C_a^s \right)}{\partial t} = \\ iC_d^r - iC_d^s - k_e \left(C_d^s - C_d^r \right) - \frac{e_r \theta}{\rho_b} \left(C_d^s - \lambda C_d^r \right) - \begin{cases} C_a^s (e_r + e_h), & e_h > 0, \\ C_a^s e_r + C_a^r e_h, & e_h < 0. \end{cases} \tag{4.29}$$

Source and sink components, e_r and e_h, can be defined in terms of Sect. 4.1.1. In our simplest model scenarios, considering splash erosion and gravitational deposition, soil particles are eroded and deposited following relationships:

$$e_r = a_0 r \text{ and } e_h = \frac{v}{q} (T_c - qS). \tag{4.30}$$

Equations (4.25), (4.26), (4.27), and (4.28) utilize two mass transfer coefficients: k_e, controlling mass exchange between soil solution and surface runoff water (Sect. 3.1.2), and k_s, governing the kinetics of sorption onto soil material. The system includes also two equilibrium sorption coefficients: K_d^s (sorption in the soil) and K_d^r (sorption onto suspension in the mobile phase). Erosion-related constants (Eq. 4.30) are a_0, v, and T_c (Sect. 4.1.1). The mixing layer is characterized by fixed depth, d_e, and saturated water content, $\theta = \theta_s$.

The above system of equations should be supplemented by a continuity equation for overland flow,

$$\frac{\partial h}{\partial t} + \frac{\partial q}{\partial x} = r - i, \tag{4.31}$$

and a mass balance equation for suspension flow

$$\frac{\partial hS}{\partial t} + \frac{\partial qS}{\partial x} = e_r + e_h. \tag{4.32}$$

Moreover, those two equations allow Eqs. 4.25 and 4.26 to be rewritten in a more compact form:

$$
\begin{aligned}
&h\frac{\partial C_d^r}{\partial t}+q\frac{\partial C_d^r}{\partial x}= \\
&k_e\left(C_d^s-C_d^r\right)+\frac{e_r\theta}{\rho_b}\left(C_d^s-\lambda C_d^r\right)-k_s hS\left(K_d^r C_d^r-C_a^r\right)-rC_d^r\ ,
\end{aligned}
\tag{4.33}
$$

$$
\begin{aligned}
&hS\frac{\partial C_a^r}{\partial t}+qS\frac{\partial C_a^r}{\partial x}= \\
&k_s hS\left(K_d^r C_d^r-C_a^r\right)+\left\langle\begin{array}{ll}\left(C_a^s-C_a^r\right)\left(e_r+e_h\right), & e_h>0, \\ \left(C_a^s-C_a^r\right)e_r, & e_h<0.\end{array}\right.
\end{aligned}
\tag{4.34}
$$

Thus, the mathematical formulation of the problem leads to a system of four coupled differential Eqs. (4.33), (4.34), (4.27) and (4.28), including four concentration functions C_d^r, C_d^s, C_a^r, C_a^s, and two additional equations, determining potential-controlling functions – the concentration, S (4.32), and hydrodynamic, h, q ($q=\alpha h^n$) (4.31). Given appropriate initial and boundary conditions, this system of equations possesses a solution:

$$
C_d^s(x,0)=C_{d0}^s,\ C_a^s(x,0)=K_d^s C_{d0}^s,\ C_d^s(0,t)=0,\ C_d^r(0,t)=0,\ S(0,t)=0,
\tag{4.35a}
$$

$$
h(x,0)=0, h(0,t)=0, q(x,0)=0, q(0,t)=0.
\tag{4.35b}
$$

If we consider sorption interaction between solutes and suspended particles under equilibrium condition, such that $C_a^r=K_d^r C_d^r$, the addition of (4.33) and (4.34) yields:

$$
\begin{aligned}
h\left(1+K_d^r S\right)\frac{\partial C_d^r}{\partial t}+q\left(1+K_d^r S\right)\frac{\partial C_d^r}{\partial x}&=k_e\left(C_d^s-C_d^r\right)+\frac{e_r\theta}{\rho_b}\left(C_d^s-\lambda C_d^r\right)-rC_d^r+ \\
&+\left\langle\begin{array}{ll}\left(C_a^s-C_a^r\right)\left(e_r+e_h\right), & e_h>0, \\ \left(C_a^s-C_a^r\right)e_r, & e_h<0,\end{array}\right.
\end{aligned}
\tag{4.36}
$$

i.e., the number of equations and variables decreases.

4.3 Analysis of Behavior of Suspended Sediments and Adsorbed Solutes in a Runoff System

As it was repeatedly mentioned above, rainfall onto polluted soil layers causes its erosion, such that chemicals can undergo rapid transport in a particulate form: the particles of erosion origin become carriers for contaminants. In the development of the model, we suppose that soil erosion is governed by raindrop impact and water runoff. The solid particles thus released from the surface are involved in downslope transport with water; however, their net deposition onto soil surface is taking place, so the concentration of suspension in water decreases.

4.3.1 Special-Case Solutions for a Suspended Sediment Transport Problem

In this section, for simplicity sake, all models considered bellow are restricted by the use of a single representative sediment-size class. Then, the derivation of analytic solutions describing sediment transport with overland flow is based on the following coupled kinematic-wave and erosion equations

$$\frac{\partial h}{\partial t}+\frac{\partial q}{\partial x}=r, \tag{4.37}$$

$$\frac{\partial hS}{\partial t}+\frac{\partial qS}{\partial x}=e_r+e_h, \tag{4.38}$$

where sink and source functions e_r (rainsplash erosion rate) and e_h (flow erosion/ gravitational sedimentation) are defined by (4.4) and (4.10)/(4.11).

To describe the erosion/deposition processes occurring in an idealized slope under constant precipitation, we consider two limiting scenarios. The first scenario implies a joint effect of two processes with different directions, namely, rainsplash erosion of the soil surface (4.4) and the gravitational settling of the eroded particles (4.11) carried by water flowing over the surface in the absence of flow-driven erosion or, in other words, when flow does not exceed the threshold for particle entrainment. The second scenario implies that the rainsplash erosion is accompanied by flow erosion induced by the soil bed shear stress (4.10); however, in this case, the model does not take into account the gravitational settling of particles onto the surface.

The choice of the model is determined by a series of natural factors, including hillslope and rainfall characteristics, soil erodibility, suspended sediment transport potential in the surface water and some others. To simplify the analytical considerations, only the case of constant rainfall excess and the rising stage of runoff period are considered.

Let the sediment transport capacity, T_c, be described by a generalized expression (4.14), rewritten as

$$T_c=\varphi_i q^{\beta}, \tag{4.39}$$

where coefficient β can take values 1 or 2; $\varphi_i=\varepsilon S_0^{\gamma}$ (the dimension depends on the exponent β). Thus, we consider the extreme values in the range, which contains the majority of experimental data (Sect. 4.1).

With the above reasoning, combining (4.38), (4.37), and (4.39) leads to the model

$$h\frac{\partial S}{\partial t}+q\frac{\partial S}{\partial x}=a_0 r-rS+\left\langle\begin{array}{ll} v\left(\varphi_i q^{\beta-1}-S\right)<0 & -\ \text{deposition,}\\ \sigma q\left(\varphi_i q^{\beta-1}-S\right)>0 & -\ \text{flow erosion,}\end{array}\right. \tag{4.40}$$

which adopts both scenarios; here, v is the velocity of particle settling under gravitational forces (ms^{-1}); σ is a coefficient (m^{-1}), determining the ratio of detachment potential of the soil (D_c) to sediment transport capacity (T_c), a_0 is the detachability coefficient.

4.3.1.1 First Scenario

Case $\beta = 1$ At the initial stage of overland flow formation ($t < t_e$, Eq. 2.12), when its depth linearly grows with time ($h = rt$, $q = \alpha(rt)^n$), Eq. (4.40) becomes

$$\frac{\partial \overline{S}}{\partial z} + z^{n-1}\frac{\partial \overline{S}}{\partial y} = \frac{1}{z}\left[1 + k_1\overline{\varphi}_1 - (1 + k_1)\overline{S}\right], \tag{4.41}$$

$\overline{S} = S/a_0$, $z = rt$, $y = rx/\alpha$, $k_1 = v/r$, $\overline{\varphi}_1 = \varphi_1/a_0$. Here φ_1 serves as the parameter S_m, which limits the saturation of runoff by disperse material ($\varphi_1 = S_m$), where $\overline{\varphi}_1$ is to be less than 1 ($S_m < a_0$), because otherwise we have $e_h > 0$, which contradicts to the initial problem formulation; k_1 is the relative settling velocity characteristic. Equation 4.41 corresponds to an ordinary differential equation

$$z\frac{d\overline{S}}{dz} = 1 + k_1\overline{\varphi}_1 - (1 + k_1)\overline{S}, \tag{4.42}$$

describing the process along characteristic lines (3.30)/(3.32).

In the domain of linear growth of flow discharge in the direction x ($q = rx$), which is valid at $t \geq t_e$, Eq. (4.40) becomes

$$\frac{\partial \overline{S}}{\partial y} + \frac{y^{1/n}}{y}\frac{\partial \overline{S}}{\partial z} = \frac{1}{y}\left[1 + k_1\overline{\varphi}_1 - (1 + k_1)\overline{S}\right]. \tag{4.43}$$

It corresponds to the ordinary differential equation

$$y\frac{d\overline{S}}{dy} = 1 + k_1\overline{\varphi}_1 - (1 + k_1)\overline{S}, \tag{4.44}$$

which describes the process along the characteristic (3.38); $y = rx/\alpha$.

When solving Eqs. (4.42) and (4.44), one faces the problem of singularity, which can be avoided by introducing an initial water depth in the problem (see Sect. 3.2.2). With this done, we obtain a single solution of Eqs. (4.42) and (4.44):

$$\overline{S} = \frac{S}{a_0} = \frac{1 + k_1\overline{\varphi}_1}{1 + k_1}, \tag{4.45}$$

At $k_1 >> 1$, we obtain the limiting value $S = S_m$; at $k_1 << 1$ $S = a_0$.

Case $\beta = 2$ At the initial moment, in the flow domain where h is a linear function of time ($h = rt$, $q = \alpha(rt)^n$), Eq. (4.38) becomes

$$z\frac{\partial \overline{S}}{\partial z} + z^n \frac{\partial \overline{S}}{\partial y} = 1 + k_1 k z^n - (1 + k_1)\overline{S}, \tag{4.56}$$

$\overline{S} = S/a_0, z = rt, y = rx/\alpha, k_1 = v/r, k = \varphi_2\alpha/a_0$. The equation for characteristics for (4.44) has the form of (3.30) or (3.32). Concentration changes along those characteristics are described by an equation

$$z\frac{d\overline{S}}{dz} = 1 + k_1 k z^n - (1 + k_1)\overline{S}. \tag{4.57}$$

The general solution of this equation is

$$\overline{S} = \frac{1}{1 + k_1} + \frac{k_1 k z^n}{1 + k_1 + n} + A_1(z)^{-(1+k_1)}, \tag{4.58}$$

where A_1 is an integration constant. It must be equal to 0, because otherwise the solution (4.58) is not determined at $t \to 0$. Thus, at the initial stage of the process, function $\overline{S}$ is determined by the expression

$$\overline{S}(z) = \frac{1}{1 + k_1} + \frac{k_1 k z^n}{1 + k_1 + n}, z < y^{1/n} = (rx/\alpha)^{1/n}. \tag{4.59}$$

In the domains were functions q and h are stationary ($q = rx$), Eq. (4.40) becomes

$$\frac{h}{y}\frac{\partial \overline{S}}{\partial z} + \frac{\partial \overline{S}}{\partial y} = \frac{1}{y}\left(1 + k_1 k y - (1 + k_1)\overline{S}\right), y = \frac{rx}{\alpha}. \tag{4.60}$$

The equation for characteristics for (4.60) has the form of (3.38). Variations in the concentration along those characteristics are described by an ordinary differential equation

$$y\frac{d\overline{S}}{dy} = 1 + k_1 k y - (1 + k_1)\overline{S}. \tag{4.61}$$

Its solution is

$$\overline{S} = \frac{1}{1 + k_1} + \frac{k_1 k}{2 + k_1} y + A_2 y^{-(1+k_1)}, \tag{4.62}$$

where A_2 is an integration constant.

The constant A_2 can be found from the condition that the concentrations determined by solutions (4.59) and (4.62) are equal in the point of intersection of characteristics (2.12) and (3.32) (Fig. 3.3a). In solution (4.59), time t is determined from the equation $rt = (rx'/\alpha)^{1/n}$, in the solution (4.62), $x = x'$. From here, we have:

$$A_2 = \frac{1-n}{(1+n+k_1)(2+k_1)} k_1 k (y')^{2+k_1}, y' = \frac{rx'}{\alpha}. \tag{4.63}$$

The solution (4.62) becomes

$$\overline{S} = \frac{1}{1+k_1} + \frac{k_1 ky}{2+k_1}\left[1 + \frac{n-1}{1+n+k_1}\left(\frac{x'}{x}\right)^{2+k_1}\right]. \tag{4.64}$$

Coordinate x' determines the relationship between the current coordinates x and t on the characteristic (3.38) in the form (3.43d), therefore, the solution (4.64) becomes:

$$\overline{S}(y,z) = \frac{1}{1+k_1} + \frac{k_1\overline{\varphi}_2}{2+k_1}\left[1 + \frac{1-n}{1+n+k_1}\left(\frac{ny^{1/n} - z}{(n-1)y^{1/n}}\right)^{n(2+k_1)}\right], y^{1/n} < z \leq ny^{1/n}, \tag{4.65}$$

$$\overline{\varphi}_2 = \frac{\varphi_2 \alpha h_x^n}{a_0} = \frac{\varphi_2 rx}{a_0}. \tag{4.66}$$

Finally, the asymptotic solution of the problem takes the form

$$\overline{S}(y) = \frac{1}{1+k_1} + \frac{k_1\overline{\varphi}_2}{2+k_1}, z > ny^{1/n}. \tag{4.67}$$

The solution of the problem can be plotted as a function $\overline{S}$ of a dimensionless complex variable, $\tau = rt/h_x$ ($h_x^n = rx/\alpha$), and dimensionless coefficients $\overline{\varphi}_2$ and k_1, which control soil erosion (Fig. 4.5). The plots show two plateaus: an initial one, where particle concentration does not depend on the position of point x on the slope, and a final one, where the steady-state value S is determined by the current value of x (the greater the distance from the water divide, $x = 0$, the larger S).

As can be seen from Fig. 4.5, the concentration of solid particles in water increases with increasing sediment transport capacity, T_c (~ φ_2). An increase in k_1 leads to an inverse result, because of an increase in the role of their gravitational deposition.

The solution of the problem at $\varphi_2 = 0$ (i.e., when $e_h = -Sv$) becomes one-parameter, corresponding to the formula

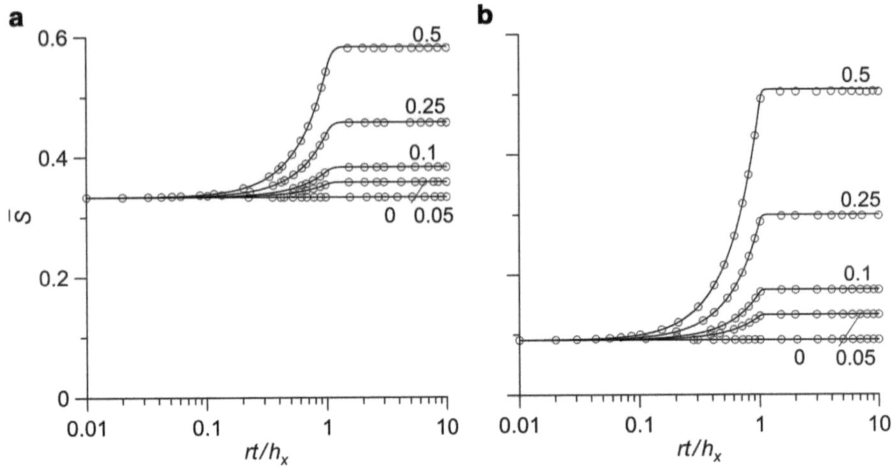

Fig. 4.5 Function $\overline{S}(\tau)$ at different values of dimensionless group $\overline{\varphi}_2$ (numbers at curves). The *full lines* show the analytical solution (4.59)/(4.65)/(4.67); the *circles* show a numerical solution of water erosion of the surface (4.40). (**a**) $k_1 = 2$, (**b**) $k_1 = 10$. $n = 5/3$, $a_0 = 10\ \text{kg/m}^3$

$$\overline{S} = \frac{S}{a_0} = \frac{1}{1 + k_1}, \tag{4.68}$$

which is valid throughout the range of values of y and τ.

The obtained formulas (4.59), (4.65), (4.67), and (4.68) are in complete agreement with the numerical solution of the initial partial differential equation (4.40) – Fig. 4.5. The applicability of the formulas is limited by the condition $e_h < 0$ (no hydraulic erosion of the surface), or, given the flow transport capacity (4.37), $\varphi_2 q < S$, which holds when the criterion

$$\overline{\varphi}_2 < \frac{2 + k_1}{2(1 + k_1)} \tag{4.69}$$

is valid.

4.3.1.2 Second Scenario

This model includes mechanisms of soil detachment by raindrops and runoff, implying the situation when, because of the high transport capacity of the flow, the gravity settling of suspended particles can be neglected.

Case $\beta = 1$ In the domain where flow depth is a linear function of time ($h = rt$, $q = \alpha(rt)^n$) Eq. (4.40) becomes

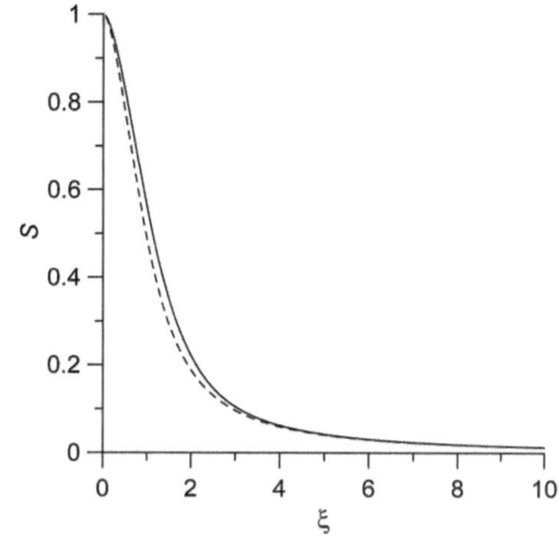

Fig. 4.6 Function $F(\xi)$ (4.72) (*full line*) and its approximation by (4.72a) (*broken line*)

$$\frac{\partial \overline{S}}{\partial z} + z^{n-1}\frac{\partial \overline{S}}{\partial y} = \frac{1}{z}\left[1 + \frac{\sigma\alpha}{r}z^n(\overline{\varphi}_1 - \overline{S}) - \overline{S}\right], \tag{4.70}$$

$\overline{S} = S/a_0,\ z = rt,\ y = rx/\alpha,\ \overline{\varphi}_1 = \varphi_1/a_0$. This corresponds to a linear differential equation

$$\frac{d\overline{S}}{dz} + \frac{1}{z}(1 + kz^n)\overline{S} = \frac{1}{z}(1 + \overline{\varphi}_1 kz^n), k = \frac{\alpha\sigma}{r}, \tag{4.71}$$

which describes the process along characteristics (3.30).

The solution of the ordinary differential Eq. (4.71) can be written as (Hjelmfelt et al. 1975)

$$\frac{\overline{S} - \overline{\varphi}_1}{1 - \overline{\varphi}_1} = F(\xi) = \frac{1}{\xi}\exp(-\xi^n)\int_0^{\xi}\exp(\lambda^n)d\lambda, \tag{4.72}$$

where $\xi^n = (rt)^n k/n$. The solution (4.72) is valid at $rt < h_x = rx/\alpha$. Function $F(\xi)$ is plotted in Fig. 4.6. The same figure gives a curve calculated by the formula

$$F(\xi) \approx \frac{1 - \exp(-n\xi^n)}{n\xi^n}, \tag{4.72a}$$

approximating this function.

In the domain where flow discharge is a linear function of coordinate x ($q = rx$), Eq. (4.40) becomes

$$\frac{\partial \overline{S}}{\partial y} + \frac{y^{1/n}}{y}\frac{\partial \overline{S}}{\partial z} = \frac{1}{y}\left(1 - \overline{S}\right) + k\left(\overline{\varphi}_1 - \overline{S}\right). \tag{4.73}$$

It corresponds to the ordinary differential equation

$$\frac{d\overline{S}}{dy} + \left(\frac{1}{y} + k\right)\overline{S} = \frac{1}{y} + k\overline{\varphi}_1, \tag{4.74}$$

which describes the process along the characteristic (3.38).

The general solution of Eq. (4.74) is given by

$$\frac{\overline{S} - \overline{\varphi}_1}{1 - \overline{\varphi}_1} = \frac{1}{ky} + \frac{A_2}{(1 - \overline{\varphi}_1)ky}e^{-ky}, \tag{4.75}$$

where A_2 is an integration constant.

The constant A_2 can be found from the condition of equality of concentrations given by solutions (4.59) and (4.62) at the intersection point of the characteristics (2.12) and (3.32) (Fig. 3.3a)

$$F(\xi') = \frac{1}{ky'}\left[1 + \frac{A_2}{(1 - \overline{\varphi}_1)}e^{-ky'}\right], \xi'^n = [\xi(y')]^n = \frac{\sigma x'}{n}, y' = x'r/\alpha. \tag{4.76}$$

Thence we have

$$A_2 = (1 - \overline{\varphi}_1)[ky'F(\xi') - 1]\exp(-ky'). \tag{4.77}$$

Since

$$\frac{rx'}{\alpha} = \left(\frac{nh_x - rt}{n - 1}\right)^n = h_x^n\left(\frac{n - \tau}{n - 1}\right)^n \tag{4.78}$$

(see Eq. 3.43), and

$$\xi'^n = \frac{h_x^n}{n}\left(\frac{\sigma\alpha}{r}\right)\left(\frac{n - \tau}{n - 1}\right)^n = \frac{kh_x^n}{n}\left(\frac{n - \tau}{n - 1}\right)^n, \tau = \frac{rt}{h_x}, h_x^n = \frac{rx}{\alpha}, \tag{4.78a}$$

the final solution becomes

$$\frac{\overline{S} - \overline{\varphi}_1}{1 - \overline{\varphi}_1} = \frac{1}{kh_x^n} + \frac{1}{kh_x^n}\left[kh_x^nF(\xi') - 1\right]\exp\left\{kh_x^n\left[\left(\frac{n - \tau}{n - 1}\right)^n - 1\right]\right\}. \tag{4.79}$$

The solution describes the process in the domain $1 < \tau \leq n$.

The stationary solution ($\tau > n$) is

$$\frac{\overline{S} - \overline{\varphi}_1}{1 - \overline{\varphi}_1} = \frac{1 - \exp\left(-kh_x^n\right)}{kh_x^n} = \frac{1 - e^{-\sigma x}}{\sigma x}. \tag{4.80}$$

The plots in Fig. 4.7 show that at the beginning of rainfall, the suspension concentration in point x is minimal $S = a_0$, after which it increases and stabilizes according to (4.80). With an increase in x, the concentration of suspension tends to a limit value $S = \varphi_1$, i.e., at some distance from the water divide point ($x = 0$), the concentration stabilizes and stops increasing.

Case $\beta = 2$ The solution (4.40) for the case where flow transport capacity is a quadratic function of its discharge (4.39) can be obtained numerically. The plots in Fig. 4.8 can be interpreted as a family of curves $\overline{S}$, obtained for a fixed point x at a given ratio $\varphi_2 r/a_0$, but at different σ. It can be seen that an increase in σ (soil erodibility) leads to an increase in particle concentration in water. As can be seen from the analysis of numerical solutions, function $\overline{S}$ ($\beta = 2$) shows steeper transient branches of the appropriate plot (Fig. 4.8), compared with the case $\beta = 1$ (Fig. 4.7), but similar steady-state values of $\overline{S}$ for the same values of the dimensionless group φ_i. At the initial moments, as before, we have $S = a_0$.

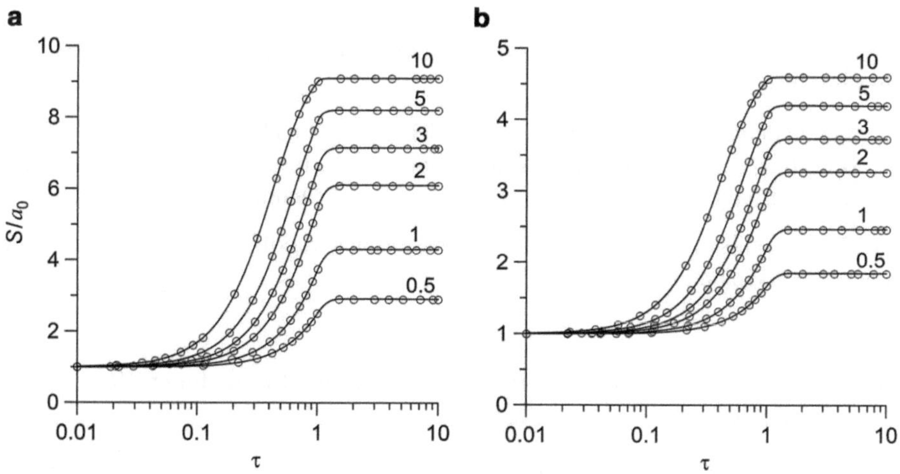

Fig. 4.7 Function $\overline{S}(\tau_x)$ at different values of dimensionless group $kh_x^n = \sigma x$ (numbers at *curves*) – the case $T_c = \varphi_1 q$. The *full lines* give the analytical solution (4.72)/(4.79)/(4.80); the *circles* give the numerical solution for water erosion of the surface (4.40). (**a**) $\overline{\varphi}_1 = \varphi_1/a_0 = 10$, (**b**) $\overline{\varphi}_1 = \varphi_1/a_0 = 5$. $n = 5/3$

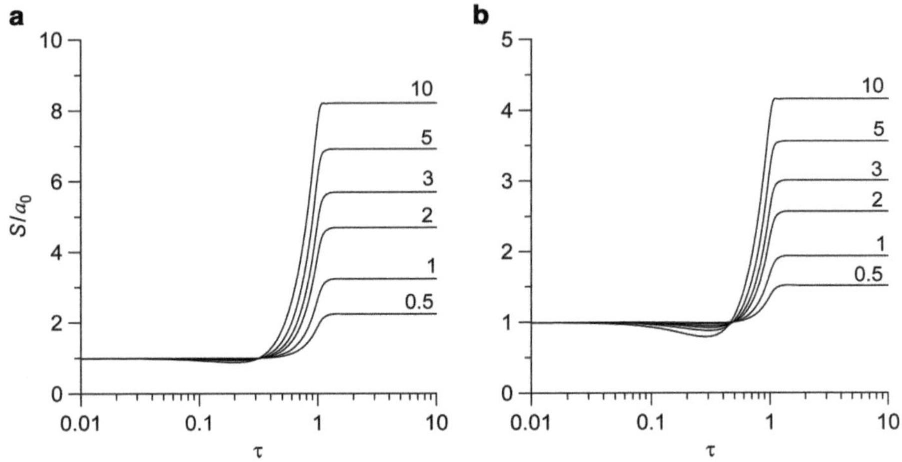

Fig. 4.8 Function $\overline{S}(\tau_x)$, a numerical solution of surface water erosion Eq. (4.40) ($T_c = \varphi_2 q^2$), at different values of dimensionless group $kh_x^n = \sigma x$ (numbers at *curves*). (**a**) $\overline{\varphi}_2 = \varphi_2 rx/a_0 = 10$, (**b**) $\overline{\varphi}_2 = \varphi_2 rx/a_0 = 10$. $n = 5/3$

4.3.2 *Approximate Solutions for Multi-size Class Soil Erosion and Deposition along a Hillslope Profile*

As an illustration, we present the mathematical problem setup restricted by two limiting conditions relating to:

- rainfall detachment, re-detachment, and deposition of soil particles of different size when flow entrainment from the original soil and the deposited layer can be ignored;
- runoff-driven erosion when detachment and re-detachment can be ignored.

Thus, in this section rainfall-driven erosion and runoff-driven erosion will be considered separately.

4.3.2.1 Dynamic Erosion of Soil Caused by Rainfall Detachment/Re-detachment

Analysis of Full Problem Solution The accepted limitation, $r_i = r_{ri} = 0$, allows the system of Eqs. (4.18), and (4.19) to be rewritten as (Hairsine et al. 1999; Hogarth et al. 2004b; Le et al. 2013):

$$\frac{\partial hS_i}{\partial t} + \frac{\partial qS_i}{\partial x} = e_i + e_{ri} - d_i, \tag{4.81}$$

$$\frac{\partial m_{di}}{\partial t} = -(e_{ri} - d_i). \tag{4.82}$$

Obviously, the system of Eqs. (4.81)–(4.82) will best describe soil erosion under the conditions of shallow rain-impacted flow that occur in interrill areas. For steady-state overland flow, when discharge is defined by a relation $q = rx$ (2.14), Eq. (4.81) can be rewritten in a form

$$\frac{\partial S_i}{\partial t} + \frac{q}{h}\frac{\partial S_i}{\partial x} = \frac{1}{h}(e_i + e_{ri} - d_i - rS_i). \tag{4.83}$$

Substituting the definition of e_i, e_{ri} (4.20) and d_i (4.22) into Eqs. (4.83) and (4.82) and assuming that each size class of the sediment contains an equal mass of sediment, i.e. $p_i = 1/I$, a system of $2I$ partial differential equations for determining $2I$ unknowns ($S_1, \ldots, S_I$ and $m_1, \ldots, m_I$) can be obtained:

$$\frac{\partial S_i}{\partial t} + \frac{q}{h}\frac{\partial S_i}{\partial x} = \frac{1}{h}\left[(1 - H)\frac{a_0 r}{I} + \frac{m_{di}}{m_{dt}^{*}}a_d r - S_i(v_i + r)\right], \tag{4.84}$$

$$\frac{\partial m_{di}}{\partial t} = S_i v_i - \frac{m_{di}}{m_{dt}^{*}}a_d r. \tag{4.85}$$

An approximate analytical solution of (4.84) and (4.85) was developed by Hogarth with co-workers (2004b) for "zero" initial and boundary conditions (treating inflow of clear water at the top of the eroding slope). It was supported by comparison with a numerical solution obtained using an upwinding finite difference method (Hogarth et al. 2004a).

The obtained analytical solution can be used to predict the spatial and temporal behavior of the system in terms of total concentration as a function of x and t, $S(x, t)$, for different particle size classes, I. The main conclusions, following from the particular solutions of the problem, can be formulated as follows (Hogarth et al. 2004b).

1. All profiles $S(x)$ at different time, t, and for fixed value of I show an increase in S from the inflow boundary down the eroding soil surface due to sediment added to water layer by rainfall dislodgment. The gradient, dS/dx, which characterizes this growth, decreases with time because of the removal of fine sediment fraction from the soil.
2. However, in the initial moments, S can be of non-monotonic (wave-like) character: the value of S abruptly increases near $x = 0$ and drops in the middle part of the flow. This decline of $S(x)$ curve beyond its maximum is most likely due to the dilution of particles dispersed in runoff by rainfall-added water. Much of the sediment concentration at this early time is due to fine sediment, whose settling velocity is so low that it will settle only very slowly if at all, and so will be transported down the slope essentially at the velocity of overland flowing water. In this case, the concentration S near the boundary $x = 0$ from the side of the slope very rapidly reaches the value S_0, which remains nearly unchanged throughout the process of overland flow formation. The equilibrium profile $S(x, \infty)$ also tends to this value.

3. As the number of soil classes adopted, I, increases, a better description is given to the distribution of the fine slow-settling fraction of the soil. It was found that the dependence of the predicted results of analysis on the value of I adopted is much greater at the early time, and, contrary, at longer times, the choice of the value of I is of much less importance.

The latter conclusion is important with a viewpoint of accuracy of contaminant transport forecasting. Since finer constituents of the sediment are more readily transported and chemically enriched, their migration determines the major adverse effect on the quality of water ecosystems and chemical enrichment of the eroded soils (Rose et al. 2007).

Approximate Solution of the Problem (A Multi-class Solution, $i = 1, \ldots, I$) Let h and q be constants with respect to x and t. In this case, HR model (4.81)–(4.82) becomes

$$h\frac{\partial S_i}{\partial t} + q\frac{\partial S_i}{\partial x} = e_i + e_{ri} - d_i, \tag{4.86}$$

$$\frac{\partial m_{di}}{\partial t} = -(e_{ri} - d_i). \tag{4.87}$$

It can be supposed that the system of sediment/suspended solids under steady flow conditions will rapidly attain a dynamic equilibrium, which can be characterized by an equality between the rate of particle re-detachment from the newly formed sediment and the rate of gravitational settling of particles, a fact implying that, during erosion, the terms e_{ri} (Eq. 4.20) and d_i (Eq. 4.22) are equal (Proffitt and Rose 1992; Hairsine and Rose 1992a; Marshall et al. 1999). Now, considering (4.24), we have

$$v_i S_i = A\frac{m_{di}}{m^*_{dt}}, \text{ or } \frac{\partial m_{di}}{\partial t} = \frac{v_i m^*_{dt}}{A}\frac{\partial S_i}{\partial t}, A = ra_d. \tag{4.88}$$

The assumption made has no strict theoretical grounds, but it does not contradict to any postulate of physicochemical hydrodynamics, which defines the dynamic equilibrium as the equality of the rates of forward and reverse reactions, in this context, rainfall detachment and deposition.

The equilibrium condition (4.88) simplifies the search of analytical solution of the formulated problem. Adding the left and right sides of Eqs. (4.86) and (4.87), and, considering (4.88), we obtain (Rose et al. 2007)

$$\frac{\partial S_i}{\partial t} + \frac{1}{E_i}\frac{\partial S_i}{\partial x} = \frac{(1 - H)B}{qE_i}, \tag{4.89}$$

$$E_i = \frac{h}{q} + \frac{v_i m^*_{dt}}{Aq}, B = ra_0. \tag{4.89a}$$

Rose et al. (2007) demonstrated the possibility to convert the partial differential Eq. (4.89) into the following ordinary differential equation for $x = L$ (measurement site length):

$$\frac{dS_i}{dt} + \frac{S_i}{E_i L} = \frac{(1-H)B}{qE_i}. \tag{4.90}$$

Equation (4.90) was used to derive two major characteristics of erosion process (Sect. 4.1.2): (1) the fraction of shielding of the original soil matrix resulting from depositing sediment as function of time, $H(t)$, and (2) the concentration of eroded products in water and sediment.

Thus, an approximate expression for H is given as a function of time:

$$H = H_\infty\left[1 - \exp\left(-t/t^*\right)\right], \tag{4.91}$$

where asymptotic expression for H_∞, determining the fraction of shielding of soil from raindrop impact after a steady state is attained (when $\partial S_i/\partial t = \partial m_i/\partial t = 0$), is given by

$$H_\infty = \frac{L}{L + L^*}, \quad L^* = \frac{q}{fIv_a}, \tag{4.92}$$

$t^* = H_\infty Eq/fIv_a$, $f = a_0/a_d$, $E = \sum v_i/\sum v_i/E_i$, $v_a = \sum v_i/I$ is the average value of settling velocity. The total steady-state concentration of suspended solids will be:

$$S(L) = \frac{a_d r}{v_a} H_\infty. \tag{4.93}$$

Substituting Eq. 4.91 for $H(t)$ into Eq. 4.90 (for $L = x$) leads to the following solution for sediment concentration for size class i at point x and time t (Rose et al. 2007):

$$\begin{gathered} S_i(x,t) = \frac{(1-H_\infty)Bx}{q} \\ \left[1 - \exp\left(-\frac{t}{E_i x}\right)\right] + \frac{H_\infty B x t^*}{q(E_i x - t^*)}\left[\exp\left(-\frac{t}{E_i x}\right) - \exp\left(-\frac{t}{t^*}\right)\right], \\ S(x,t) = \sum_i S_i(x,t). \end{gathered} \tag{4.94}$$

Approximate Solution of the Problem (A Single-Class Solution, $i = 1$) For a particular scenario $i = 1$ (single-size class sediment), the original system of Eqs. (4.81)–(4.82) (steady-state runoff) becomes:

$$\frac{\partial hS}{\partial t}+\frac{\partial qS}{\partial x}=-vS+\frac{m_d}{m_d^*}(a_d-a_0)r+a_0r, \tag{4.95}$$

$$\frac{\partial m_d}{\partial t}=vS-\frac{m_d}{m_d^*}a_dr, \tag{4.96}$$

$$\frac{\partial q}{\partial x}=r. \tag{4.97}$$

Let us rewrite (4.95) and (4.96) in new dependent variables $m_s \equiv m_s(x,t) = S(x,t)h(x)$ (mass per unit area of sediment in the water column, kgm^{-2}) and m_d (mass per unit area of sediment in the deposited layer, kgm^{-2}) (Lisle et al. 1998; Barry et al. 2010), taking into account that the water flux varies with distance downslope according to Eq. 4.97:

$$\frac{\partial m_s}{\partial t}+\frac{\partial m_s}{\partial t_0}=-\frac{m_s}{h}(v+r)+\frac{m_d}{m_d^*}(a_d-a_0)r+a_0r, \tag{4.98}$$

$$\frac{\partial m_d}{\partial t}=\frac{m_s}{h}v-\frac{m_d}{m_d^*}a_dr, \tag{4.99}$$

where

$$t_0=\int\limits_{x_0}^{x}(h/rx)dx=(n/r)(r/\alpha)^{1/n}\left(x^{1/n}-x_0^{1/n}\right)$$

for relationship (2.3) between flow discharge and flow depth. Such problem formulation, unlike the case of $q=\text{const}$, accounts for the dilution of suspension by rainwater. Normalizing m_s and m_d by m_d^* and introducing an averaged value of flow depth h_a in the determination of the first terms in the right sides of Eqs. 4.98 and 4.99, we come to the system of partial differential equations:

$$\frac{\partial p}{\partial t}+\frac{\partial p}{\partial t_0}=-\lambda(1+k_1)p+\left(a'_d-a'_0\right)g+a'_0, \tag{4.100}$$

$$\frac{\partial g}{\partial t}=\lambda k_1 p-a'_d g, \tag{4.101}$$

$p=m_s/m_d^*$ is the normalized mass m_s [−], $g=m_d/m_d^*$ is the normalized mass m_d [−], $a'_0=a_0r/m_d^*$ [s^{-1}], $a'_d=a_dr/m_d^*$ [s^{-1}], $\lambda=r/h_a$ [s^{-1}], $k_1=v/r$ (relative settling velocity characteristics, Sect. 4.3.1) [−].

The system of Eqs. (4.100), (4.101) is solved at zero initial and boundary conditions:

$$p(0,t) = 0,\ p(x,0) = 0,\ g(x,0) = 0. \tag{4.102}$$

The solution of (4.100), (4.101) in the Laplace space is:

$$\widetilde{p}(x,s) = \frac{a'_0}{s\,y(s)}[1 - \exp[-t_0 y(s)]], \tag{4.103}$$

$$\widetilde{g}(x,s) = \frac{k_1}{s + a'_d}\widetilde{p}(x,s), \tag{4.104}$$

where $\widetilde{p}(x,s)$ and $\widetilde{g}(x,s)$ are the transformed function of $p(x, t)$ and $g(x, t)$; s is the Laplace transform variable (with respect to t),

$$y(s) = s + \lambda(1 + k_1) - \frac{(a'_d - a'_0)\lambda k_1}{s + a'_d}. \tag{4.105}$$

The inverse transform of functions $\widetilde{p}$ and $\widetilde{g}$ were taken using formulas given in Barry et al. (2010). For the relative density of soil particles in runoff water, $p = m_s/m^*_d$, the solution is

$$p(x,t) = \overline{w}(\tau) - H(\tau - \tau_0)\exp[-(1 + k_1)\tau_0]\times \left\{ \begin{array}{l} \overline{w}(\tau - \tau_0) + \\ \int\limits_{\tau_0}^{\tau} \overline{w}(\tau - \xi)\exp[-\overline{a}_d(\xi - \tau_0)]\sqrt{\dfrac{\overline{a}_d(1 - f)k_1\tau_0}{\xi - \tau_0}}\, I_1\left[2\sqrt{\overline{a}_d(1 - f)k_1\tau_0(\xi - \tau_0)}\,\right] d\xi \end{array} \right\}, \tag{4.106}$$

$$\overline{w}(\tau) = \frac{f\overline{a}_d}{1 + k_1 f}\times \left\{1 + \exp\left[-\frac{\tau}{2}(1 + k_1 + \overline{a}_d)\right]\left[\frac{(1 + k_1)^2 - \overline{a}_d^2 - \psi}{2\overline{a}_d\psi^{1/2}}\sinh\left(\frac{\tau\psi^{1/2}}{2}\right) - \cosh\left(\frac{\tau\psi^{1/2}}{2}\right)\right]\right\}. \tag{4.106a}$$

The solution for the relative density of particles on soil surface, $g = m_d/m^*_d$, has the same representation (4.106), where

$$\overline{w}(\tau) = \frac{k_1 f}{2(1 + k_1 f)\psi^{1/2}} \left\{ \begin{array}{l} 2\psi^{1/2} + (1 + k_1 + \overline{a}_d - \psi^{1/2})\exp\left[-\frac{\tau}{2}\left(1 + k_1 + \overline{a}_d + \psi^{1/2}\right)\right] \\ -(1 + k_1 + \overline{a}_d + \psi^{1/2})\exp\left[-\frac{\tau}{2}\left(1 + k_1 + \overline{a}_d - \psi^{1/2}\right)\right] \end{array} \right\}, \tag{4.106b}$$

$H(\tau)$ is the Heaviside function; $\psi = (1 + k_1 - \overline{a}_d)^2 + 4k_1\overline{a}_d(1 - f)$, $\tau = \lambda t = rt/h_a$, $\tau_0 = \lambda t_0$, $\overline{a}_0 = a_0 h_a/m^*_d$, $\overline{a}_d = a_d h_a/m^*_d$, $f = \overline{a}_0/\overline{a}_d = a_0/a_d$.

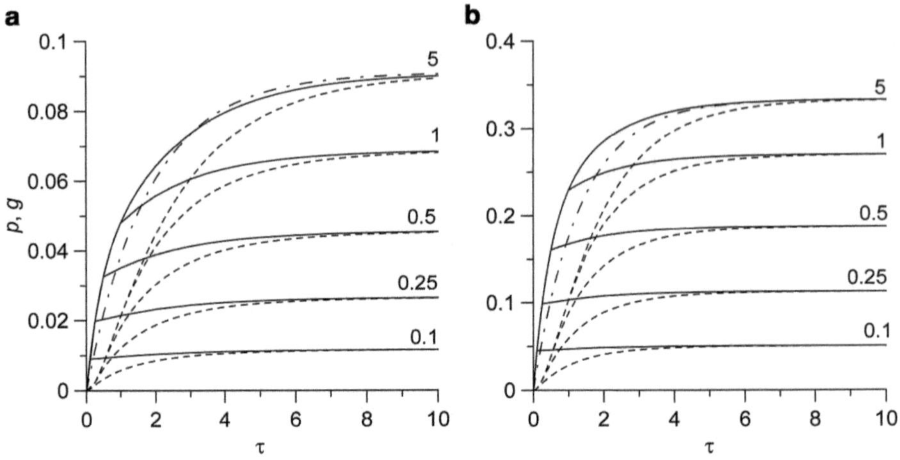

Fig. 4.9 An illustration of variation of sediment concentrations p (*solid curves*) and g (*dashed curves*) predicted by Eq. 4.106. Parameters are $\overline{a}_d = 1, k_1 = 1, f = 0.1$ (**a**), $f = 0.5$ (**b**) Predictions are shown for a range of $\tau_0 = \lambda t_0$ (numbers at the curves). *Dash*-and-*dot curves* correspond to approximate solution for p (4.110)

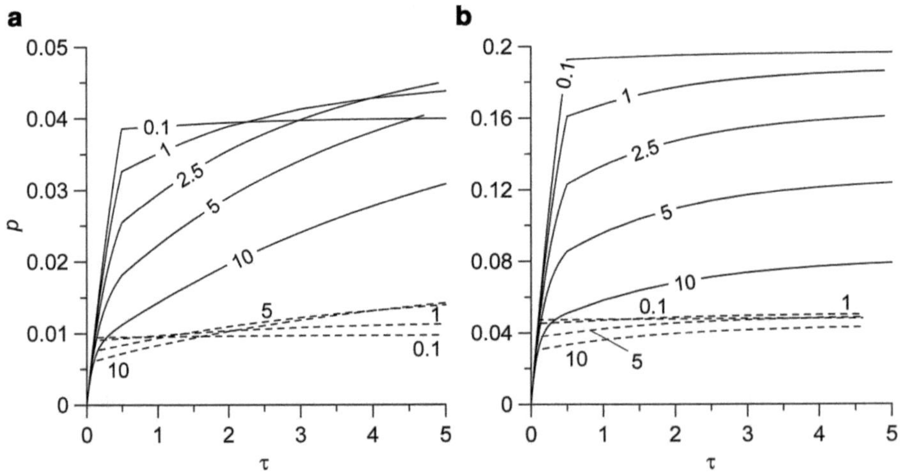

Fig. 4.10 An illustration of variation of sediment concentration p predicted by Eq. 4.106. The parameters are $\overline{a}_d = 1, f = 0.1$ (**a**), $f = 0.5$ (**b**). Predictions are shown for a range of k_1 (numbers at the *curves*) and $\tau_0 = \lambda t_0$ (0.5 – *solid curves*, 0.1 – *dashed curves*)

An illustration of the functional form of Eq. 4.106 is given in Figs. 4.9, 4.10, and 4.11. The figures present an evaluation of the rainfall-driven soil erosion process in terms of HR model. The behavior of breakthrough curves is largely determined by the space and time scale at which the process is considered. The erosion involves simultaneous detachment, deposition, and re-detachment of particles, which, as can be seen from Fig. 4.9, leads to sediment accumulation on soil surface. The amount

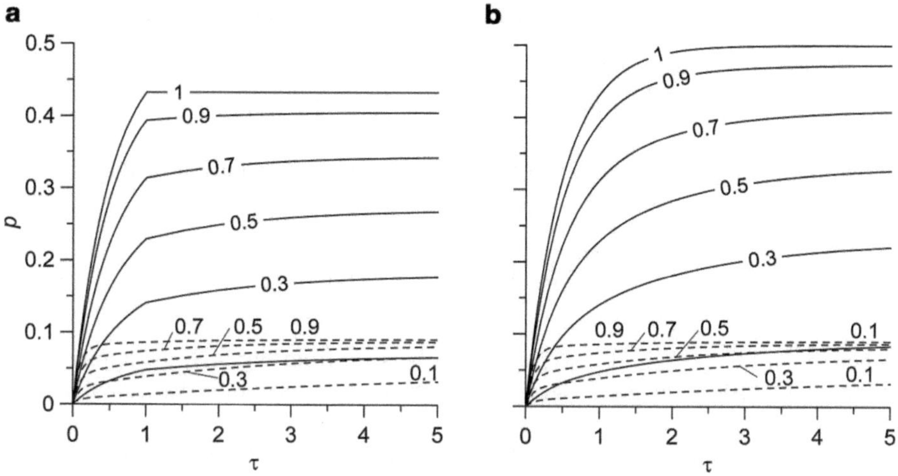

Fig. 4.11 An illustration of variation of sediment concentration p predicted by Eq. 4.106. Parameters are $\overline{a}_d = 1$, $\tau_0 = 1$ (**a**), $\tau_0 = 5$ (**b**). Predictions are shown for a range of f (numbers at the *curves*) and k_1 (1 – *solid curves*, 10 – *dashed curves*)

of sediment on the surface and the concentration of suspension in runoff is larger, the higher the value of the characteristic parameter $\tau_0 = (n/h_a)(rx/\alpha)^{1/n}$, i.e., the further the observation point (x) from the water divide line ($x = 0$). When $\tau = rt/h_a >> 1$, the relationship between g and p concentrations is governed by settling velocity characteristics, $k_1 = v/r$,

$$g/p = k_1 = v/r. \tag{4.107}$$

At large τ_0, the functions p and g attain their maximums:

$$p = \frac{f\overline{a}_d}{1 + k_1 f}, g = \frac{k_1 f\overline{a}_d}{1 + k_1 f}. \tag{4.108}$$

Assuming $S = m_s/h_a$, we can rewrite (4.108) as

$$S = \frac{a_0}{1 + k_1 f}, \tag{4.108a}$$

which at $f = 1$ ($a_0 = a_d$) coincides with solution (4.68). Thus, within a time–space range, which corresponds to dynamic equilibrium of detachment and deposition, and with equal characteristics of the detachability of the original soil and the newly formed sediment, the HR model degenerates into the classic one-component model of rainsplash soil erosion (Sect. 4.1.1). At the same time, it is clear that, since in fact $a_0 < a_d$ ($f < 1$), the classical model, which ignores re-detachment, should be expected to steadily underestimate the concentration of disperse particles in runoff.

Formulas (4.108) can be rewritten in terms of the fraction of shielding, H_∞,

$$p = \frac{H_\infty}{k_1}\overline{a}_d,\ g = H_\infty \overline{a}_d, H_\infty = \frac{k_1 f}{1 + k_1 f}. \tag{4.109}$$

Now, by analogy with solution (4.91), an approximate solution of the problem ($\tau_0 >> 1$) is given by

$$p = \frac{H_\infty}{k_1}\overline{a}_d\left[1 - \exp\left(-\frac{\tau}{\tau^*}\right)\right], g = k_1 p, \tag{4.110}$$

where $\tau^* = (1 + k_1\overline{a}_d)/(1 + k_1 f)$. As can be seen from Fig. 4.9, the approximate solution (4.110) is in good agreement with curves $p(\tau, \tau_0 = 5)$, calculated by Eq. 4.106.

The exhibited unsteady-state behavior of the detachment–deposition–re-detachment process near the water divide point (when $\tau_0 < 1$), complicates the character of curves $p(\tau)$ (Figs. 4.10 and 4.11), making them less predictable at the intuitive level. For example, the analysis of Figs. 4.10 and 4.11 suggests that, against the background of the general trend toward a decrease in function $p(\tau)$ with increasing settling velocity characteristics, k_1, the inverse tendency can also materialize. Thus, in Fig. 4.10a, one can see that curves $p(\tau)$, calculated at different values of k_1, may intersect, i.e., after some moment, curves $p(\tau)$, calculated at larger k_1, may lay above curves $p(\tau)$, calculated at lower k_1. This suggests the ambiguous role of settling velocity characteristics: on the one hand, an increase in k_1 contributes to a decrease in suspension concentration in the runoff water, while, on the other hand, the same increase in k_1 leads to the formation of a sediment layer which show lower cohesive properties and, accordingly, greater susceptibility to erosion, resulting in a greater concentration of suspension in runoff water. Finally, Figs. 4.10 and 4.11 show that a steady value of $p(\tau)$ can be attained at lower values of k_1 and τ.

4.3.2.2 Dynamic Erosion of Soil Caused by Runoff Entrainment/Re-entrainment (Approximate Solution)

A limitation $e_i = e_{ri} = 0$, allows the system of Eqs. (4.18)–(4.19) to be rewritten for the case of steady-state overland flow with averaged characteristics $h = \text{const}$ and $q = \text{const}$ to become:

$$h\frac{\partial S_i}{\partial t} + q\frac{\partial S_i}{\partial x} = r_i + r_{ri} - d_i, \tag{4.111}$$

$$\frac{\partial m_i}{\partial t} = -(r_{ri} - d_i). \tag{4.112}$$

As in the case of rainfall-driven erosion (Sect. 4.3.2.1), we assume that, during erosion, the terms r_{ri} (Eq. 4.21) and d_i (Eq. 4.22) are very closely equal. Now it is easy to show that the system of Eqs. 4.111 and 4.112, with (4.24) taken into account, reduces to the form (4.89), where constants A and B are defined as

$$A = \frac{F}{gh}\left(\frac{\rho_s}{\rho_s - \rho_w}\right)(\Omega - \Omega_c), B = \frac{F}{IJ}(\Omega - \Omega_c). \quad (4.113)$$

In this case, we come to the validity of the earlier constructions, reflected in the resulting solutions for the fraction of shielding of the original soil matrix due to depositing sediment, H, and concentration of eroded products in water, S (or S_i) (Eqs. 4.91, 4.92, 4.93, and 4.94).

4.3.3 *Solutions Describing Contaminant Transport with Sorption onto Suspended Particles*

4.3.3.1 An Equilibrium (Asymptotic) Transport Model with Fast Sorption Kinetics

The asymptotic solution of the problem given below can be obtained based on the assumption that the physical system under consideration can have local sorption equilibria and conditions allowing the mass exchange between the surface of soil and the mixing layer in the soil column to be ignored. Such exchange will be taken into account in the model considered below in Sect. 4.3.3.2.

Suppose that chemical sorption onto the soil matrix and the suspended particles is fast enough compared with changes in flow dynamics (theoretically instantaneous), and the sorption can be described by equilibrium linear sorption reactions, implying

$$C_a^s = K_d^s C_d^s, \ C_a^r = K_d^r C_d^r. \quad (4.114)$$

Now, the system of Eqs. (4.25), (4.26), (4.27), (4.28), and (4.29) for the case $e_h < 0$ (the gravitational settling of suspended particles) and $i = 0$ (the absence of the water loss due to infiltration) reduces to a system of two partial differential equations

$$\begin{aligned} &h\left(1 + SK_d^r\right)\frac{\partial C_d^r}{\partial t} + q\left(1 + SK_d^r\right)\frac{\partial C_d^r}{\partial x} = \\ &k_e\left(C_d^s - C_d^r\right) + \frac{e_r\theta}{\rho_b}\left(C_d^s - \lambda C_d^r\right) + e_r K_d^r\left(\overline{K}C_d^s - C_d^r\right) - rC_d^r, \end{aligned} \quad (4.115)$$

$$\begin{aligned} &d_e\frac{\partial\left(\theta C_d^s + \rho_b K_d^s C_d^s\right)}{\partial t} = \\ &-k_e\left(C_d^s - C_d^r\right) - \frac{e_r\theta}{\rho_b}\left(C_d^s - \lambda C_d^r\right) - K_d^r\left(\overline{K}e_r C_d^s + e_h C_d^r\right), \end{aligned} \quad (4.116)$$

where $\overline{K} = K_d^s/K_d^r$. Adding the right and left parts of Eqs. (4.115) and (4.116), we obtain:

$$h\left(1+SK_d^r\right)\frac{\partial C_d^r}{\partial t}+d_e\frac{\partial\left(\theta C_d^s+\rho_b K_d^s C_d^s\right)}{\partial t}+q\left(1+SK_d^r\right)\frac{\partial C_d^r}{\partial x}= \\ -K_d^r C_d^r(e_r+e_h)-rC_d^r\ . \tag{4.117}$$

Next, we assume that e_r is a linear function of r and the sediment transport capacity is very low, $T_c \to 0$ (see Eqs. 4.30), i.e.,

$$e_r = a_0 r,\ e_h = -\nu S,\ S = \frac{a_0}{1+k_1},\ k_1 = \frac{\nu}{r}. \tag{4.118}$$

Now, if we suppose that the system rapidly reaches chemical equilibrium between soil water in the mixing layer and runoff, i.e., $C_d^r = C_d^s = C_d$, then

$$\left[h(1+\psi)+d_e\left(\theta+\rho_b K_d^s\right)\right]\frac{\partial C_d}{\partial t}+q(1+\psi)\frac{\partial C_d}{\partial x}=-(1+\psi)rC_d, \tag{4.119}$$

or

$$\left(h+\frac{\theta d_e R^s}{1+\psi}\right)\frac{\partial C_d}{\partial t}+q\frac{\partial C_d}{\partial x}=-rC_d, \tag{4.120}$$

i.e., we have an equation similar to (3.11), which, however, contains a combined parameter characterizing additional effective capacity of the system: it takes into account sorption processes and the erosion character of pollutant washout

$$R^s = 1+\frac{\rho_b K_d^s}{\theta},\ \psi = \frac{K_d^r a_0}{1+k_1}. \tag{4.121}$$

The total concentration of a chemical in the surface flow is:

$$C = C_d + SC_a = C_d\left(1+K_d^r S\right) = C_d(1+\psi). \tag{4.122}$$

Thus, solving Eq. (4.120) and using the relationship (4.122) allows us to obtain the required solution of the problem. Since similar problem has been solved in Sect. 3.3.3, then, with the above said taken into consideration, the relationships can be represented in the following dimensionless form:

$$\frac{C(rt)}{C_0}=\frac{\theta d_e R^s}{rt+\theta d_e R^s/(1+\psi)}\ \text{at}\ rt/h_x \le 1, \tag{4.123}$$

$$\frac{C(rt,h_x)}{C_0}=\frac{\theta d_e R^s \overline{h}^n}{h_x\overline{h}+\theta d_e R^s/(1+\psi)}\ \text{at}\ rt/h_x > 1, \tag{4.123a}$$

where function $\overline{h}(\tau)$ can be found from the solution of the system of equations

$$\overline{h} = \frac{n\overline{\Delta}_e}{1-n}\ln\left(\frac{\overline{h}}{\overline{h}_0}\right) - \frac{n}{1-n}\overline{h}_0, \tag{4.124a}$$

$$\tau = -n\overline{\Delta}_e \ln \overline{h}_0 + n\left(1 - \overline{h}_0\right), \tag{4.124b}$$

given dimensionless time $\tau = rt/h_x$ ($\overline{h}_0$ is a parameter), C_0 is component concentration in pore water, $\overline{\Delta}_e = \Delta_e/h_x$, $h_x = (rx/\alpha)^{1/n}$, $\Delta_e = \theta d_e R^s$.

Unlike the particular solutions considered earlier (Sect. 3.3.3), the relationships in (4.123) can give the values of concentrations many times greater than the concentration of a chemical in soil water because of its transfer from the soil in adsorbed state with suspended particles of erosion origin.

The proposed solutions are standardized by the initial porous concentration, C_0. If a contaminant is sorbable, then, strictly speaking, the value C_0 is determined (under equilibrium conditions) by the distribution coefficient K_d^s (or the retardation coefficient, R^s). Therefore, in the practical formulation, concentration C_0 should be evaluated basing either on the density of the initial contamination of the soil surface, N_0 or data on pollutant concentration in a unit soil mass M_0:

$$C_0 = \frac{N_0}{\theta d_e R^s}, \text{ or, } C_0 = \frac{M_0 \rho_b}{\theta R^s}. \tag{4.125}$$

Now solution (4.123) can be given in the generalized form:

$$C = \begin{cases} \dfrac{P_0}{\tau + R^s/(1+\psi)} & \text{at } \tau \leq 1, \\ \dfrac{P_0 \overline{h}^n}{\overline{h}_x \overline{h} + R^s/(1+\psi)} & \text{at } \tau > 1, \end{cases} \tag{4.126}$$

where P_0, depending on problem formulation, is evaluated from $P_0 = N_0/\theta d_e$ or $P_0 = M_0\rho_b/\theta$; $\overline{h}_x = h_x/\theta d_e$; $\tau = rt/\theta d_e$.

When precipitation falls on soils polluted by a highly absorbable component ($R^s >> 1, a_0 K_d^r >> 1$), such as a radionuclide or a heavy metal, the concentration in runoff, C, in the initial period weakly depends on time and can be evaluated from simple relationships

$$C \approx \frac{M_0 a_0}{(1+k_1)\overline{K}}, \text{ or, } C \approx \frac{N_0 a_0}{\rho_b d_e (1+k_1)\overline{K}}. \tag{4.126a}$$

We see that concentration C increases with increasing ratio K_d^r/K_d^r ($1/\overline{K}$). An increase in the contribution of the gravity precipitation of particles (governed by an increase in k_1) leads to an inverse tendency. With $\overline{K} = 1$ and $k_1 << 1$, the concentration is determined by the degree of soil pollution and the value of detachment coefficient a_0. As an example, let us take characteristic values: $M_0 = 10$ mg/kg (0.01 mg/mg), $a_0 = 10$ g/L (10,000 mg/L). Now, basing on (4.126), we obtain $C = 100$ mg/L.

4.3.3.2 A Nonequilibrium Transport Model with Slow Sorption and Erosion Kinetics

The nonequilibrium character of the process in this case is due to (1) the relatively slow exchange rate characterizing solute transfer from the mixing layer into the surface runoff and (2) soil erosion kinetics. We suppose that the sorption processes are equilibrium, described by linear sorption isotherms (4.114).

Mathematical (Numerical) Model The original system of governing partial differential equations, (4.31), (4.32), (4.36) and (4.29), can be conveniently represented in a dimensionless form:

$$\frac{\partial \overline{h}}{\partial z} + n\overline{h}^{n-1}\frac{\partial \overline{h}}{\partial y} = 1 - \gamma, \tag{4.127}$$

$$\overline{h}\frac{\partial \overline{S}}{\partial z} + \overline{h}^{n}\frac{\partial \overline{S}}{\partial y} = 1 + \overline{e}_r + \overline{e}_h - (1-\gamma)\overline{S}, \tag{4.128}$$

$$\begin{aligned} &\overline{h}\left(1 + a_0^r\overline{S}\right)\frac{\partial C_d^r}{\partial z} + \overline{h}^{n}\left(1 + a_0^r\overline{S}\right)\frac{\partial C_d^r}{\partial y} = \\ &k_2\left(C_d^s - C_d^r\right) + \overline{a}_0\left(C_d^s - \lambda C_d^r\right) - C_d^r + a_0^r\left(\overline{K}C_d^s - C_d^r\right)\left\langle \begin{array}{l} \overline{e}_r + \overline{e}_h,\ \overline{e}_h > 0, \\ \overline{e}_r,\ \overline{e}_h < 0, \end{array}\right. \end{aligned} \tag{4.129}$$

$$\begin{aligned} &\overline{d}_e\frac{\partial C_d^r}{\partial z} = \\ &\gamma\left(C_d^r - C_d^s\right) - k_2\left(C_d^s - C_d^r\right) - \overline{a}_0\left(C_d^s - \lambda C_d^r\right) - a_0^r\left\langle \begin{array}{ll} \overline{K}C_d^s(\overline{e}_r + \overline{e}_h), & \overline{e}_h > 0, \\ \overline{K}C_d^s\overline{e}_r + \overline{e}_h C_d^r, & \overline{e}_h < 0, \end{array}\right. \end{aligned} \tag{4.130}$$

where $z = rt/h_x$; $y = (rx/\alpha)/h_x^n$; $\overline{h} = h/h_x$; $\overline{S} = S/a_0$; $\overline{d}_e = d_e\theta R^s/h_x$; $a_0^r = a_0 K_d^r$; $\overline{K} = K_d^s/K_d^r$; $\overline{a}_0 = a_0\theta/\rho_b$; $\overline{a}_0 = a_0\theta/\rho_b$; $k_1 = v/r$; $k_2 = k_e/r$; $\gamma = i/r$; $\overline{e}_r = e_r/a_0 r$ (4.4); $\overline{e}_h = e_h/a_0 r$ (4.10/4.11); λ is analogous to that in Eq. (3.21).

The independent variables are normalized by a factor $h_x = (rx/\alpha)^{1/n}$, representing the maximal flow depth in point x. The boundary conditions are further considered as homogeneous, i.e., invariable over both space and time:

$$\begin{aligned} &\overline{h}(y,0) = 0,\ \overline{S}(y,0) = 0,\ C_d^r(y,0) = 0,\ C_d^s(y,0) = C_{d0}^s; \\ &\overline{h}(0,z) = 0,\ \overline{S}(0,z) = 0,\ C_d^r(0,z) = 0,\ C_d^s(0,z) = C_{d0}^s. \end{aligned} \tag{4.131}$$

Description of Modeling Results The main results of calculations will be given as plots of dimensionless total concentration $\overline{C}(\tau) = C/C_0$, including the concentrations of dissolved components and components adsorbed by suspended particles: $C = C_d^r + SC_a^r$, C_0 is the initial concentration of a chemical in soil water in the mixing layer. Additionally, the plots will present curves $\overline{C}(K_d^s = 0)$, corresponding to the case of migration of a nonreactive component under equivalent conditions.

First, we consider the results of modeling chemical transport under conditions of rainfall detachment and deposition sediment in the absence of flow-driven erosion processes with assumptions of (a) linear dependence of sediment transport capacity, T_c, on the discharge q (see Eq. 4.39 with $\beta = 1$) and (b) quadratic dependence ($\beta = 2$). However, first, we consider the variant $T_c = 0$, for which an asymptotic analytical solution is available. Next, we discuss numerical examples illustrating the behavior of a concentration function calculated for soil erosion due to combined influence of rainfall detachment and flow entrainment of soil particles. Also, two variants, $\beta = 1$ and $\beta = 2$, will be analyzed.

All calculations were carried out for a moderately absorbable component $K_d^s = 0.1\ \text{m}^3/\text{kg} = 100\ \text{cm}^3/\text{g}$ (the order of magnitude of $10^2\ \text{cm}^3/\text{g}$ is typical of many chemical pollutants, i.e., heavy metals, radionuclides, and pesticides). In the calculations, it was taken into account that the silty material that forms in the erosion process shows higher sorption capacity: $K_d^r/K_d^s = 10-100$. It should be mentioned that all (four) model variants, cases 1, 2, 3, and 4 (Figs. 4.12, 4.13, 4.14, 4.15, and 4.16), show the concentration of the solute in runoff, which is being absorbed, to be noticeably higher than its concentration in soil water filling the pore space of the mixing layer ($\overline{C} >> 1$). This is due to the significant role of the transport of solutes in runoff in particulate form. Also, all breakthrough curves are characterized by long tails due to dilution effects in the mixing layer. That is why the model results are presented in form of the semi-logarithmic scale graphs.

Case 1 As can be seen from Fig. 4.12, the slow transfer kinetics, controlling chemical exchange between soil water and runoff, can be a factor that appreciably hampers the pollution of surface runoff. Represented results (Fig. 4.12) correspond to the case of a relatively low capacity of the mixing layer (the solutes are concentrated in a very thin surface soil layer), resulting in its relatively rapid depletion and an appreciable drop in its concentration in the flow. This tendency is the stronger, the higher the sorption capacity of suspended particles. At a high intensity of the soil water/runoff exchange process ($k_2 = k_e/r > 100$), the total concentration tends to equilibrium asymptotics ($C_d^r \approx C_d^s$), while the numerical solution is in agreement with the analytical solution (4.123) and (4.124), where $\psi = 1 + [a_0(R^s - 1)]/[\overline{K}(1 + k_1)]$ (in terms of (4.126a), (4.127), (4.128), (4.129), and (4.130) model).

Case 2 At $T_c > 0$ (model variant $\beta = 1$), the values of $\overline{C}$ at initial moments (Fig. 4.13) are in excess of the appropriate equilibrium value obtained at $T_c = 0$ ($\overline{C} \approx 1 + \psi = 1 + 10/(1+2) = 4.33$). The variant $\beta = 2$ (Fig. 4.14), unlike the previous one, at initial times, is characterized by a single value $\overline{C}$, not depending on T_c (φ_2) and coinciding with the variant $\beta = 1$, $\varphi_1 = 0$. However, the variants $\beta = 1$ (Fig. 4.13) and $\beta = 2$ (Fig. 4.14) show qualitative differences between the character of behavior of breakthrough curves at later times: the linear function ($\beta = 1$) yields concentration curves, which monotonically decrease with time (Fig. 4.13), while the quadratic function yields curves which have an extremum (Fig. 4.14).

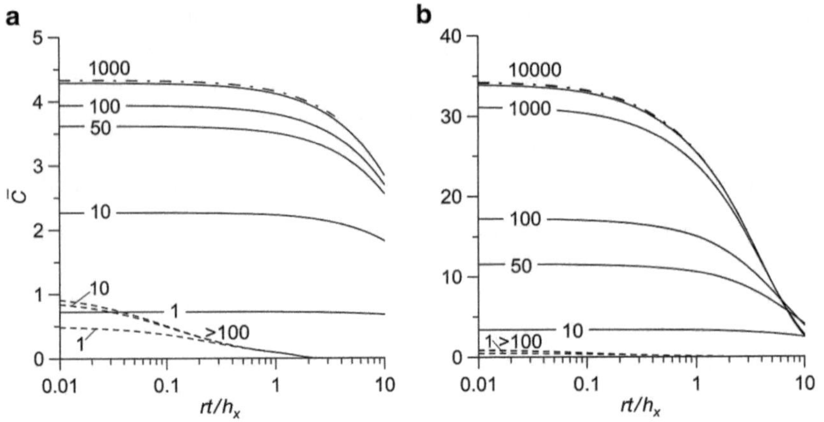

Fig. 4.12 The effect of mass exchange kinetics, $k_2 = k_e/r$ (numbers at the *curves*), on the relative concentration of the solute being absorbed in the surface runoff water for the case of $T_c = 0\,(\varphi_i = 0)$, corresponding to the maximal sedimentation intensity of disperse particles. The *solid lines* are for the numerical model, the *dash*-and-*dot line* is for the analytical solution (4.123)–(4.124) for an equilibrium system, the broken lines are for an unsorbable component. (**a**) $K_d^r/K_d^s = 1/\overline{K} = 10$; (**b**) $K_d^r/K_d^s = 1/\overline{K} = 100$. $a_0 = 0.01$ g/cm^3, $\rho_b/\theta = 10$ g/cm^3, $k_1 = 2, d_e\theta/h_x = 0.1, K_d^s = 100$ cm^3/g, $n = 5/3, \gamma = 0, e_h < 0, \lambda = 1$

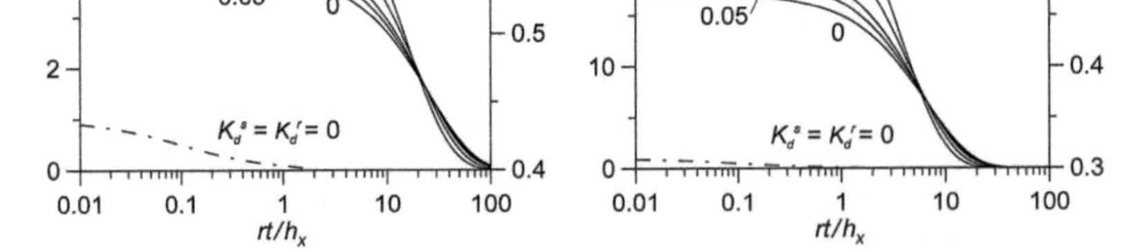

Fig. 4.13 The effect of dimensionless group $\overline{\varphi}_1 = \varphi_1/a_0$ (numbers at the *curves*) on the concentration distributions ($e_h < 0$). The *broken line* is $\overline{S}$ concentration, the *dash*-and-*dot line* is for the unsorbable component. (**a**) $K_d^r/K_d^s = 10$; (**b**) $K_d^r/K_d^s = 100$. $a_0 = 0.01$ g/cm^3, $k_2 = 100$, $d_e\theta/h_x = 0.1$, $K_d^s = 100$ cm^3/g, $k_1 = 2$, $\rho_b/\theta = 10$ g/cm^3, $n = 5/3, \gamma = 0, \lambda = 1$

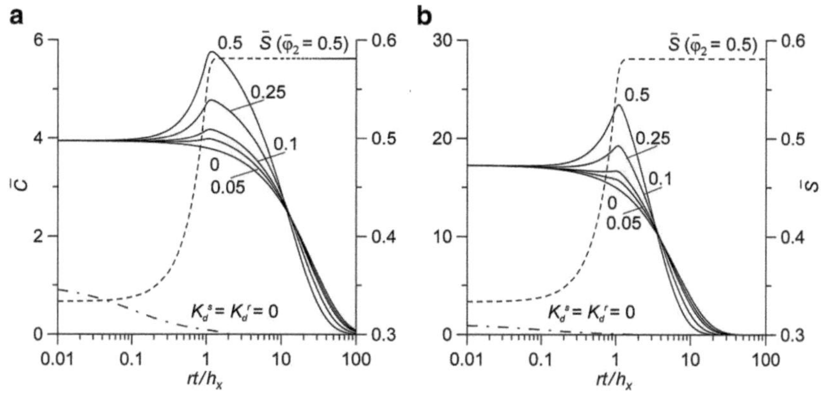

Fig. 4.14 The effect of dimensionless group $\overline{\varphi}_2 = \varphi_2 rx/a_0$ (numbers at the *curves*) on the concentration distributions. The *broken line* is $\overline{S}$ concentration, the *dash*-and-*dot line* is for the unsorbable component. (**a**) $K_d^r/K_d^s = 10$; (**b**) $K_d^r/K_d^s = 100$. $a_0 = 0.01$ g/cm^3, $k_2 = 100$, $d_e\theta/h_x = 0.1$, $K_d^s = 100$ cm^3/g, $k_1 = 2$, $\rho_b/\theta = 10$ g/cm, $n = 5/3$, $\gamma = 0$, $e_h < 0$, $\lambda = 1$

All plots considered here (Figs. 4.13 and 4.14) reflect the tendency toward an increase in the concentration with increasing parameter φ_i. This is due to the higher flow transportation capacity, which reduces the rate of gravitational settling of the suspended particles, thus contributing to the retention of larger amount of chemicals, adsorbed on those particles, in water layer.

Case 3 The plots that reflect the process of rainfall detachment and flow entrainment of soil particles from the surface when gravitational settling is neglected ($e_h > 0$), are qualitatively similar to the plots described above (Figs. 4.15 and 4.16). The concentration at the initial stage is determined by the interaction between the components being absorbed and the particles that release from soil due to drop interaction: $\overline{C} \approx K_d^s \overline{K} a_0$. One can see that an increase in the dimensionless group σx (see Sects. 4.1.1 and 4.3.1) is accompanied by an increase in the total solute concentration in water, which is due to the greater concentration of disperse particles, which are carriers for solutes in runoff (Figs. 4.15 and 4.16). Those plots also show the effect of the value of dimensionless group $d_e\theta/h_x$ on the character of breakthrough concentration curves, similar to that described before: with an increase in the mixing layer capacity ($d_e\theta/h_x$), the descending branches of the curves become flatter and a second plateau appears.

Essentially, the maximum concentration of solute at point x is attained at the moment when the formation of steady-state flow profile on the eroded hillslope is completed, i.e., when $h = h_x$ ($\tau = 1$).

Case 4 The example in Fig. 4.17 illustrates graphically rather high sensitivity of a numerical solution of the system (4.126a)–(4.129), presented for the dimensionless mass flux,

$$\overline{q}_s = \frac{Cq}{a_0 rx} = \overline{C}\,\overline{h}^n, \tag{4.132}$$

to the ratio $\gamma = i/r$. As one may expect, an increase in infiltration velocity, i, reduces the flux of dissolved solids due to the net solute loss (from the upper mixing layer) via vertical infiltration. Obviously, such effect can be attributed to reduction of the pollution potential

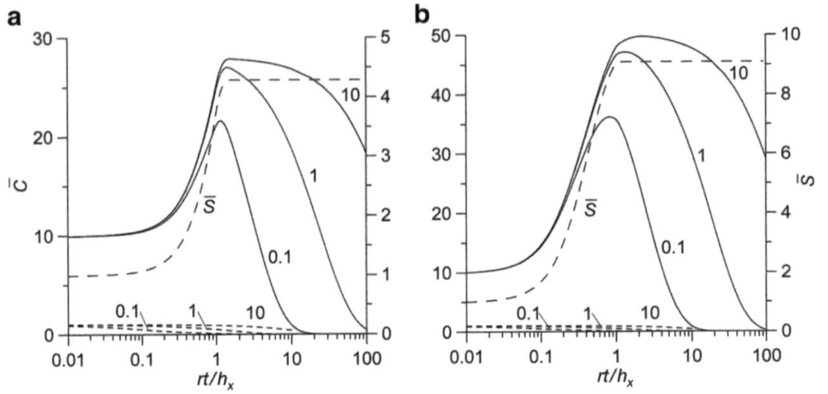

Fig. 4.15 The character of concentration functions $\overline{C}(rt/h_x)$ (the *full lines* are for the component being sorbed, the fine *dashes* are for the nonreactive component) and $\overline{S}(rt/h_x)$ (*long dashes*) for the scenario of raindrop and hydraulic soil erosion ($e_h > 0,\ \beta = 1$) at different values of mixing layer capacity ($d_e\theta/h_x$ are given by figures at the curves). (**a**) $\sigma x = 1$; (**b**) $\sigma x = 10$. $\overline{\varphi}_1 = \varphi_1/a_0 = 10, K_d^s = 100\ \text{cm}^3/\text{g}, \overline{K} = 0.1, a_0 = 0.01\ \text{g/cm}^3$, $k_2 = 100$, $\rho_b/\theta = 10\ \text{g/cm}^3$, $n = 5/3$, $\gamma = 0$, $\lambda = 1$

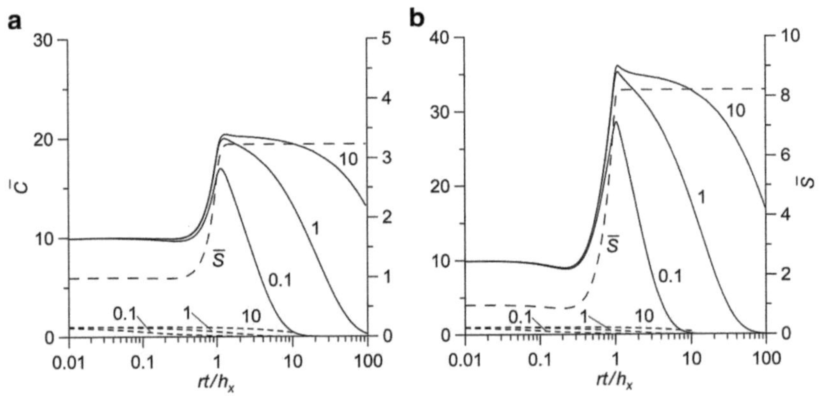

Fig. 4.16 The character of concentration functions $\overline{C}(rt/h_x)$ (*full lines* are for the component being sorbed, *fine dashes* are for the nonreactive component) and $\overline{S}(rt/h_x)$ (*long dashes*) for the scenario of raindrop and hydraulic soil erosion ($e_h > 0,\ \beta = 2$) at different values of mixing layer capacity ($d_e\theta/h_x$, given by numbers at the *curves*). (**a**) $\sigma x = 1$; (**b**) $\sigma x = 10$. $\overline{\varphi}_2 = \varphi_2 rx/a_0 = 10$, $K_d^s = 100\ \text{cm}^3/\text{g}$, $\overline{K} = 0.1$, $a_0 = 0.01\ \text{g/cm}^3$, $k_2 = 100$, $\rho_b/\theta = 10\ \text{g/cm}^3$, $n = 5/3$, $\gamma = 0$, $\lambda = 1$

of runoff from hillslopes and catchment areas. The plots in Fig 4.17 also reflect the tendency toward an increase in the dimensionless mass flux with increasing the sorption capacity of suspended solids.

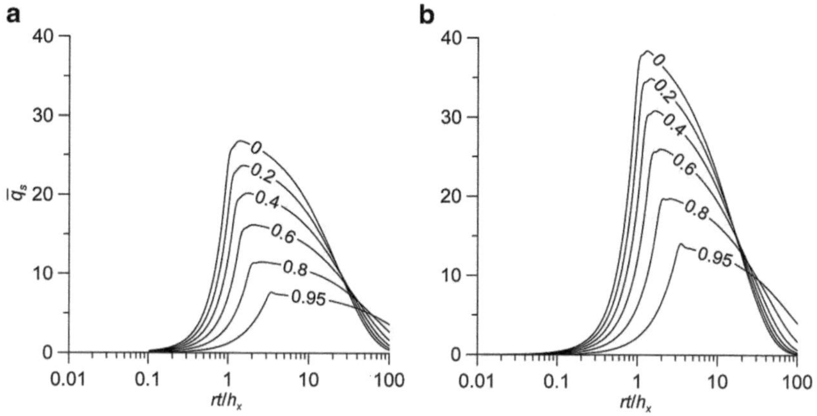

Fig. 4.17 Influence of the infiltration/raifall ratio, i/r (numbers at the *curves*), on the relative mass flux. Parameters for the scenario of raindrop and hydraulic soil erosion ($e_h > 0$, $\beta = 1$) are assumed as been fixed in Fig. 4.15a ; $d_e\theta/h_x = 1$; (**a**) $\overline{K} = 0.1$, (**b**) $\overline{K} = 0.05$

The above examples illustrate importance of the coupled hydrological-mechanical-chemical approach (the flow, dissolved solids transport, suspension solids generation and transport, chemical reaction equations are solved simultaneously) to examine the transport potential of runoff and the system behavior at the catchment scale. Despite the simplicity of the conceptual scheme, it helps to identify a suite of specific parameters to be chosen to describe the hydrological system and must be obtained by hydrologists in field studies.

References

Al-Hamdan OZ, Pierson FB, Nearing MA (2012) Concentrated flow erodibility for physically based erosion models: Temporal variability in disturbed and undisturbed rangelands. Water Resour Res. doi:10.1029/2011WR011464

Barry DA, Jomaa S, Sander GC et al (2010) Exact solutions of the Hairsine–Rose precipitation-driven erosion model for a uniform grain-sized soil. J Hydrol 389:217–230

Bryan RB (2000) Soil erodibility and processes of water erosion on hillslope. Geomorphology 32:385–415

Deng Z-O, de Lima JLMP, Jung H-S (2008) Sediment transport rate-based model for rainfall-induced soil erosion. Catena 76:54–62

Downer CW, Ogden FL (2004) GSSHA: A model for simulating diverse streamflow generating processes. J Hydrol Eng 9(3):161–174

Finkner SC, Nearing MA, Foster GR (1989) A simplified equation for modeling sediment transport capacity. Trans ASAE 32:1545–1550

Flanagan DC, Nearing MA (1995) Water erosion prediction project: Hillslope profile and watershed documentation. NSERL Report N 10, USDA-ARS National Soil Erosion Research laboratory. West Lafayette. Indiana, USA

Foster GR, Flanagan DC, Nearing MA (1995) Chapter 11. Hillslope erosion component. In: Flanagan DC, Nearing MA (eds) Technical Documentation. USDA – Water Erosion Prediction Project (WEPP), NSERL. Report N 10. National Soil Erosion Research Laboratory, West Lafayette

Foster GR, Meyer LD (1975) Mathematical simulation of upland erosion by fundamental erosion mechanics. In: Present and prospective technology for predicting sediment yields and sources, ARS-S-40. Agricultural Research Service, US Department of Agriculture. Washington, DC, p 190

Gabet EJ, Dunne T (2003) Sediment detachment by rain power. Water Resour Res. doi:10.1029/2001WR000656

Hairsine P, Beuaelinck L, Sander G (2002) Sediment transport through an area of net deposition. Water Resour Res. doi:10.1029/2001WR000265

Hairsine P, Rose C (1991) Rainfall detachment and deposition: sediment transport in the absence of flow-driven processes. J Soil Sci Soc Am 55(2):320–324

Hairsine PB, Rose CW (1992a) Modeling water erosion due to overland flow using physical principles: Sheet flow. Water Resour Res 28:237–243

Hairsine PB, Rose CW (1992b) Modeling water erosion due to overland flow using physical principles: Rill flow. Water Resour Res 28:245–250

Hairsine PB, Sander GC, Rose CW et al (1999) Unsteady soil erosion due to rainfall impact: a model of sediment sorting on the hillslope. J Hydrol 220:115–128

Hjelmfelt AT, Piest RP, Saxton KE (1975) Mathematical modeling of erosion on upland areas. In: Proceedings of the XVI congress of the international association for Hydraulic research 2, Sao Paulo, Brazil, pp 40–47

Hogarth WL, Parlange J-Y, Rose CW (2004a) Soil erosion due to rainfall impact with inflow: an analytical solution with spatial and temporal effects. J Hydrol 295:140–148

Hogarth WL, Rose CW, Parlange J-Y (2004b) Soil erosion due to rainfall impact with no inflow: a numerical solution with spatial and temporal effects of sediment settling velocity characteristics. J Hydrol 294:229–240

Johnson BE, Zhang Z (2005) Development of a distributed source contaminant transport model for ARAMS. Engineer Research and Development Center, Vicksburg

Julien PY, Simons DB (1985) Sediment transport capacity of overland flow. Am Soc Agric Eng 28(3):755–762

Kinnell PIA (2005) Raindrop-impact-induced erosion processes and prediction: a review. Hydrol Process 19:2815–2844

Le M-H, Cordier S, Lucas C et al (2013) An improved numerical scheme for a coupled system to model soil erosion and polydispersed sediments transport. Report N hal-00839681. http://citeseerx.ist.psu.edu/viewdoc/download?doi=10.1.1.405.8419&rep=rep1&type=pdf

Lisle IG, Rose CW, Hogarth WL et al (1998) Stochastic sediment transport in soil erosion. J Hydrol 204:217–230

Marshall TJ, Holmes JW, Rose CW (1999) Soil Physics. Cambridge University Press, Cambridge, p 457

Meyer LD (1981) How rainfall intensity affects interrill erosion. Trans ASAE 24(4):1472–1475

Morgan RPC, Quinton JN, Smith RE (1998) The European soil erosion model (EUROSEM): A dynamic approach for predicting sediment transport from fields and small catchments. Earth Surf Process Landf 23:527–544

Planchon O, Mouche E (2010) A Physical Model for the Action of Raindrop Erosion on Soil Microtopography. Soil Sci Soc Am J 74:1092–1103. doi:10.2136/sssaj2009.0063

Proffitt A, Rose C (1992) Relative contributions to soil loss by rainfall detachment and runoff entrainment. In: Hurni H, Tato K (eds) Erosion, conservation and small-scale farming. Geographica Bernensia, Bern, pp 75–89

Rose CW, Yu B, Ghadiri H et al (2007) Dynamic erosion of soil in steady sheet flow. J Hydrol 333:449–458

Rumynin VG (2011) Subsurface solute transport models and case histories (with applications to radionuclide migration), vol 25, Series: Theory and applications of transport in porous media. Springer, Dordrecht, p 815
Sander GC, Hairsine PB, Beuselinck L (2002) Steady state sediment transport through an area of net deposition: Multisize class solutions. Water Resour Res 38(6):1086. doi:10.1029/2001WR000265
Sander GC, Parlange J-Y, Barry DA et al (2007) Limitation of the transport capacity approach in sediment transport modeling. Water Resour Res. doi:10.1029/2006WR005177
Shaw SB, Walter MT, Steenhuis TS (2006) A physical model of particulate wash-off from rough impervious surfaces. J Hydrol 327:618–626
Trimble SW (2007) Encyclopedia of water science, 2nd edn. CRC Press, Boca Raton, p 1586
Woolhiser DA, Smith RE, Goodrich DC (1990) A kinematic runoff and erosion model: documentation and user manual. US Department of Agriculture, Series: Agricultural Research Service, vol 77, p 130
Yalin YS (1963) An expression for bed-load transportation. J Hydraul Div Am Soc Civ Eng 89:221–250
Zhang GZ, Liu Y, Han Y et al (2009) Sediment Transport and Soil Detachment on Steep Slopes: I. Transport Capacity Estimation. Soil Sci Soc Am J 73(4):1291–1297
Zhang XC, Nearing MA, Miller WP et al (1998) Modeling interrill sediment delivery. Soil Sci Soc Am J 62:438–444

Part II
Water Flow and Solute Transport Models at the Catchment Scale

The aim of this, second part of the book, is to demonstrate: (1) how elemental processes as presented in the first part of the book can manifest at the mezzo-(regional) scale, and (2) how rational approaches from the elementary theory of the overland flow dynamics can be meaningfully applied to analyzing mezzo-scale hydrological processes. Although these issues are related to the global upscaling problem, the latter is not particularly relevant to the present analysis.

The emphasize will be put on the lumped-conceptual approach, explicitly or implicitly. According to that the catchment (watershed) is considered as a single homogeneous unit (e.g., a whole river basin) considering as a time-invariant system with a limited number of average/effective parameters, i.e. spatial variability in physical properties is ignored. Also the rainfall is considered to be spatially uniform over the catchment. In other word, a catchment is considered here as a control volume which is characterized by a well-defined hydrological structure (Chap. 5). It is worth mentioning that, as noted earlier, the terms catchment and watershed are used interchangeably without defining the distinctions between them to refer to a drainage area (basin) characterized by all runoff being conveyed to the same outlet.

According to viewpoints of some authors, errors in runoff predictions that are based on lumped hydrological models are not due to neglecting the special variability of hydrological parameters and input data within a particular catchment, but mostly result from errors related to a selection of model structure. It was shown that if the structure of a lumped model is adequate to describe the catchment nature and a model is properly calibrated, it can produce predictions that are almost identical to those generated by a distributed model (Das et al. 2008; Kling and Gupta 2009). Lumped-parameter models can be simply validated with the available hydrological data and global tracer concentration measurements (Maloszewski et al. 2000).

Finally, such lumped-concept may provide a basis for semi-distributed (spatially lumped continuous) models (Karnieli et al. 1994; Sheffer et al. 2010), where the catchment area is considered as a series of isolated or interconnected cell units/subcatchments (each cell produces the surface runoff, evapotranspiration and seepage losses as a response to rainfall input based on simple infiltration or saturation excess concept). The lateral flow component between the spatially lumped cells is ignored.

The lumped-parameter formulation of the water budget and flow problems significantly simplifies the solving of solute transport problem and makes practically possible the long-range prediction of chemical component distributions between the near-surface domains relying on a restricted number of hydrogeological parameters (Chap. 6).

Physically based, distributed-parameter models that are based on rigorous mathematical formulations of physical laws governing the coupled surface and subsurface flow and solute transport will be introduced as well, and practical application of a numerical simulator from this software family will be given in the conclusion Chap. 7 of the book.

References

Das T, Bárdossy A, Zehe E et al (2008) Comparison of conceptual model performance using different representations of spatial variability. J Hydrol 356:106–118

Karnieli AM, Diskin MH, Lane LJ (1994) CELMOD5–a semi-distributed model for conversion of rainfall into runoff in semi-arid watersheds. J Hydrol 157:61–85

Kling H, Gupta H (2009) On the development of regionalization relationships for lumped watershed models: the impact of ignoring sub-basin scale variability. J Hydrol 373:337–351

Maloszewski P, Stichler W, Rank D (2000) Combined application of black box models to environmental tracer data for determination of transport and hydraulic parameters in karstic aquifer of Schneealpe (Austria). In: Use of isotopes for analyses of flow and transport dynamics in groundwater systems. Results of a co-ordinated research project 1996–1999, IAEA

Sheffer NA, Dafny E, Gvirtzman H et al (2010) Hydrometeorological daily recharge assessment model (DREAM) for the Western Mountain Aquifer, Israel: Model application and effects of temporal patterns. Water Resour Res. doi:10.1029/2008WR007607

Chapter 5
A Short Review of Water Budget and Flow Models for a Lumped Catchment

A rigorous and unambiguous classification of the lumped catchment (or watershed) models commonly faces objective difficulties, because similar approaches to mathematical formalization of processes with different physical nature can be used in different models. That is one of the main reasons why there is no universal method to characterize this category of the catchment models. For this review, four categories of models can be provisionally identified: (1) balance (budget) models, (2) reservoir models, (3) soil moisture accounting models (approach), and (4) combined models which synthesize some properties of the above models and empirical features of hydrological processes description. All of them, of course, are simplifications of reality and have a high degree of empiricism.

Catchment (watershed) runoff and associated transport processes are controlled by a number of parameters characterizing current and antecedent conditions of a drainage area. A water budget is a quantitative summation of many of these characteristics, and also the inputs, outputs and net changes to a particular catchment over a fixed period. Since then, it would be logically to begin our analysis with description of the macroscopic water balance (water fluxes) components based on lumped-parameter approach which has been widely used in the analysis of stream flow discharge in response to some precipitation event, evapotranspiration, surface runoff and the impact of the land use over a catchment or its hydrologic response units (Duffy and Gelhar 1985; Burnash 1995; Maloszewski et al. 2000; Dingman 2002; Gupta et al. 2003).

Then, it makes sense to focus on the reservoir-like representation of a catchment as the simplest way for mathematical formulation of hydrological phenomena at the regional scale. In such a reservoir, the change in water discharge induced by precipitation is defined as a linear or non-linear function of the change in the reservoir storage. In case of fragmentation of a catchment basin into several hydrological units (sub-catchments), a system of interconnected reservoirs put in series or parallel. The actual temporal variability of the rainfall can be included in the reservoir model. The antecedent rainfall can be also included, because the time variation of the storage in the system is taken into account.

V.G. Rumynin, *Overland Flow Dynamics and Solute Transport*, Theory and Applications of Transport in Porous Media 26, DOI 10.1007/978-3-319-21801-4_5

Also, it will be shown that prediction capacity of the physically-based models may be improved by using soil moisture accounting procedures according to which soil moisture dynamics and water fluxes are coupled and updated continuously. Models that utilize these procedures relay on the *threshold soil moisture concept* (Bergström 1995; Kling and Gupta 2009). According to that the flow components related to surface/subsurface runoff, percolation flux and groundwater recharge/loss do not occur until soil moisture content is lower or higher than the set limits. This approach also can be used in combination with empirical models.

Finally, it should be mentioned that at the present the increased interest to watershed flow modes is due to their efficiency for assessing hydrological responses to global climate change accompanying the changes in regional water availability (Farmer et al. 2003; Xu and Singh 2004).

5.1 Water Cycle Components and Their Representation in Water Budget Models

The basic concept of water budget for a lumped catchment considered as a closed system is represented by the formula:

$$\Delta S = P - ET - OR, \tag{5.1}$$

where P is the total precipitation depth; OR is total depth of overall runoff (surface, subsurface and groundwater); ET is the total actual evapotranspiration depth (sometimes this member includes interception); ΔS is the change in total storage. All components in this equation correspond to the thickness of a water layer accumulated or lost within the given period. The unit for all measuring water budget components is [L] (usually mm).

In the context of a time continuum the quantities from Eq. 5.1 represent the integration of dynamic fluxes over some timescale (Farmer et al. 2003):

$$\frac{dS}{dt} = r - e - q, \tag{5.2}$$

where r is the rate of rainfall; e is the evapotranspiration rate; q is the rate of runoff generation; dS/dt represents the rate of change of stored water; the unit for all rate components is [LT^{-1}] (usually mm/d).

The assessment of the runoff depth components (Eq. 5.1) and runoff flow components (Eq. 5.2), is a target of water budget modeling considered as a tool for studying the hydrological cycle of catchments and solving many environmental problems.

Thornthwaite (1948) and Thornthwaite and Mather (1955) developed some of the first water budget models, and since then many other models with varying degrees of complexity have been formulated. The existing verified models used to assess water balance within catchments number many dozens, if not hundreds. Their comprehensive review and classification can be found in the literature (Lascano 1991; Alley 1984; Beven 2006). In this section, we consider only the basic components that are common for most water balance models. Particular attention will be paid to water retention mechanisms.

The majority of water balance models are used to estimate the monthly and annual runoff from a daily rainfall record (*P*) and potential evapotranspiration (*PET*) estimates (Sect. 1.5.3). They have been developed for different conditions and purposes. In the recent years, special attention was given to the use of balance models for studying the impact of land use and climate changes on the hydrological regime of river basins. Some models are able to calculate surface runoff and recharge (as base flow component of streamflow) with daily time steps. There are models that use hourly time steps for modeling flood runoff.

As a measure of excess precipitation, the runoff includes several components; two of them are basic ones (Sects. 1.1 and 1.2): surface flow (direct or overland flow) and base flow (ground water flow). The direct runoff, which enters surface streams immediately after a rainfall, is the main contributor to the peak discharge, while the base flow response in discharge areas is detected with a noticeable delay after the rainfall. Several other (minor) components of the subsurface runoff, such as stormwater and supplementary base flow, can be incorporated in the model formulation as well. The interflow is the portion of the runoff represented by infiltrated water that moves laterally on top of the subsoil composed of consolidated sediment. The interflow is a slower process than surface runoff but a quicker than groundwater runoff.

Many of the water balance models are based on the assumption that runoff occurs, when precipitation exceeds potential evapotranspiration and the soil has reached its field capacity. Any additional water applied to the soil runs off. When soil moisture is deficient, as in the case when potential evapotranspiration exceeds actual evapotranspiration ($PET > ET$), the formation of runoff becomes impossible. Thus, water balance models generate runoff by *saturation excess* in the soil moisture storage: water excess of saturation becomes runoff.

Major problems of balance models are (Alley 1984): (1) selecting an appropriate procedure for distributing the overall runoff between direct flow and base flow, and (2) accounting for the lag of the runoff in its transfer to the surface stream. The latter requires incorporating calculation procedures allowing runoff to be generated in periods without precipitation.

A considerable portion of the balance models includes two storage components/stores (Figs. 5.1 and 5.2): one, associated with the short-term storage capacity, for soil water flow, and one, related to the long-term storage capacity, for groundwater flow. The *soil store* plays an important role in determining the runoff and the amount of actual evapotranspiration. It requires initial values of the soil water storage contents. *Groundwater (aquifer) store* is important for determining base flow (Ibrahim and Cordery 1995).

The usual approach to splitting the overall runoff into two components, *R* and *Q* (Fig. 5.1), is to specify some fraction of the rainfall excess, *f*, that remains in the soil as part of the groundwater storage (Alley 1984; Dingman 2002). Water budget in the groundwater store is calculated for each time step (e.g., month), which allows the recharge

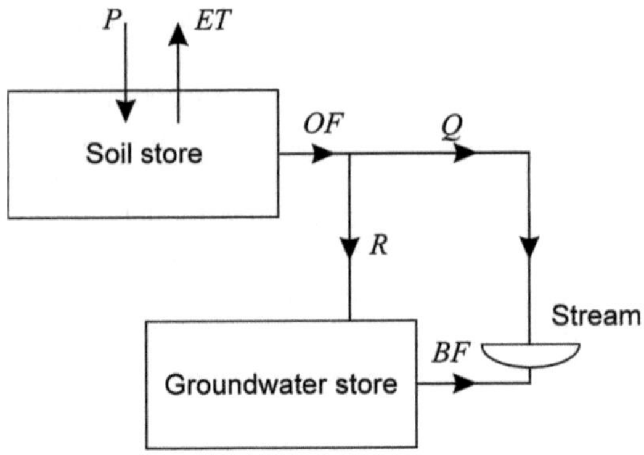

Fig. 5.1 Schematic diagram of the conceptual two-store distribution model. *P* precipitation, *ET* evapotranspiration, *OF* overall flow = rainfall excess, *Q* surface flow, *R* recharge, *BF* base flow

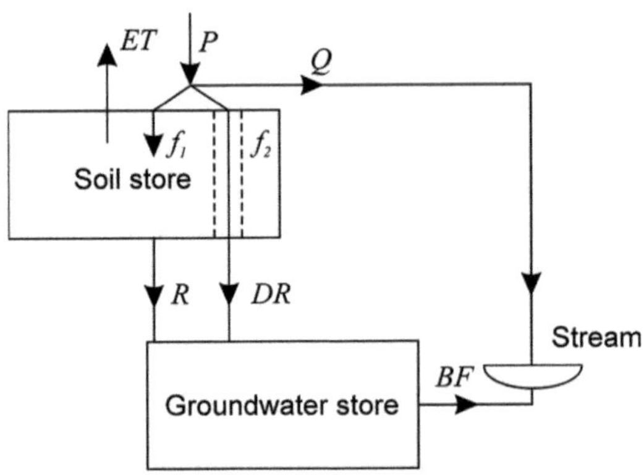

Fig. 5.2 Schematic diagram of the conceptual two-store two-component infiltration model. *P* precipitation, *ET* evapotranspiration, *Q* surface flow, f_1 and f_2 two infiltration components, *R* recharge, *DR* direct recharge, *BF* base flow

at a given time step to contribute to discharge at the subsequent time step. The aquifer store, having unlimited capacity, drains as a function of the base flow recession constant with the remainder being the initial aquifer storage for the next time step (Dingman 2002). The other approach conceptualizes that all saturation excess runoff is routed to aquifer store, which is allowed to fill until the maximum capacity, *S*, is reached. If runoff at the current step summed with the previous aquifer storage exceeds *S*, then the difference becomes surface runoff, and the remaining storage becomes a subject for the subsurface discharge controlled by the recession constant (Dingman 2002).

The above concept can be used for quantifying some particular components of the hydrological budget, for example, groundwater recharge. Thus, let the precipitation minus the surface runoff (*P*, *Q*) be accumulated infiltration (*F*), which can be estimated independently using some methods discussed in Sect. 1.3. Then, using the water budget approach for soil moisture store associated with the root zone (moisture storage), one can estimate the annual groundwater recharge on daily basis as it is proposed in the FAO model (Allen et al. 1998; Erickson and Stefan 2007). In this model, as in the previous ones, it is assumed that in the dry seasons, when soil moisture content is below the field capacity, θ_{fc}, pore water is not allowed to move under gravitational force and moisture is held in the soil.

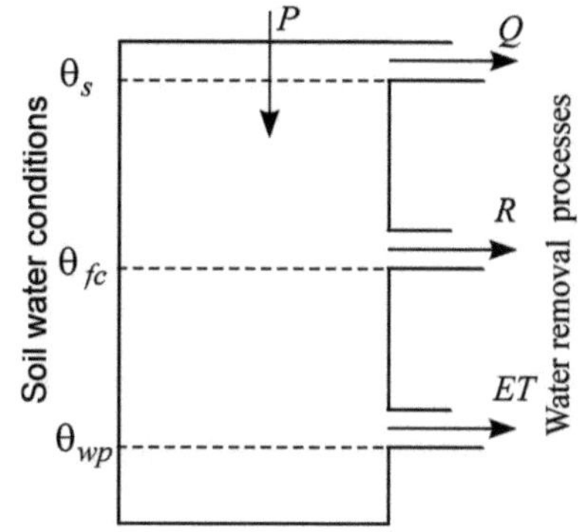

Fig. 5.3 A conceptual single-store model of a daily water budget (Sheffer et al. 2010). *P* precipitation, *ET* evapotranspiration, *Q* runoff, *R* recharge

Oppositely, in the heavier precipitation periods, when soil water content exceeds θ_{fc}, all soil water that is above the field capacity percolates through the root zone under gravity and becomes groundwater recharge.

All models presented above have a common drawback: the above described water budget technique is unable to model recharge when soil water content is near field moisture capacity, while field observations show that phreatic aquifers yield noticeable recharge even in periods of soil moisture deficiency.

To overcome this drawback, some balance models were modified by incorporating, along with traditional infiltration component, represented by water filling the soil moisture zone (f_1) and reducing any soil moisture deficit, a certain portion of water (f_2) which can directly recharge the groundwater (Khan and Mawdsley 1982; Moor and Bell 2001; Vaze et al. 2012). This feature of the model (Fig. 5.2) allows the recharge to be generated even in the case of soil moisture deficiency. Thus, the percolation entering the groundwater is represented by direct recharge and indirect recharge, which drains the partly saturated soil moisture storage zone.

To support the above analysis in terms of soil moisture accounting, we refer to a diagram in Fig. 5.3. In this diagram, the soil profile is conceptualized as a single-store with one inlet and three outlets, representing activation of different output contributions as a function of soil moisture content (Sheffer 2009; Sheffer et al. 2010). The inlet represents the daily precipitation over the area (*P*). The three outlets represent: (1) daily evapotranspiration (*ET*), which produces output only when the water content in the store is higher than the soil wilting point, θ_{wp} (the minimal point of soil moisture the plant requires not to wilt), (2) daily recharge (deep percolation or recharge, *R*), which generates output only when the moisture content in the store exceeds the soil field capacity, θ_{fc}, and (3) daily runoff (*Q*) which generates output only when the water content in the store reaches saturation (porosity), θ_s. As seen, the model includes three key soil water storage capacity thresholds: θ_{wp}, θ_{fc} and θ_s.

5.2 Conceptual Reservoir Models

Reservoir models are simplest among the models designed to describe the transformation of rainfall excess to runoff. They reflect the limiting degree of formalization of the hydrological process where a set of physical mechanisms is taken into

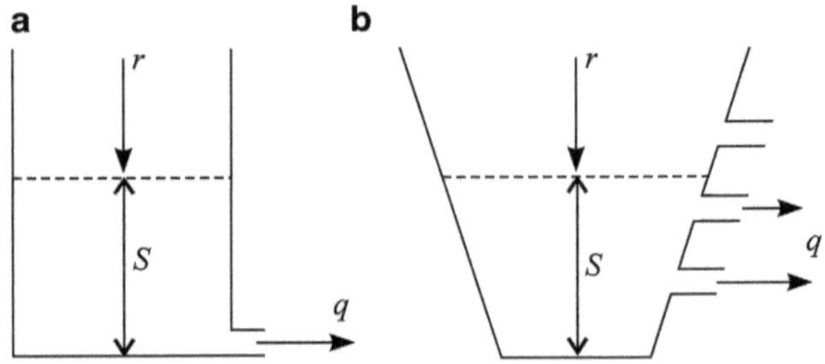

Fig. 5.4 (**a**) A linear and (**b**) nonlinear (with multiple outlets) reservoirs

account in a generalized form by the introduction into the model of empirical coefficients taken from algebraic or differential equations. That is why this type of models is often referred to "black box" models. Unlike the water budget (balance) models, considered above, which deal with soil moisture dynamics, the only input data required by the reservoir models are rainfall excess and some empirical rate constants characterizing the speed at which a reservoir fills or drains.

To produce the surface runoff hydrograph as a response to a rainfall onto a small catchment with a short response time, one may use a single linear reservoir model (Chow et al. 1988). The use of such models can be justified, for example, in the analysis of hydrological processes in intermountain basins, were groundwater plays a subordinate role in the total water balance. For catchments with variable source area hydrology, several capacity- or flow-limited processes, and complex geometry, multi-linear or non-linear reservoir models can be considered as an alternative to the single linear reservoir models.

The linear reservoir model is based on the concept that catchment behaves as a reservoir (Fig. 5.4a) in which outflow (runoff) rate, q [LT^{-1}], is a linear function of storage, S [L], thus:

$$q = \lambda S, \tag{5.3}$$

where λ is a rate constant (catchment discharge coefficient) [T^{-1}] characterizing the overall retention (buffering) capacity of a catchment or its water release readiness. Note that in a linear model, S is an unbounded function (its value can be arbitrarily large).

Water balance equation accounting for the basin storage change can be expressed as follows

$$r - q = \frac{dS}{dt}, \tag{5.4}$$

where the left part is the potential of water accumulation written as the difference between *rainfall excess* (effective rainfall), $r = r(t)$ [LT^{-1}], and *outflow rate*,

$q = q(t)$ [LT^{-1}]. For evaluating the effective rainfall, a "pre-reservoir" procedure should be provided, allowing one to determine what part of the precipitation is intercepted by the plants and evapotranspirates from soil, and what part is available for runoff generation. Equation (5.4) can also be termed as lumped continuity equation.

The solution of Eqs. (5.3) and (5.4) can be given in finite-difference form, making it possible to determine the discharge in any time moment, given its value in the previous moment and the average inflow for the given interval of time (Overton 1970; Pedersen et al. 1980).

On the other hand, combining (5.3) and (5.4) yields the linear differential equation:

$$\frac{dq}{dt} = \lambda(r - q). \tag{5.5}$$

Integrating (5.5) with the initial condition $q(t = 0) = 0$ yields (O'Donnell 1960):

$$q(t) = \int_0^t r(\xi)U(t - \xi)\, d\xi, \tag{5.6}$$

where ξ is a dummy variable. The solution (5.6) for an arbitrarily varying function $r(t)$ describes both the rising and falling (recession) limbs of hydrograph. The functions r and q account for time variations in the precipitation and runoff depths, respectively. Clearly, volumetric rates can be used to characterize appropriate flows; the dimension of such functions will be [L^3T^{-1}].

Equation 5.6 is called convolution integral or Duhamel integral, in which an input-independent kernel,

$$U(t) = \lambda e^{-\lambda t}, \tag{5.7}$$

can be associated with the *instantaneous unit hydrograph* (IUH) [T^{-1}], a linear transfer function assumed constant for a particular catchment (Chow et al. 1988; Wood et al. 1990) which represents the runoff from the catchment due to instantaneous precipitation of the rainfall excess of fixed volume (ΔS), normally, of 1 cm (Subramanya 2008). Indeed, such result yields the solution Eq. 5.5, provided that, at the initial moment $t = 0$, the inflow fills the reservoir storage up to ΔS instantaneously:

$$q(t) = \Delta S\, U(t). \tag{5.8}$$

Formally Eq. 5.8 describes the falling (*recession*) limbs of a hydrograph $\Delta S = q^* / \lambda$, where q^* is the outflow at time $t = t^*$ when the rainfall excess ceases.

The solution (5.6) can be normalized with respect to an arbitrary linear parameter (e.g., ΔS):

$$h(t) = \int_0^t r'(\xi)\,U(t-\xi)\,d\xi, \tag{5.9}$$

h and r' have the dimension of $[T^{-1}]$.

For the linear model under consideration, the time to the peak of the unit hydrograph, t_p (time lag of the event), is not sensitive to the discharge peak or rainfall intensity.

Many authors use not the rate constant λ, but its reciprocal $\tau = 1/\lambda$, referred to as the time constant [T], the storage coefficient of the reservoir, or the storage delay constant. The time τ can be associated with the mean residence time (Buytaert et al. 2004). In some studies, an attempt is made to determine how the time τ can be related with measurable physical characteristics of a catchment. Such relationship can be uniquely determined for the simple case of a conceptual catchment represented by a planar surface and a constant effective rainfall intensity (Pedersen et al. 1980).

Equation 5.8 for the recession limb can be linearized in a semilogarithmic plot:

$$\ln q = \ln q^* - t/\tau, \tag{5.10}$$

thus, constant τ can be determined directly from the slope of a particular recession limb of the drainage hydrograph. If the excess rainfall–direct runoff process is actually linear, the value of τ thus determined would be a constant for all storm events.

The integral (5.6) reflects the superposition principle, implying that the additional response of the model to a rainfall event is the same irrespective of the antecedent condition on the catchment (Jowitt 1999). The basic solution is (5.6), which for $r = \text{const}$ becomes:

$$q(t) = rF(t), \quad F = \left(1 - e^{-\lambda t}\right). \tag{5.11}$$

In the case of a short ($0 \le t \le t_1$) rain, the superposition (5.11) yields the solution:

$$q(t) = \begin{cases} r\left(1 - e^{-\lambda t}\right), & 0 \le t \le t_1, \\ r\left(e^{\lambda t_1} - 1\right)e^{-\lambda t}, & t > t_1. \end{cases} \tag{5.12}$$

Based on the superposition principle, a solution for discharge function, $q(t)$, can be constructed for any input function of rainfall excess, $r(t)$, allowing piecewise-homogeneous (step-wise) representation

$$q_i = \sum_{j=1}^{i} \left(r_j - r_{j-1}\right)F\left(t_i - t_{j-1}\right) \text{ at } t_i > t_{j-1}, \tag{5.13}$$

by analogy with the solutions of linear problems of solute transport in porous media (Rumynin 2011, p. 133); here $i = 1, 2 \ldots$ are the numbers of time intervals within which it is assumed that rainfall excess $r = r_i = \text{const}$; $r_0 = 0$, $t_0 = 0$.

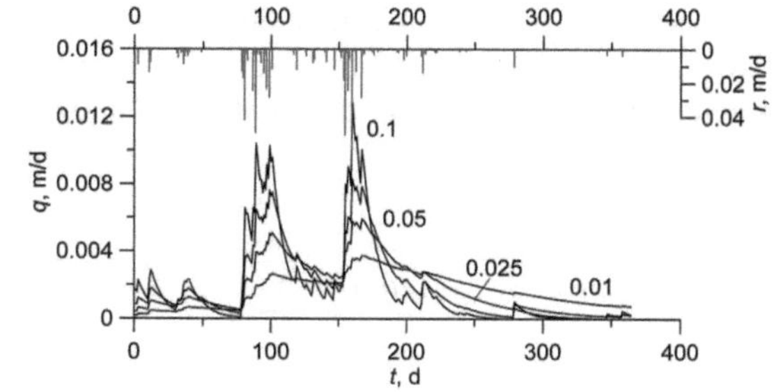

Fig. 5.5 Influence of the rate constant λ (figures at the curves, d^{-1}) on the hydrograph behavior

A response to a synthesized time-discretized rainfall excess (1 year cycle) input is illustrated by the curves in Fig. 5.5, which are calculated using formula (5.13) for different kinetic coefficients. It can be clearly seen that, the less the values of λ, the flatter the recession limbs of hydrographs. In this case, the maximal values of discharge functions increase with increasing λ.

The linear reservoir concept considered above has many limitations for its practical application. The use of a combination of two or more linear reservoirs makes it possible to eliminate those limitations and improve the predictive ability of reservoir models.

The multilinear modeling approach has been used by many researchers. The best known multi-reservoir model is the Nash cascade model representing a number of identical linear *reservoirs in a series* (Nash 1960). In this *cascade model*, the output of each reservoir constitutes the input into the next downstream reservoir. Suppose that the response of the first (uppermost) reservoir to a unit rainfall excess at the initial moment ($t=0$) is described by function (5.7). Then the output flow for the second reservoir can be obtained by substituting (5.7) into the solution (5.9):

$$h_2(t) = \int_0^t \lambda e^{-\lambda(t-\xi)} \lambda e^{-\lambda\xi} d\xi = \lambda^2 t e^{-\lambda t}. \tag{5.14}$$

The corresponding impulse response of the n-th reservoir will be

$$h_n(t) = \frac{\lambda}{(n-1)!} (\lambda t)^{n-1} e^{-\lambda t}, \tag{5.15}$$

which is the IUH of the reservoir in the series model. In later studies, attempts were made to attribute some physical meaning to parameters n and λ. It was shown that n can be considered as a dimensionless shape parameter, while the ratio n/λ is a scale parameter corresponding to the mean delay time of the IUH (Shamseldin and Nash 1998).

On the other hand, if a catchment has several capacity- or flow-limited processes, it can be represented by a series of independent *parallel reservoirs*, each characterized by mean time constants τ (Kundzewicz and Napiórkowski 1986; Buytaert et al. 2004):

$$q(t) = \sum_{i=1}^{n} \int_{0}^{t} U_i(t - \xi) r_i(\xi) d\xi \tag{5.16}$$

under the condition

$$\sum_{i=1}^{n} r_i(t) = r(t). \tag{5.16a}$$

For example, the joint analysis of two linear models (Diskin 1964) makes it possible to represent two components of the runoff in (5.16), namely, surface flow and subsurface flow, as follows

$$r_1(t) = \alpha r(t),\ r_2(t) = (1 - \alpha) r(t), \tag{5.16b}$$

where α is a constant value of the distribution factor.

Buytaert with coauthors (2004), based on linear reservoirs in parallel-model concept, managed to identify two fast time constants and a third slow time constant for small catchments within a high Andes region. These constants were identified as characteristics of the overland flow and throughflow occurring in the soil layer.

It is an accepted fact that, for the majority of practical situations related to hydrologic data analysis, no single IUH can be found to characterize the response of any catchment to all rainfall events. Thus, catchments often behave in a very non-linear way and both the linear and multi-linear reservoir models described above may result in crude approximations of actual hydrographs. This is the reason why alternative nonlinear models are used.

The features of nonlinear behavior of a catchment are as follows (Diskin 1973; Zondervan 1978; Broome and Spigel 1982; Jowitt 1999; Koivusalo et al. 2001; Ding 2011): (1) flow events of greater magnitude have a flashier response than smaller ones; (2) the more the catchment is saturated, the flashier its response to the rainfall excess; (3) the moment of flood peak discharge, t_p, caused by a rain excess storm is found to depend on rainfall intensity (for storms with a low average rainfall rate, an increase in this rate is associated with a considerable reduction in the lag time, while for storms with a high average rainfall intensity, the lag time approaches a constant value), thus, (4) there is a tendency towards linearity at higher runoff values. In some cases, the nonlinear behavior of the system can be attributed to the fact that an increase in the degree of reservoir filling with water causes an increase in the capacity, as it schematically shown in Fig. 5.4b. However, the physics of the process is far more complex and individual for each catchment.

Despite the nonlinearity of hydrological systems, there have been many attempts to model them via linear mathematics (Snyder et al. 1970; Ding 1974, 2011; Kundzewicz and Napiórkowski 1986). The idea is to extend the convolution procedure (5.6) to nonlinear systems. According to this approach, in the convolution integral (5.6), the kernel function (unit hydrograph) remains constant for one event but may vary from event to event. Such variable unit hydrographs are derived from storm data analysis.

Another concept is based on nonlinear extension of the storage–discharge Eq. (5.3). An example of this approach is a nonlinear reservoir with the power outflow law for storage–discharge relation:

$$q = \lambda S^m, \tag{5.17}$$

where m is the storage exponential known as a shape parameter (dimensionless), λ is the discharge coefficient known as a scale parameter [(LT^{-1})/L^{m}].

Combining (5.17) and (5.4) yields

$$\frac{dq}{dt} = \lambda^{1/m} m \frac{r - q}{q^{(1-m)/m}}. \tag{5.18}$$

For some inflow pattern, $r(t)$, the outflow can be calculated with a numerical technique.

The model (5.18) has been used by many authors, with greater or lesser success, for the analysis of hydrographs obtained for catchments under different natural conditions (Ding 2011). The values of the shape parameter, m, vary on the average within the range from 1.2 to 3.4. The scale parameter varies within a much wider range, and the values of λ tend to decrease with increasing basin area.

5.3 A Standard SCS-CN Model

The Soil Conservation Service (SCS) curve number (CN) model (method) refers to the class of basin-scale empirical parameter models for predicting direct runoff or infiltration from rainfall excess (USDA 1986; Chow et al. 1988; Mishra and Singh 2003). It deals with two cumulative functions: the depth of precipitation, P (mm), where a rainfall is considered as an event of fixed duration; the depth of excess precipitation or direct runoff Q, (mm), established by the hydrograph curve in the outlet section over the same period. The application domain is limited to individual rainfall events assumed mutually independent.

The depth of potential runoff is defined as the difference $P - I_a$, where I_a is the initial abstraction (mm) – a characteristic which accounts for the fact that falling of some amount of precipitation in the beginning of an event does not lead to runoff formation (i.e., while $P \leq I_a$, $Q = 0$). The depth of excess precipitation or direct runoff, Q, is always less than or equal to the depth of precipitation, P. After the runoff started to form on a watershed, the additional amount of water retained in the watershed is the cumulative abstraction, F_a (mm), which can be associated with the cumulative infiltration (Sect. 1.3). The value of F_a is less than or equal to some potential maximum retention, S (mm).

According to the original postulate, the ratios of real values of Q and F_a, to the potentially possible values, $P - I_a$ and S, should be equal to one another, i.e.,

$$\frac{Q}{P - I_a} = \frac{F_a}{S}. \tag{5.19}$$

Solving this equation in combination with the balance equation

$$P = Q + I_a + F_a \tag{5.20}$$

with respect to Q, we obtain

$$Q = \frac{(P - I_a)^2}{P + S - I_a}, \text{ when } P > I_a, \tag{5.21a}$$

$$Q = 0, \text{otherwise } (P \leq I_a). \tag{5.21b}$$

In some cases, it is of use to analyze the expression for cumulative abstraction:

$$F_a = \frac{S(P - I_a)}{P + S - I_a}, \text{ when } P > I_a. \tag{5.22}$$

In its original formulation, the SCS-CN method implies that Eqs. 5.21a, 5.21b and 5.22 are satisfied at the end of the event – the period of shower, i.e., the time scale is taken equal to the duration of the event, commonly the period of shower. The sum $F_a + I_a$ is the part of precipitation not transformed into surface runoff.

It is clear that, since the balance items (5.20) do not include evapotranspiration and deep percolation, the model (5.21) should describe the flood periods with minimal errors. When the model (5.21) is used for long observation periods including recessions, when significant components of moisture flow are evapotranspiration and recharge, this model needs to be modified.

The generalization of a vast body of field data (mostly on drainage basins in USA) suggested the conclusion that I_a may be defined implicitly as a fraction of the potential storage depth in the soil:

$$I_a = \lambda S, \tag{5.23}$$

where $\lambda = 0.2$ can be taken as the first approximation.

Empirical studies made it possible to obtain the following formula for assessing the maximal retaining capacity of water in a watershed (USDA 1986):

$$S = 25.4\left(\frac{1000}{\text{CN}} - 10\right), \tag{5.24}$$

where CN is a characteristic referred to as the number of runoff curve, which characterizes the type of soil, its vegetation cover, and land use conditions, as well as some other features, which determine the ability of the catchment to retain water (Erickson and Stefan 2007).

Various procedures are known to allow one to determine the number of the runoff curve for different types of soils covered by vegetation. CN varies from 0 (the soil absorbs all precipitation falling onto it) to 100 (no absorption at all). Shestakov and Pozdniakov (2003), with references to previous studies, give the following ranges of CN for four typical hydrologic soil groups:

Soil	CN	Soil	CN
Sand	40–60	Silty loam	75–85
Sandy loam	60–75	Silty clay loam	85–98

A major weakness of the SCS-CN method is the very high sensitivity of estimated runoff to errors in the selection of the CN. Changes of about 15–20 % in the curve number double or halve the total estimated runoff (Boughton 1989). Usually the standard SCS-CN method underestimates Q and overestimates infiltration for large runoff events (Erickson and Stefan 2007; Shi et al. 2009).

In its *original* form, the method is mostly aimed at the estimation of the hydrological consequences of short precipitation events (with intensity large enough), with the consequences of an event assumed having nothing in common with the past, i.e., it is a priori supposed that between rains the soil returns to its initial state described by parameter S. However, the statistical character of function P determines the strong time variations in soil moisture content; therefore, the potential maximum retention, strictly speaking, is not constant in a catchment. It varies at both annual and many-year scale, so the response of the system in any moment depends on the history. This limits the applicability of the method.

Indeed, many authors note that the value of S is related to catchment features and to antecedent moisture condition. For example, the greater the soil moisture content, the larger the portion of precipitation converted into surface runoff. Formally speaking, CN is not a constant, but varies from event to event. Therefore, some researchers propose a correction of S, taking into account the moisture regime preceding the event (Chow et al. 1988). For dry period, the values $S = S_{\mathrm{I}}$ are about 2.4 times greater than the "normal value of S_{II}; conversely, for the wet period, $S = S_{\mathrm{III}} = S_{\mathrm{II}}/2.3$. However, the abrupt changes from one value of S to another hamper the calibration of hydrological models.

Recently, a CN methodology of continuous hydrogeological simulation models was developed based on some analytical relationships, which ensure smooth time variations of parameter S in the model (5.21). With this in view, parameter S is linked to some soil moisture depletion coefficients or soil available water capacity. In such approach, referred to as soil moisture accounting (SMA) procedure (taking into account changes in soil moisture content), the value of parameter S at moment t depends on S (in some cases, on P and Q, as well) at the previous moment (Kannan et al. 2007).

5.4 A Lumped Continuous SMA Model

5.4.1 A Basic Hypothesis for Correction of the Standard SCS-NS Model

Michel et al. (2005) suggested not to limit the applicability domain of formula (5.21) to a unit event, characterized by the total volume of precipitation onto the watershed (total precipitation depth P) and runoff volume (the total runoff depth Q), but to consider this formula as an analytical model valid for any moment t. As a consequence, P and Q in (5.21) should be considered as *continuous* time functions. Their time derivatives are characteristics of flow rate, $r = dP/dt$ (rainfall rate) and $q = dQ/dt$ (runoff rate). Thus, the new model is unsteady-state. Note that in the classic SCS-CN model, P and Q are *discrete* functions, which can be determined only for the end of an event, with which a period of shower is commonly associated.

In a new model proposed by Michel et al. (2005), which takes into account soil moisture dynamics, i.e., SMA procedure, its authors introduce the concept of soil moisture (accounting) store (SMS), which can accumulate the part of liquid precipitation that does not transform into surface runoff. This SMA model is based on the assumption that the greater soil moisture content, the larger portion of precipitation will transform into surface runoff. If soil moisture content has reached its maximal possible value, i.e., SMS is fully filled, all precipitation becomes surface runoff. Later, the SMA model (Michel et al. 2005) was revised for initial soil moisture store level and improved by Sahu et al. (2007).

Suppose V_0 is the initial soil moisture store level, mm; V is the soil moisture store level at time t, i.e. when the accumulated rainfall is equal to P, mm. Then the balance equation becomes

$$V = V_0 + P - Q. \tag{5.25}$$

The characteristic V_0 is ignored in the SCS-CN model, however at the end of the event, the difference $V - V_0$ corresponds to $I_a + F_a$ $(P > I_a)$.

If we replace Q in (5.25) by an expression for this function from the main SCS-CN Eq. (5.21), we can come to problem solution in terms of accumulative functions:

$$V = V_0 + \frac{(S + I_a)P - I_a^2}{P + S - I_a}. \tag{5.26}$$

The assumption that functions P and Q are continuous and differentiable allows us to differentiate the basic Eq. (5.25) and to derive a relationship between the runoff rate, q, and rainfall rate, r

$$q = r\frac{(P - I_a)(P + 2S - I_a)}{(P + S - I_a)^2}, \quad P > I_a \tag{5.27}$$

($r = dP/dt$ and $q = dQ/dt$); otherwise, $P \le I_a$, $q = 0$. Now, if we express P in (5.27) in terms of V using (5.26), we will have

$$q = r\frac{V - (V_0 + I_a)}{S}\left(2 - \frac{V - (V_0 + I_a)}{S}\right), \text{ if } V > V_0 + I_a, \tag{5.28}$$

when $V \le V_0 + I_a$, $q = 0$. We can see that, in such representation, the degree of precipitation fractionation depends on the current filling of SMS, V, rather than on the arbitrarily chosen state of the system. The Eq. (5.28) contains the sum

$$S_a = V_0 + I_a, \tag{5.29}$$

which is an intrinsic parameter of the SMA модели. Thus, the initial abstraction, I_a, is excluded from the further analysis as an independent characteristic, and the generalized parameter S_a is used instead. With this new parameter, (5.28) becomes:

$$q = r\frac{V - S_a}{S}\left(2 - \frac{V - S_a}{S}\right), \text{ if } V > S_a, \tag{5.30a}$$

$$q = 0, \text{ if } V \le S_a. \tag{5.30b}$$

Model (5.30) allows a vivid physical interpretation (Fig. 5.6). In the beginning of an event, all rainfall penetrates into the soil layer and fills the soil moisture store (SMS), so there is no surface runoff (Fig. 5.6b). Once the amount of water accumulated in SMS reaches the value of S_a, surface runoff starts forming (Fig 5.6a). At $q = r$, i.e., when all precipitation transforms into surface runoff, the entire soil moisture store capacity is filled: $V = S + S_a$.

Finally, the continuity equation following from the balance Eq. (5.25) can be represented in the form

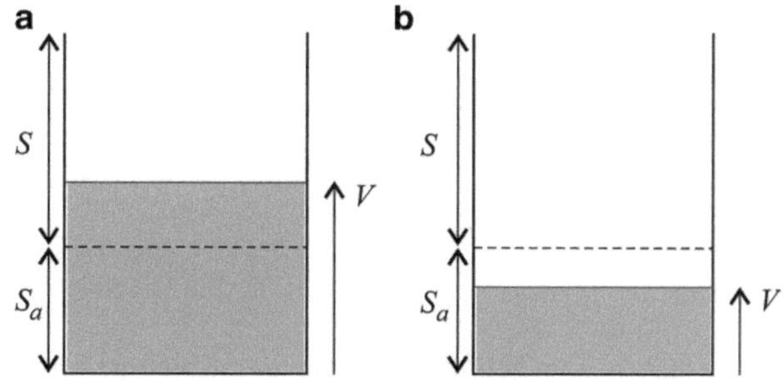

Fig. 5.6 Conceptual representation of behavior of an SMA store at different phases of the fallout–infiltration–runoff process: (**a**) $V > S_a$, Eq. 5.30a, (**b**) $V \le S_a$, Eq. 5.30b

$$\frac{dV}{dt} = r - q, \tag{5.31}$$

written for the scenario $Q > 0$. Obviously, this equation is valid for describing short-time processes, in which the effect of seasonal factors, which determine the soil moisture regime, can be neglected (see below).

The formalization of the process in the form of Eqs. (5.30a), (5.30b) and (5.31), corresponding to the SMA model, is more general as compared with the SCS-CN model (Michel et al. 2005). The SMA model includes cases, such as (1) $S_a > S$, where $V_0 < S_a - S$, the case ignored by the SCS-CN model, because here we have $I_a > S$; (2) $V_0 > S_a$, which corresponds to the situation of $I_a < 0$, not covered by the SCS-CN model; $I_a < 0$ implies the existence of flow in the beginning of the simulated event.

The return from (5.30), (5.31) to problem formulation in terms of Q, P allows a more complete system of equations to be written for those cumulative characteristics (Michel et al. 2005). It is shown that the calibration of runoff models from watersheds can be simplified with a linear relationship similar to (5.23) assumed to exist between S_a and S, e.g., $S_a = S/3$.

5.4.2 A Further Extension of the SMA Model

As we have already mentioned in Sect. 5.3, the prediction capacity of the standard SCS-NC model can be improved by introducing some correction procedures in the evaluation of its parameter S, which relate S with the characteristics of moisture regime, showing seasonal trends. The extension of the application domain of this approach to studying within-year hydrological cycles implies the inclusion into the model of mechanisms responsible for the description of depletion of soil moisture reserves under the effect of evaporation, evapotranspiration, and percolation of water into the zones lying in vertical section below the soil layer, i.e., the zone of aeration and aquifers. For long-range forecasts, especially in arid and semiarid regions, of particular significance are evaporation and evapotranspiration (Kannan et al. 2007). In this case, the continuity Eq. (5.31), rewritten for two scenarios of behavior of a hydrogeological system, receives additional terms:

$$\frac{dV}{dt} = r - q - ET - w, \ V > S_a, \tag{5.32a}$$

$$\frac{dV}{dt} = r - ET - w, \ V \leq S_a, \tag{5.32b}$$

where ET is the actual evapotranspiration rate, mm/d; w is the recharge rate, mm/d. The flow components ET and w, which are time functions, are responsible for soil moisture depletion during interstorm periods.

The lumped model (5.32a), (5.32b) is supposed to take daily (mean) precipitation and evaporation as input. In the general case, *ET* and *w* are functions of soil water conditions and soil hydraulic properties. If the soil saturation profile is assumed to be vertically uniform, then the soil water content averaged over the soil depth $\theta = \theta_{avr}$ is a function controlling the behavior of *ET* and *w*. On the other hand, the dimensionless function

$$s(t) = \frac{V(t)}{S + S_a}, \tag{5.33}$$

which characterizes the degree of SMS filling, can be regarded as an analogue of the soil moisture, θ. Basing on this analogy, we can write relationships for *ET* (*s*) and *w*(*s*), which are convenient to use.

Thus, a soil water parameterization method is commonly used to determine the actual evaporation, *ET*, based on potential evaporation, *PET*, and soil moisture conditions, $ET = f(\theta) \cdot PET$, or, using the mentioned analogy, one can write:

$$ET = f(s) \cdot PET. \tag{5.34}$$

In a more complete setup (Bergström 1992; Kling and Gupta 2009):

$$ET = f(s)PET \text{ when } V < (S + S_a)\eta_{cr}, \tag{5.34a}$$

$$ET = PT \text{ when } V \geq (S + S_a)\eta_{cr}, \tag{5.34b}$$

where η_{cr} is a critical soil moisture for actual evapotranspiration.

Following (Schaake and Chunzhen 1989; Brutsaert 1991; Kim et al. 1996; Michel et al. 2005), we can choose one of the analytical relationships:

$$f(s) = s^m, \; m = 1, 2, \text{ or} \tag{5.35a}$$

$$f(s) = [1 - (1 - s)^m], \; m = 2. \tag{5.35b}$$

Infiltration water losses from SMS can be determined as a flow under unit hydraulic gradient with this flow also related to the variable (5.33)

$$w = k(s) = k_s s^n, \tag{5.36}$$

where k_s is the hydraulic conductivity at complete filling of SMS, $s = 1$ ($V = S + S_a$).

In this case, the system of Eqs. (5.32a), (5.32b) becomes:

$$S\frac{ds}{dt} = \frac{r}{\overline{S}}(1 - s)^2 - \overline{S}\left\{PET\left[1 - (1 - s)^2\right] + k_s s^n\right\}, \quad s > 1 - \overline{S}, \tag{5.37a}$$

$$S\frac{ds}{dt} = \overline{S}\left\{r - PET\left[1 - (1 - s)^2\right] - k_s s^n\right\}, \quad s \leq 1 - \overline{S}, \tag{5.37b}$$

where $\overline{S} = S/(S + S_a)$.

5.4.3 Numerical Example

The specific features of the behavior of function $s = V/(S + S_a)$ are illustrated by an example (Fig. 5.7), which reflects the response of the system to a rain event with a limited duration. The plot was constructed for dimensionless time $\tau = rt/S$. The full lines give a finite-difference solution of the initial problem (5.37a), (5.37b). Dots in the plot are the results of integration of Eq. (5.37a). The coincidence of solutions obtained by two methods demonstrates the reliability of the numerical algorithm.

This plot in particular shows that an increase in the exponential coefficient, which controls infiltration rate, contributes to an increase in soil capacity to retain pore moisture.

5.4.4 A Case Study

Of greater interest is studying the behavior of a real hydrological system under variable precipitation and evapotranspiration conditions. A characteristic example of such system is the basin of the Limnatis R., which is a part of the Kouris catchment (including also two other river basins, Kouris and Krios), the major drainage system of the Upper Troodos Mountains in Cyprus (Fig. 5.8). At the turn of the twentieth–twenty first centuries, the Kouris catchment became the focus of detailed hydrological and hydrogeological studies carried out by the Geological Survey Department of Cyprus, supported by the European Commission and several European institutions (Boronina et al. 2003; Mederer 2005, 2009; Stadelbacher 2007; Zagana et al. 2007). Those studies were a part of a regional work aimed to assess water resources in Cyprus (Udluft et al. 2004).

The Kouris catchment is located at the southern flank of the Troodos massive, the "water tower of the island" (Afrodisis et al. 1986). The Kouris catchment encompasses an area of 300 km^2 and extends from the southern side of the Troodos Massif of Cyprus to the Mediterranean Sea (Boronina et al. 2003). The major groundwater resources are associated with the ophiolitic complex, which comprises harzburgites, gabbros, sheeted dykes, and pillow lavas. The igneous lithologies and consolidated sediments form fractured aquifers or aquifer-systems. They are characterized by strong heterogeneity in hydraulic conductivity

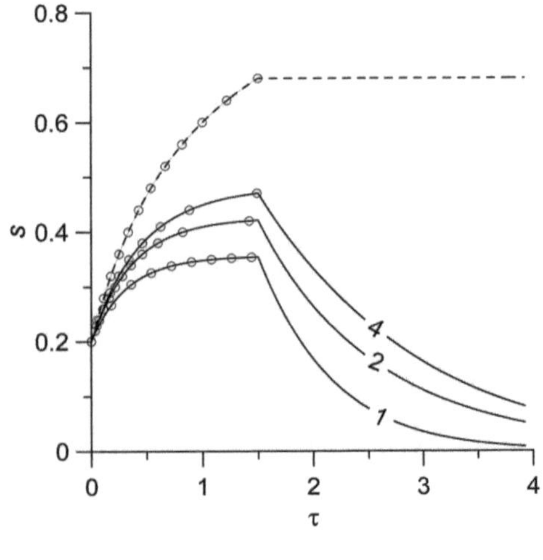

Fig. 5.7 Function $s(\tau)$, evaluated for a short-time ($\tau_0 = 1.5$) rain event. The *dashed line* is for $\overline{k} = 0$ (no infiltration). $\tau = rt/S$, $\overline{k} = k_s/r = 1, \overline{E} = PET/r = 0.5, \overline{S} = 0.8, m = 2$ (Eq. 5.35b). The *curve number* is the value of n

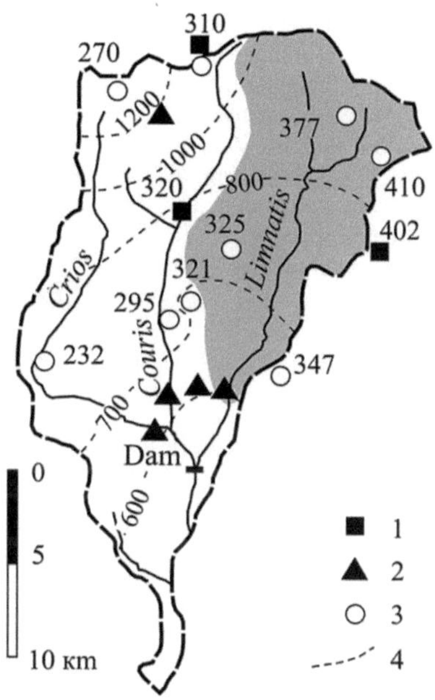

Fig. 5.8 Kouris catchment. (1) Climatological stations where observations of potential evaporation were carried out; (2) gaging stations on rivers; (3) weather stations; (4) distribution of mean annual precipitation, mm

and porosity related to tectonic fracturing and hydrothermal alteration (Boronina et al. 2003; Mederer 2009).

The climate in the study area is Mediterranean semi-arid, experiencing mild wet winters and dry hot summers. Precipitation in the Kouris catchment ranges from 580 mm/a at the lowest elevations to 1000 mm/a in the uppermost parts of the catchment (Fig. 5.8). This results in a mean annual precipitation of 725 mm/a for the whole catchment during the analyzed time series (1986–1996). Precipitation varies in a wide range both from year to year and within a year (Fig. 5.9a). The daily potential evapotranspiration rates were calculated using a version of Penman's equation. The calculated mean annual potential evapotranspiration rate for the Kouris catchment for 1986–1996 varies from 1060 to 1360 mm for the stations at different surface elevations. The plot of daily potential evaporation *PET* (Fig. 5.9a) has distinct sinusoidal character, caused by seasonal variations in climatic factors. Since the evaporation is maximal in the warm, dry season, the function *PET* is in phase opposition to function *P*. This means that the moisture accumulated in the soil in periods with intense rainfall is spent for evaporation and transpiration mostly in dry seasons. The other moisture loss from soil occurs due to the deep water percolation to the underlying aquifers.

The sub-catchment Limnatis covers 115 km^2. Limnatis River flow rate in periods of storm rainfall reaches 5–12 m^3s^{-1}, while in dry periods, it falls to tenths or hundredths of cubic meter per second (Fig. 5.9b). The elevation in the study area ranges from 1612 m in the north to 277 m at the gauging station.

Several surface-subsurface flow and balance models, such as MODFLOW, MODBIL and SWAT, have been generated to analyze groundwater dynamics and rainfall–infiltration–runoff processes within the Kouris catchment and its sub-areas (Boronina et al. 2003; Mederer 2005, 2009; Stadelbacher 2007). The dynamics of the fractured aquifers are simulated using groundwater flow model (MODFLOW). Vertical fluxes between soil, vegetation, and atmosphere are simulated and balanced with water balance models (MODBIL, SWAT). In the balance models, soil water balance is solved for each

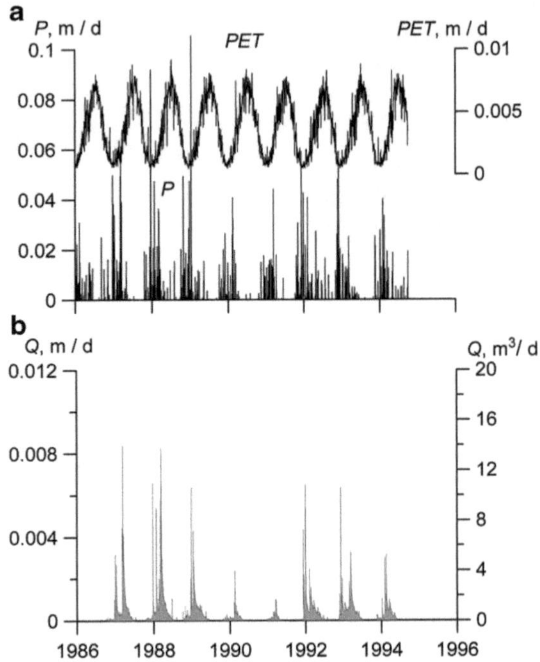

Fig. 5.9 Time distributions (daily measurements) of (**a**) daily precipitation, *P* (averaged over five weather stations: 295, 310, 321, 325, 320, see Fig. 5.8), and daily potential evapotranspiration, *PET*; (**b**) river runoff rate

single grid cell. The selected calibration and validation period for all models was approximately the same (1987–1997).

For the flow and water balance models, soil and subsoil hydraulic properties, as well as data base on morphology and landscape environment, were used as input parameters. These models have been shown to simulate quite accurately the runoff for both floods and recessions.

In this respect the earlier discussed simplified lumped parameter model cannot a priory provide a strong competition to physically based models with distributed parameters. In this study the SMA model was applied to Limnatis sub-catchment only for illustrative purpose to test the potential capacity of the lumped approach to expert assessments. This approach avoids model calibration; rather we wish to use the hydrograph observations to explore some of hydrological effects.

The catchment was considered as a single reservoir with averaged characteristics, which are coefficients of the system of Eqs. (5.37), where functions ET and w are specified according to (5.34), (5.35) and (5.36). Precipitation and potential evapotranspiration (Fig. 5.9a) were two input functions for the model. The four model parameters requiring calibration are: S, S_a, k_s, n. A daily time step was adopted to run the SMA model. The target function for the choice of the parameters was taken to be the available variations in the flow rate of the Limnatis River in 1986–1994 (Fig. 5.9b). The calibration was carried out manually by adjusting parameters by trial and error until model results match qualitatively the field observations.

The simulated hydrograph of the Limnatis is in general agreement with the monitoring data (Fig. 5.10). However, the simulated rising limbs and peak amplitudes did not fit the

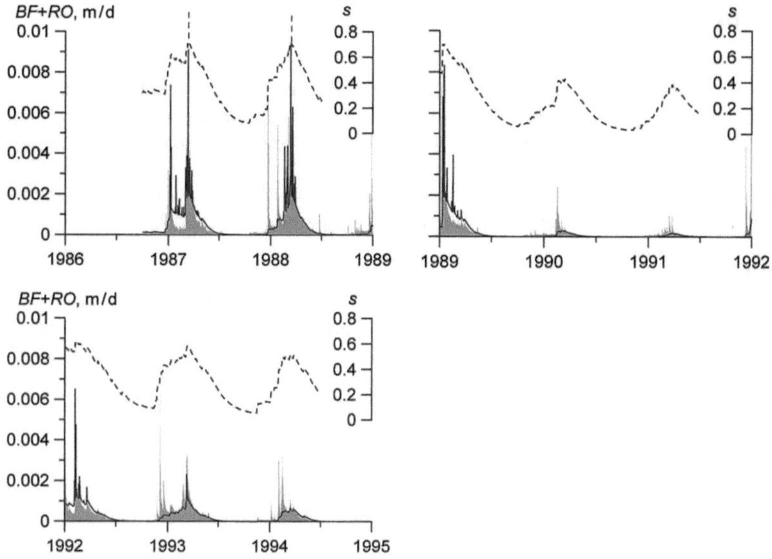

Fig. 5.10 Total (baseflow plus runoff) hydrographs of measured (shaded area) and simulated (*solid line*) discharge for Limnatis river. *Dashed line* is the *s*-function

Table 5.1 Water balance components for the Limnatis catchment

Components	MODBIL (10/1987–9/1997)	This study (10/1986–9/1995)
Precipitation, mm/a	653	689
Real evapotranspiration, mm/a	545	587
Groundwater recharge, mm/a	67	92
Surface runoff, mm/a	44	39
Groundwater recharge/Precipitation, %	10	13

observed hydrographs perfectly, moreover, the model was not able to reproduce some observed peaks in the rising limbs. The recession limbs of the hydrograph were simulated better.

The values of parameters at which the agreement was attained are $S=420$ mm, $S_a=513$ mm, $k_s=0.01$ m/d, $n=4$. The water balance components, estimated with MODBIL software for the Limnatis catchment (Mederer 2009), reach the order of magnitude of the present study's results (Table 5.1). Lumped parameter modeling of the area for period 1986–1995 indicated that recharge was around 13 % of the total annual rainfall, which is in good agreement with the results of Boronina et al. (2003) (between 12 and 16 %, or 100–150 mm for those years for the whole Kouris catchment).

In general, analysis of the obtained results suggests that the physical concepts taken as the basis of the model are in agreement with the observed tendency in river runoff formation in the catchment.

References

Afrodisis S, Avraamides C, Fischbach P et al (1986) Hydrogeological and hydrochemical studies in the Troodos region, Technical report N6 in Cyprus-German geological and pedological project N 81.2224.4. Ministry of Agriculture and Natural Resources, Geological Survey Department, Nicosia, p 101

Allen RG, Pereira LS, Raes D et al (1998) Crop evapotranspiration, guidelines for computing crop water requirements, FAO irrigation and drainage paper 56. Food and Agriculture Organization of the United Nations, Rome

Alley WM (1984) On the treatment of evapotranspiration, soil moisture accounting, and aquifer recharge in monthly water balance models. Water Resour Res 20(8):1137–1149

Bergström S (1992) The HBV model – its structure and applications. SMHI RH, Norrköping

Bergström S (1995) The HBV model. In: Singh VP (ed) Computer models of watershed hydrology. Water Resources Publication, Colorado, pp 443–476

Beven KJ (2006) Rainfall-runoff modeling: introduction. In: Anderson M (ed) Encyclopedia of hydrological sciences. Wiley, Chichester

Boronina A, Renard P, Balderer W et al (2003) Groundwater resources in the Kouris catchment (Cyprus): data analysis and numerical modelling. J Hydrol 271:130–149

Boughton WC (1989) A review of the USDA SCS curve number method. Aust J Soil Res 27(3):511–523

Broome P, Spigel RH (1982) A linear model of storm runoff from some urban catchments in New Zealand. J Hydrol 21(1):13–33

Brutsaert W (1991) The formulation of evaporation from land surfaces. In: Bowles DS, O'Connell PE (eds) Recent advances in the modelling of hydrologic systems. NATO, Series C: Mathematical and physical sciences. Boston, Kluwer Academic Publishers, 345, pp 67–84

Burnash RJC (1995) The NWS river forecast system – catchment modeling. In: Buttle JM Isotope hydrograph separations and rapid delivery of pre-event water from drainage basins. Prog Phys Hydrol 18:16–41

Buytaert W, De Bièvre B, Wyseure G et al (2004) The use of the linear reservoir concept to quantify the impact of land use changes on the hydrology of catchments in the Ecuadorian Andes. Hydrol Earth Syst Sci 8:108–114

Chow VT, Maidment DR, Mays LW (1988) Applied hydrology. McGraw-Hill, New York, p 572

Ding JY (1974) Variable unit hydrograph. J Hydrol 22:53–69

Ding JY (2011) A measure of watershed nonlinearity: interpreting a variable instantaneous unit hydrograph model on two vastly different sized watersheds. Hydrol Earth Syst Sci 15:405–423. www.hydrol-earth-syst-sci.net/15/405/2011/. doi:10.5194/hess-15-405-2011

Dingman SL (2002) Physical hydrology. Prentice-Hall Inc., Upper Saddle River

Diskin MH (1964) A basic study of the linearity of the rainfall-runoff process in watersheds. Thesis, University of Illinois, p 157

Diskin MH (1973) The role of lag in a quasi-linear analysis of the surface runoff system. In: Proceedings of the second international symposium in hydrology. Fort Collins, Colorado, pp 133–144

Duffy CJ, Gelhar LW (1985) A frequency domain approach to water quality modeling in groundwater: theory. Water Resour Res 21:1175–1184

Erickson T, Stefan HG (2007) Groundwater recharge from a changing landscape. S. Anthony Falls Laboratory, Project report 490. Minnesota Pollution Control Agency, St Paul, p 112

Farmer D, Sivapalan M, Jothityangkoon C (2003) Climate, soil, and vegetation controls upon the variability of water balance in temperate and semiarid landscapes: downward approach to water balance analysis. Water Resour Res 39(2):1035. doi:10.1029/2001WR000328

Gupta HV, Sorooshian S, Hogue TS et al (2003) Advances in automatic calibration of watershed models. Calibration of watershed models. In: Duan Q, Gupta HV, Sorooshian S, Rousseau AN, Turcotte R (eds) Water science and application. AGU, Washington DC, 6, pp 9–28

Ibrahim AB, Cordery I (1995) Estimation of recharge and runoff volumes from ungauged catchments in eastern Australia. Hydrol Sci J des Sci Hydrol 40A:499–515

Jowitt PW (1999) A conceptual systems model of rainfall-runoff on the Haast River. J Hydrol NZ 38(1):121–144

Kannan N, Santhi C, Williams JR et al (2007) Development of a continuous soil moisture accounting procedure for curve number methodology and its behaviour with different evapotranspiration methods. Hydrol Process Published online in Wiley Inter Science. www.interscience.wiley.com. doi:10.1002/hyp.6811

Khan LR, Mawdsley JA (1982) Effects of land-use changes on groundwater recharge assessed using a nonlinear catchment model. In: Improvements of methods of long term prediction of variations in groundwater resources and regimes due to human activity (Proceedings of the exeter symposium) IAHS Publ 136, pp 97–106

Kim CP, Stricker JNM, Torfs PJJF (1996) An analytical framework for the water budget of the unsaturated zone. Water Resour Res 32(12):3475–3484

Koivusalo H, Kokkonen T, Karvonen T et al (2001) Accounting for response differences in runoff events of different magnitudes. In: Ghassemi F, Post D, Sivapalan M, Vertessy R (eds) Proceedings, MODSIM: international congress on modelling and simulation-integrating models for natural resources management across disciplines, issues and scales. Canberra, Australia. [Place of publication unknown]: Modelling and Simulation Society of Australia and New Zealand Inc., pp 89–94

Kundzewicz ZW, Napiórkowski JJ (1986) Nonlinear models of dynamic hydrology. J Hydrol Sci 31(2):163–185

Lascano RJ (1991) Review of models for predicting soil water balance. Soil water balance in the Sudano-SaheUan Zone (Proceedings of the Niamey workshop). IAHS Publ 199:443–458

Maloszewski P, Stichler W, Rank D (2000) Combined application of black box models to environmental tracer data for determination of transport and hydraulic parameters in karstic aquifer of Schneealpe (Austria). In: Use of isotopes for analyses of flow and transport dynamics in groundwater systems. Results of a co-ordinated research project 1996–1999, IAEA

Mederer J (2005) Groundwater resources of the Limnatis catchment in Cyprus: Application of the water balance modelling program MODBIL in an area of irrigation. Zeitschrift des Lehr- und Forschungsbereichs Hydrogeologie und Umwelt (ISSN 09309–3757) N 33:1–13

Mederer J (2009) Water resources and dynamics of the Troodos Igneous Aquifer-system, Cyprus – balanced groundwater modelling. Doctorate Thesis. Julius-Maximilians University of Würzburg, p 146

Michel C, Vazken A, Perrin C (2005) Soil conservation service curve number method: how to mend a wrong soil moisture accounting procedure. Water Resour Res 41(2):1–6. doi:10.1029/2004WR003191

Mishra SK, Singh VP (2003) Soil Conservation Service Curve Number (SCS-CN) methodology. Kluwer Academic Publishers, Dordrecht. ISBN 1-4020-1132-6

Moor RJ, Bell VA (2001) Comparison of rainfall–Runoff models for flood forecaseting. Part 1. Literature review of models. R&D Technical report W241Environmental Agency, Bristol, p 94

Nash JE (1960) A unit hydrograph study with particular reference to British catchments. Proc Inst Civil Eng 17:249–282

O'Donnell T (1960) Instantaneous unit hydrograph derivation by harmonic analysis. Int Assoc Sci Hydrol Pub 51:546–557

Overton DE (1970) Route or convolute? Water Resour Res 6(1):43–52

Pedersen JT, Peters JC, Helweg OJ (1980) Hydrographs by single linear reservoir model. J Hydraulics 106(HY5):837–852

Rumynin VG (2011) Subsurface solute transport models and case histories (with applications to radionuclide migration), vol 25, Series: Theory and applications of transport in porous media. Springer, Dordrecht, p 133

Sahu RK, Mishra SK, Eldho TI et al (2007) An advanced soil moisture accounting procedure for SCS curve number method. Hydrol Process 21:2872–2881

Schaake JC, Chunzhen L (1989) Development and application of simple water balance models to understand the relationship between climate and water resources. In: New Directions for Surface Water Modeling Proceedings of the Baltimore Symposium IAHS Publ N181. Baltimore, MD

Shamseldin AY, Nash JE (1998) The geomorphological unit hydrograph: a critical review. Hydrol Earth Syst Sci 2(1):1–8

Sheffer NA (2009) Variable scale recharge measurement and modelling using the hydrometeorological DReAM. PhD thesis, Hebrew University of Jerusalem, p 111

Sheffer NA, Dafny E, Gvirtzman H et al (2010) Hydrometeorological daily recharge assessment model (DREAM) for the Western Mountain Aquifer, Israel: Model application and effects of temporal patterns. Water Resour Res. doi:10.1029/2008WR007607

Shestakov VM, Pozdniakov SP (2003) Geohydrology. IKC "Academkniga", p 176

Shi Z-H, Chen L-D, Fang N-F et al (2009) Research on the SCS-CN initial abstraction ratio using rainfall-runoff event analysis in the Three Gorges Area. China CATENA 77(1):1–7. doi:10.1016/j.catena.2008.11.006

Snyder WM, Mills WC, Stephens JC (1970) A method of derivation on nonconstant watershed response functions. Water Resour Res 6(1):261–274

Stadelbacher V (2007) Water balance modeling of a meso-scale mountainous catchment in Cyprus – process analysis and climate change impact. Diploma thesis. Institut fur Hydrologie Albert-Ludwigs-Universitat Freiburg i Br, p 98

Subramanya K (2008) Engineering hydorlogy. Tata McGraw-Hall Education Publ. Company Limited, New Delhi, p 435

Thornthwaite CW (1948) An approach toward a rational classification of climate. Geograph Rev 38(1):55–94

Thornthwaite CW, Mather JR (1955) The water balance. Publications in Climatology, Drexel Institute of Technology, Centerton, New Jersey, vol VIII(1)

Udluft P, Dünkeloh A, Mederer J et al (2004) Water balances for catchments and the whole island. GRC-Project Report T 6/7, Nicosia, Geological Survey Department of Cyprus, 363 p

USDA (1986) Natural Resources Conservation Service. Technical Release 55. Urban Hydrology for Small Watersheds, TR-55, Washington, DC, p 164

Vaze J, Jordan P, Beecham R (2012) Guidelines for rainfall-runoff modelling: towards best practice model application. Water Cooperative Research Centre, San Francisco, p 47

Wood EF, Sivapalan M, Beven K (1990) Similarity and scale in catchment storm response. Rev Geophys 28(1):1–18

Xu C-Y, Singh VP (2004) Review on regional water resources assessment models under stationary and changing climate, Water resource management 18. Kluwer Academic Publishers, Amsterdam, pp 591–612

Zagana E, Kuells C, Udluft P et al (2007) Methods of groundwater recharge estimation in eastern Mediterranean – a water balance model application in Greece, Cyprus and Jordan. Hydrol Process 21:2405–2414

Zondervan JG (1978) Modelling urban runoff – a quasilinear approach. Agric Res Rep 874. ISBN 90 220 0665 4. Centre for Agricultural Pub. and Documentation, Wageningen, Netherlands

Chapter 6
Lumped-Parameter Models for Solute Transport with Runoff

The use of lumped-parameter models is justified when runoff solute content at a hillslope or catchment outlet is governed by the kinetics of soil solute release (or solute removal from the contaminated soil surface) into runoff, rather than overland flow dynamics (variations in the velocities and thickness). Such models allow the infiltration (rainwater flowing downwards into the soil) and capillary effects at the interphase between the soil and water flowing over its surface to be described in detail and soil column inhomogeneity to be taken into account. In such cases, the inertia of water flow, resulting in a time lag between rainfall excess and the slope or catchment outlet discharge can be accounted for in effective parameters in linear or nonlinear flow kinetic equations or such inertia can be neglected completely with the response of the outlet discharge to the rainfall event assumed instantaneous.

The mathematical framework of such macroscopic scale approach is based on solutions of a system of two coupled equations: (1) a lumped continuity equation for water storage within a lumped hydrological unit (representing a hillslope or catchment/watershed), and (2) a lumped equation of mass balance, including a source term for chemical interaction accounting for solute yield response to rainfall impulses at the regional scale. It is assumed that all hydrological functions (associated with rainfall excess, water storage, and solute concentration) and parameters controlling overall flow and discharge, as well as chemical exchange between runoff and soil are presented in basin-scale average values.

The models discussed below address different aspects of solute behavior in watershed areas and can be implemented to predict both nonpoint contaminant discharge through runoff into surface water (and/or subsoil domains) and the rate of natural remediation of soil contaminated by agricultural chemicals, radionuclides and other anthropogenic components, which generate nonpoint-source pollution. In this chapter, special emphasis is placed on the lumped models whose development was motivated by studies of the consequences of regional-scale radioactive fallout.

V.G. Rumynin, *Overland Flow Dynamics and Solute Transport*, Theory and Applications of Transport in Porous Media 26, DOI 10.1007/978-3-319-21801-4_6

6.1 Lumped Hillslope Models for Overland Flow Dynamics

We focus here on surface runoff models, which can be regarded as an alternative to the kinematic wave model considered above (Chaps. 2, 3, and 4). Such models are approximate and based on some kinds of averaging of hydrodynamic characteristics of the water flow that forms on an idealized hillslope (Fig. 6.1).

6.1.1 Transformation of the Original Kinematic-Wave Equation

Suppose that at any moment, t, the increment (gradient) of overland flow discharge (per unit width), $Q_x(t)$, along the flow direction, x (Fig. 6.1), is a variable independent of the distance (Rose et al. 1983; Singh 1996; Agnese et al. 2001). This approximation immediately leads to a differential relation for a characteristic known as runoff rate:

$$q(t) = \frac{\partial Q_x(t)}{\partial x}. \tag{6.1}$$

As one can see, hereafter, following the tradition, for the denotation of this characteristic of the overland flow, we reserve the symbol $q = q(t)$ [LT^{-1}], which has been used to denote flow discharge, $q = q(x,t)$ [L^2T^{-1}], in the mathematical formulation of the kinematic wave problem (Chaps. 2, 3, and 4). Also, it is clear, that in this context, the runoff rate is synonymous to the outflow rate.

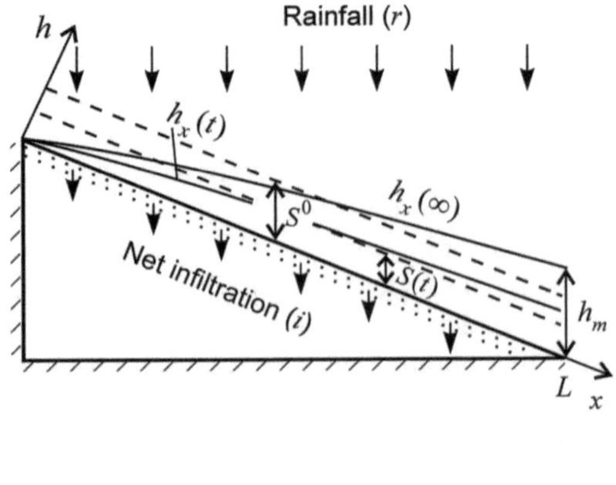

Fig. 6.1 Formation of transient water surface; characteristics $h_x(\infty)$, h_m and S^0 correspond to steady-state conditions ($r_e = r - i = \text{const}$)

With the assumption (6.1), the continuity Eq. (2.2), for arbitrary time-dependent excess rainfall rate, $r_e(t)$ (Eq. 1.1b),

$$\frac{\partial h_x(t)}{\partial t} + \frac{\partial Q_x(t)}{\partial x} = r_e(t), \tag{6.2}$$

transforms into

$$\frac{\partial h_x(t)}{\partial t} = r_e(t) - q(t). \tag{6.3}$$

where $h_x(t)$ is the depth of water flowing over the soil surface.

Integrating Eq. 6.1 yields $Q_x(t) = q(t)x$, thus the flow discharge at the bottom of the hillslope ($x = L$) is equal to $Q_L(t) = q(t)L$. Substituting this expression into formula (2.3), which relates water flux and surface water depth, we obtain

$$h_x(t) = \left(\frac{Q_x(t)}{\alpha}\right)^{1/n} = \left(\frac{q(t)x}{\alpha}\right)^{1/n}. \tag{6.4}$$

Obviously, under steady-state condition ($r_e = \text{const}$), we have $q(\infty) = r_e$ and $h_x(\infty) = (r_e x/\alpha)^{1/n}$, which corresponds to kinematic flow approximation (Chap. 2).

Suppose that there exists an effective water depth, $S(t)$, which characterizes the amount of water that is temporarily stored on the soil surface per unit area (Fig. 6.1) and can be regarded as the mean value of $h_x(t)$,

$$S(t) = \frac{1}{L}\int_0^L h_x(t)dx = \left(\frac{q(t)}{\lambda}\right)^{1/n}, \tag{6.5}$$

$$\lambda = \frac{\alpha}{L_e}, \quad L_e = \left(\frac{n}{1+n}\right)^n L, \tag{6.6}$$

where λ can be considered as a rate constant, characterizing the dynamics of delayed slope discharge $[\mathrm{L}^{1\text{-}n}\mathrm{T}^{-1}]$; L_e is the effective length of the slope [L]. The relationship (6.5) can be readily obtained by multiplying the left and right sides of Eq. 6.4 by dx and integrating from 0 to L. Now, Eq. 6.3 becomes

$$\frac{dS}{dt} = r_e - \lambda S^n. \tag{6.7}$$

As seen, a peculiar characteristic of Eq. 6.7 is the non-linear dependence of the lumped hillslope response to the rainfall excess intensity, similar to relationship (Eq. 5.17) which was introduced for a nonlinear model of reservoir type (Sect. 5.2).

A useful characteristic for the analysis of solute transport by runoff is the *mean residence time* of runoff water on the slope. This characteristic can be easily obtained for a steady-state water flow profile. Therefore, in the following analysis, a constant rainfall excess and steady-state condition is assumed. With this assumption, we have

$$u_x h_x = r_e x. \tag{6.8}$$

With h_x expressed from this equality and substituted into Manning's Eq. (2.3), rewritten as

$$u_x = Q_x(t)/h_x = \alpha h_x^{n-1}, \tag{6.8a}$$

we obtain

$$u_x = \alpha^{1/n}(r_e x)^{1-1/n}. \tag{6.8b}$$

Now the time that takes water to travel from a point, x, on the slope to the outlet of the slope, $x = L$, can be determined:

$$t_x = \int_x^L \frac{dx}{u_x} = n\alpha^{-1/n} r_e^{1/n-1}\left(L^{1/n} - x^{1/n}\right). \tag{6.9}$$

The time lag or the mean residence time, τ^0, can be determined as the average travel time over the hillslope (Yu et al. 2000):

$$\tau^0 = \frac{1}{L}\int_0^L t_x dx = \frac{1}{r_e}\left(\frac{r_e}{\lambda}\right)^{1/n} = \frac{n}{n+1} t_e^*, \tag{6.10}$$

where t_e^* is the time of concentration of overland flow at $x = L$ obtained earlier for the kinematic flow approximation (see Eq. 2.13). From differential Eq. 6.7 it follows that, under steady-state condition, the mean runoff water depth (Fig. 6.1) is

$$S^0 = (r_e/\lambda)^{1/n}, \tag{6.11a}$$

now Eq. 6.10 becomes

$$\tau^0 = S^0/r_e. \tag{6.11b}$$

Comparing Eqs. 6.10 and 6.11b and taking into account the relationship for the time of concentration, it can be shown that, under steady-state condition, mean water depth, S^0, as the basic characteristic of the lumped model is related with the maximum water depth on the sloped surface, h_m (at $x = L$), considered in the kinematic wave model, through the ratio:

$$S^0 = \frac{n}{n+1} h_m. \tag{6.11c}$$

The assumption that runoff flow is steady and the runoff water depth is constant allows us to use in the further description of solute transport the residence time distribution (RTD) function (Wallach et al. 1988; Wallach 1991):

$$E(t) = \frac{1}{\tau^0} \exp\left(-\frac{t}{\tau^0}\right). \tag{6.12}$$

An idealistic model (6.12) is based on the assumption that the rainfall excess is mixed, completely and instantly, into the bulk of the water stored on the soil surface. The stored and the outlet water have identical, homogeneous compositions at all times.

6.1.2 *Basic Differential Equation for Unsteady Overland Flow*

Differentiating Eq. 6.4 with respect to time, we come to the expression

$$\frac{\partial h_x(t)}{\partial t} = \frac{1}{n} \left(\frac{x}{\alpha}\right)^{1/n} q(t)^{(1-n)/n} \frac{dq(t)}{dt}. \tag{6.13}$$

Now Eq. 6.3 becomes

$$\frac{1}{n} \left(\frac{x}{\alpha}\right)^{1/n} q(t)^{(1-n)/n} \frac{dq(t)}{dt} = r_e(t) - q(t). \tag{6.14}$$

Multiplying the right and left sides of Eq. 6.14 by dx and integrating both sides from $x = 0$ to L (Agnese et al. 2001)

$$\frac{1}{n} q(t)^{(1-n)/n} \frac{dq(t)}{dt} \int_0^L \left(\frac{x}{\alpha}\right)^{1/n} dx = \left[r_e(t) - q(t)\right] \int_0^L dx, \tag{6.15}$$

we come to a differential identity for outflow rate:

$$\frac{q^{1/n}}{(r_e - q)} \frac{dq}{q} = n\lambda^{1/n} dt, \tag{6.16}$$

where λ is the effective rate constant (see Eq. 6.6); $q = q(t)$; $r_e = r_e(t)$. Equation 6.16 can be directly obtained by combining Eqs. 6.5 and 6.7. In the case of $r_e =$ const, Eq. 6.16 can be represented in the dimensionless form:

$$\frac{\overline{q}^{1/n}}{(1-\overline{q})}\frac{d\overline{q}}{\overline{q}} = nd\tau, \quad \overline{q} = q/r_e, \quad \tau = t/\tau^0, \tag{6.16a}$$

where τ^0 is as in Eq. 6.11b.

Integrating the ordinary differential Eq. (6.16a) with separable variables in fixed intervals [0, q] and [0, τ] yields

$$\tau = \frac{1}{n}\overline{q}^{1/n}\text{Lerch Phi}\,(\overline{q}, 1, 1/n), \tag{6.17}$$

where Lerch Phi (z, s, a) is a special function determined by the formula

$$\text{Lerch Phi}\,(z, s, a) = \sum_{k=0}^{\infty} \frac{z^k}{(a+k)^s}. \tag{6.17a}$$

Relationship (6.17) can be represented in a graphical form (Fig. 6.2).

Essentially, the above transformations reduce the equation (model) of one-dimensional flow (6.2) to a "0-dimensional" reservoir-type equation, i.e., to a lumped-parameter model in terms of kinematic wave equations (Chap. 2). Equation (6.16) is a basic equation for the analysis of hydrograph behavior at the hillslope scale.

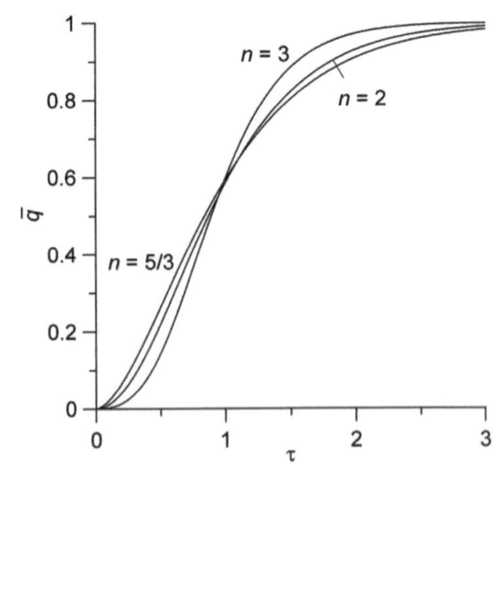

Fig. 6.2 Function $\overline{q} = \overline{q}(\tau, n), \overline{q} = q/r_e$, $\tau = t/\tau^0$

A solution for Eq. 6.16a for arbitrary n can be obtained numerically. For $n = 2$ and r_e = const, integrating (6.16a) yields an analytical solution to the problem with respect to the required function:

$$\overline{q} = \tanh^2\left[\operatorname{arctanh}\sqrt{\overline{q}_0} + \sqrt{\lambda r_e}(t - t_0)\right] \quad r_e > q_0\,(\overline{q}_0 < 1),\ \overline{q} = q/r_e,\ q_0 = q(t_0); \tag{6.18a}$$

$$\overline{q} = \coth^2\left[\operatorname{arccoth}\sqrt{\overline{q}_0} + \sqrt{\lambda r_e}(t - t_0)\right] \quad r_e > q_0\,(\overline{q}_0 < 1),\ r_e \neq 0,\ \overline{q} = q/r_e; \tag{6.18b}$$

$$\overline{q} = \left[1 + \sqrt{\lambda q_0}(t - t_0)\right]^{-2} \quad r_e > q_0\,(\overline{q}_0 < 1),\ r_e = 0,\ \overline{q} = q/q_0. \tag{6.18c}$$

For recession period ($r = 0$), integration of Eq. 6.16a also results in a closed-form solution:

$$\overline{q} = \tan^2\left[\arctan\sqrt{\overline{q}} - \sqrt{\lambda i}(t - t_0)\right] \quad i > 0,\ \overline{q} = q/i,\ \overline{q}_0 = q_0/i; \tag{6.18d}$$

$$\overline{q} = \left[1 + \sqrt{\lambda q_0}(t - t_0)\right]^{-2} \quad i = 0,\ \overline{q} = q/q_0. \tag{6.18e}$$

6.1.3 Comparative Analysis (Case $i = 0$*)*

To verify the applicability of the lumped-parameter model for overland flow discussed in the section above, the basic differential identity for dimensionless outflow rate (6.16a) was re-written in terms of the kinematic wave model (Sect. 2.2.1), specifically, in terms of the dimensionless overland flow discharge function, $\overline{q}(x = L, \tau)$ (Eqs. 2.22 and 2.22a, $h_0 \equiv h_m$), using relationships (6.11a), (6.11b) and (6.11c) and assuming $\overline{q} = q/r$ $(i = 0)$.

As can be seen from several computation experiments for step-wise input function $r(t)$, $0 < t \le t_1 = T$ (Fig. 6.3), the results obtained by different approaches are in good agreement in what regards the formation of an equilibrium profile of water flow (Sect. 2.2.1), i.e., while $\tau \ge 1$ (Figs. 6.3a, b). However, the lumped-parameter model is less accurate in describing the recession stage of inequilibrium hydrodynamic profile $\tau < 1$ (Figs. 6.3c, d). Because of its integral character, the model always satisfies the condition of water balance in the system, and, in this aspect, it can be regarded as reliable for the use in water dynamic and solute transport calculations.

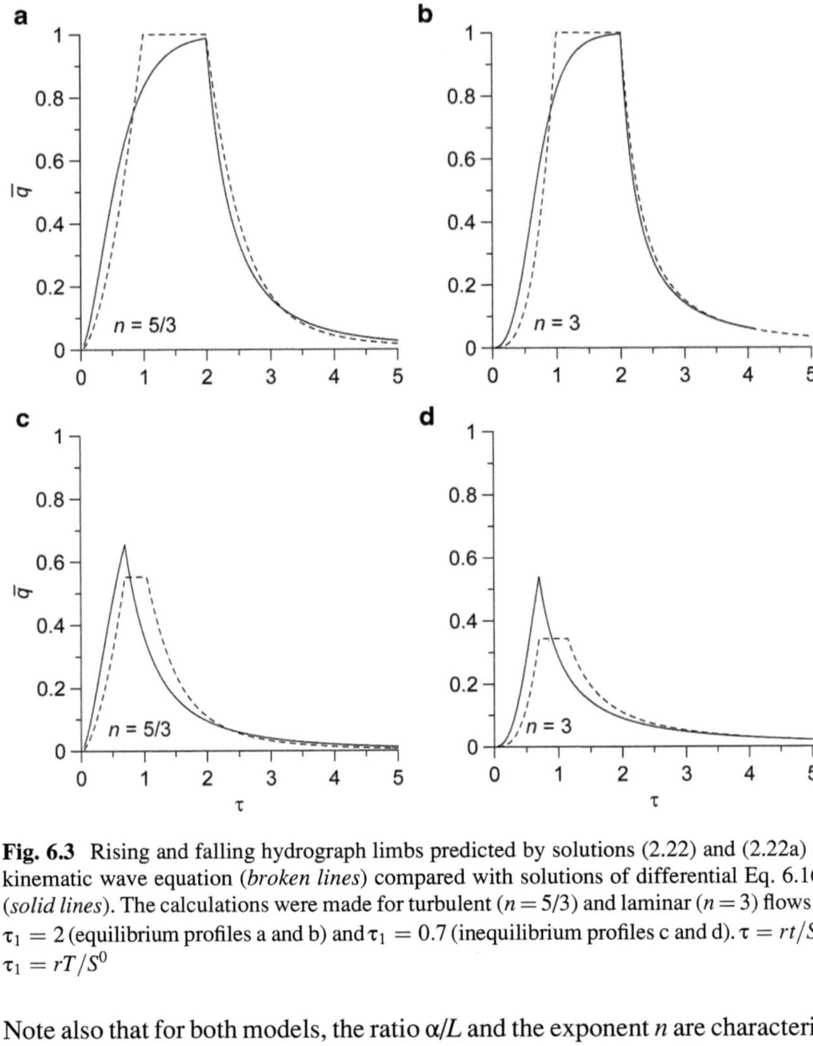

Fig. 6.3 Rising and falling hydrograph limbs predicted by solutions (2.22) and (2.22a) of kinematic wave equation (*broken lines*) compared with solutions of differential Eq. 6.16a (*solid lines*). The calculations were made for turbulent ($n = 5/3$) and laminar ($n = 3$) flows at $\tau_1 = 2$ (equilibrium profiles a and b) and $\tau_1 = 0.7$ (inequilibrium profiles c and d). $\tau = rt/S^0$; $\tau_1 = rT/S^0$

Note also that for both models, the ratio α/L and the exponent n are characteristics that control the dynamics of surface runoff. The ratio α/L is a kind of discharge rate coefficient (for hillslope scale flow process), depending on the domain size and the roughness of the surface, parameter n is responsible for water flow regime.

6.1.4 *General Solution for a Piecewise-Homogeneous Function $r_e(t)$*

Because of its homogeneity, the model (6.16), unlike kinematic-wave model, does not require the identification of characteristic space–time intervals, within which the behavior of functions $h(x, t)$ and $q(x, t)$ is described by different relationships. With $r_e(t)$ considered as a piecewise-homogeneous function, a generalized solution for variable-intensity fallout can be constructed by integrating the differential Eq. (6.16):

$$\int_{q_0}^{q} \frac{q^{(1-n)/n}}{r_e - q} dq = n\lambda^{1/n} \int_{t_0}^{t} dt \tag{6.19}$$

for fixing the initial condition $q(t = t_0) = q_0$, assuming that, during the closed interval $[t_0, t]$, the rainfall excess, r_e, is constant (Agnese et al. 2001).

The integral identity (6.19) can be represented as an equality, including the summation of infinite series. Three different cases for (6.19) can be distinguished, depending on the relationship between r_e and q_0 (Agnese et al. 2001).

1. In the case $r_e > q_0$, it is reasonable to introduce a new variable $q_* = q/r_e$ in the left-hand integral in Eq. 6.19

$$r_e^{\frac{1-n}{n}} \int_{q_{*0}}^{q_*} q_*^{\frac{1-n}{n}} (1 - q_*)^{-1} dq_*, \tag{6.20a}$$

where $q_{*0} = q_0/r_e$. Because $0 \le q_* < 1$, $(1 - q_*)^{-1} = \sum_{j=0}^{\infty} q_*^{\,j}$, formula (6.20a) becomes

$$r_e^{\frac{1-n}{n}} \int_{q_{*0}}^{q_*} \sum_{j=0}^{\infty} q_*^{\frac{1+n(j-1)}{n}} dq_* = n r_e^{\frac{1-n}{n}} \left[\sum_{j=0}^{\infty} \frac{q_*^{\frac{1+nj}{n}}}{1+nj} - \sum_{j=0}^{\infty} \frac{q_{*0}^{\frac{1+nj}{n}}}{1+nj} \right]. \tag{6.20b}$$

Thus, the solution of Eq. (6.20a) can be given in an implicit form

$$\left[\sum_{j=0}^{\infty} \frac{q_*^{\frac{1+nj}{n}}}{1+nj} - \sum_{j=0}^{\infty} \frac{q_{*0}^{\frac{1+nj}{n}}}{1+nj} \right] = \frac{1}{\tau^0} (t - t_0). \tag{6.20c}$$

2. In the case $r_e < q_0$ and $r_e \ne 0$, another new variable $r_* = r_e/q$ transforms the left-hand integral in the Eq. 6.19 as follows

$$r_e^{\frac{1-n}{n}} \int_{r_{*0}}^{r_*} r_*^{-\frac{1}{n}} (1 - r_*)^{-1} dr_*, \tag{6.20d}$$

where $r_{*0} = r_e/q_0$. Because $0 \le r_* < 1$, $(1 - r_*)^{-1} = \sum_{j=0}^{\infty} r_*^{\,j}$, formula (6.20d) becomes

$$r_e^{\frac{1-n}{n}} \int_{r_{*0}}^{r_*} \sum_{j=0}^{\infty} r_*^{\frac{nj-1}{n}} dr_* = n r^{\frac{1-n}{n}} \left[\sum_{j=0}^{\infty} \frac{r_*^{\frac{n(j+1)-1}{n}}}{n(j+1)-1} - \sum_{j=0}^{\infty} \frac{r_{*0}^{\frac{n(j+1)-1}{n}}}{n(j+1)-1} \right]. \tag{6.20e}$$

The solution of Eq. (6.20d) for this relationship between r_e and q_0 can be given in a form

$$\left[\sum_{j=0}^{\infty}\frac{r_*^{\frac{n(j+1)-1}{n}}}{n(j+1)-1}-\sum_{j=0}^{\infty}\frac{r_{*0}^{\frac{n(j+1)-1}{n}}}{n(j+1)-1}\right]=\frac{1}{\tau^0}(t-t_0). \quad (6.20f)$$

3. In the case $r_e < q_0$ and $r_e = 0$, q can be found directly from the following expression (Agnese et al. 2001)

$$q = q_0\left[1+(n-1)\lambda^{1/n}q_0^{1-1/n}(t-t_0)\right]^{-\frac{n}{n-1}}; \quad (6.20g)$$

here $\tau^0 = S^0(r_e)/r_e$; $r_e = r(t_k) - i(t_k)$ is rainfall excess rate in interval $t - t_{k-1}$; q_0 is the outflow rate by the end of period $k-1$, preceding the current period, k; τ^0 is the mean residence time determined by Eq. 6.11b. Function $r_e(t)$ can be generated using appropriate infiltration model (Sect. 1.3).

As an example, we will use solution (6.16), (6.20) to construct the hydrograph $q(t)$, being the response to an artificially generated function $r(t)$, assuming that some precipitation soaks into the soil with a constant rate $i = 5$ mmh^{-1}. Suppose also that the slope length is $L = 1000$ m and the slope itself can be characterized by a generalized coefficient of hydraulic resistance $\alpha = S_0^{1/2}/m$, equal to 5 m$^{1/3}$/s $= 1.8\cdot10^5$ mm$^{1/3}$h^{-1}. With those parameters, we have $\lambda = \alpha/L_e = 0.39$ mm$^{-2/3}$h^{-1} (L_e is the effective slope length which is determined by Eq. 6.6, assuming $n = 5/3$). Further, assume that the rainfall can be approximated by a step-wise input function (Fig. 6.4): it characterizes a 5-h precipitation period including three rain events with irregular (time-variable) rate. One can readily see that the relationship between characteristics q and q_0 within the cycles was different; therefore, all three cases mentioned above (Eqs. 6.20) are realized. The obtained hydrograph shows a quick response to precipitation events; however, function $q(t)$ can be generally characterized as a long-tail hydrograph: runoff over slope surface continues for a considerable time after rain cessation.

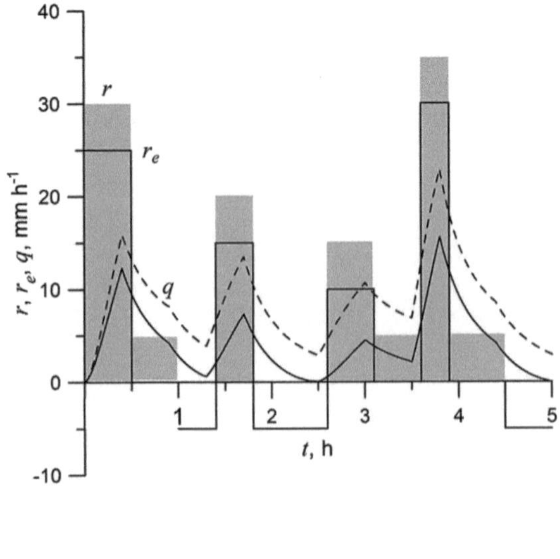

Fig. 6.4 Hydrograph of the outflow rate, $q(t)$ (*solid curve*), for a multistep input function, $r(t)$ (*shaded bar*), and the relevant rainfall excess function, $r_e(t) = r(t) - i$ (*solid bare lines*); the *dashed curve* is the hydrograph for $i = 0$

6.2 Solutions for Solute Transport in Runoff Over Lumped Hillslope

Previously, we described the transport of solutes in water flowing over the soil surface using solutions of the generalized system of Eqs. 3.1–3.2 obtained by the method of characteristics (Sects. 3.2–3.3). In that case, we took into account the differences between the behavior of functions $h(x,t)$ and $q(x,t)$ in different zones of the flow and in different time intervals, i.e., the concentration function $C(x, t)$ was determined by several partial solutions, each relating to the spatial and temporal features of flow dynamics on the slope. Of greater convenience for practical application can be relationships obtained using lumped-parameter models, dealing with the mean flow dynamic characteristics (Sect. 6.1). In particular, the water flow is characterized by the average overland flow depth, $S(t)$ (Eq. 6.5). With this formulation, the concentration of a solute, $C(t)$, in water in contact with the surface is an averaged function as well.

As previously (Chap. 3), we will analyze two mathematical frameworks conceptualizing the solute exchange at the soil runoff interface: (1) kinetically controlled transfer of chemicals from uniformly contaminated soil water to runoff; (2) equilibrium solute exchange between the soil surface instantaneously contaminated from fallout and runoff.

6.2.1 *A Model of Kinetic Mass Exchange at the Water–Soil Interface*

The original system of equations incorporates a flow dynamic equation and a mass balance equation in a water layer, accounting for the kinetics of the chemical's release from contaminated land surface or emission from internal pore volume:

$$\begin{aligned} \frac{dS}{dt} &= r_e - q, \\ \frac{dSC}{dt} &= J_s - qC + rC_r, \end{aligned} \tag{6.21}$$

where the source term, J_s, is defined as in Eq. 3.5a; other variables are introduced in the previous sections.

Simple transformations allow the system of Eq. (6.21) to be reduced to a first-order ordinary differential equation

$$\left(\frac{q}{\lambda}\right)^{1/n}\frac{dC}{dt} + (r + k_e)C - rC_r - kC_s = 0, \tag{6.22}$$

which takes into account the functional relationship between runoff (outflow) rate, q (Eq. 6.16), and average water depth, S, through Eq. 6.5; here k_e is the mass transfer coefficient (as in Eqs. 3.5 and 3.5a); λ is the rate constant for slope discharge (Eqs. 6.5 and 6.7); C_s is the concentration of saturation or solute content in soil water.

6.2.1.1 Step-Wise Input Function

Let us consider the solution of the problem relating the response of a contaminated slope to a rain event having limited duration, $t = t_1$ ($t_1 \equiv T$). The Eq. (6.22) for period $t_0 \leq t \leq t_1$, when $r_e > 0$, has a dimensionless representation

$$\left(\frac{q}{\lambda}\right)^{1/n} \frac{d\hat{C}}{dt} = r - r(1+\beta)\hat{C}\,, \tag{6.23}$$

where $\hat{C} = (C_s - C)/(C_s - C_r)$; $\beta = k_e/r$; outflow rate, q, is determined from the solution of the ordinary differential Eq. (6.16) or from its series representations (Eqs. 6.20).

As the differential dt of the variable t is connected with the differential dq of q through Eq. 6.16, the solute balance Eq. (6.23) can be transformed such that the concentration function $\hat{C}(t)$ becomes a function of outflow rate, $\hat{C}(q)$:

$$nq(r_e - q)\frac{d\hat{C}}{dq} = r\left[1 - (1+\beta)\hat{C}\right]. \tag{6.24}$$

Separation of variables results in the equation

$$\frac{dq}{nq(r_e - q)} = \frac{d\hat{C}}{r\left[1 - (1+\beta)\hat{C}\right]}. \tag{6.24a}$$

Integrating the left part of Eq. 6.24a from q_0 to q and its right part from $\hat{C}_0$ to $\hat{C}$ we get a solution of the problem

$$\hat{C} \equiv \frac{C_s - C}{C_s - C_r} = \frac{1}{1+\beta} - \left[\frac{1}{1+\beta} - \frac{C_s - C_0}{C_s - C_r}\right]\left[\frac{q_0(1-q)}{q(1-q_0)}\right]^{(1+\beta)/n\overline{r}}. \tag{6.25}$$

If $C_r = 0$, Eq. 6.25 can be simplified

$$\overline{C} = \frac{C}{C_s} = \frac{\beta}{1+\beta} + \left(\overline{C}_0 - \frac{\beta}{1+\beta}\right)\left[\frac{q_0(r_e - q)}{q(r_e - q_0)}\right]^{(1+\beta)/n\overline{r}}; \tag{6.26}$$

here $\overline{r} = r_e/r$; $\overline{C}_0 = C_0/C_s$; C_0 is the solute concentration at time t_0, when $q = q_0$.

At $q_0 = 0$, which corresponds to rain falling onto an initially dry surface,

$$\overline{C} = \frac{\beta}{1+\beta}, \quad 0 < t < t_1, \tag{6.27}$$

the result derived before from a kinetic wave model, see Eq. 3.64; t_1 is the rainfall duration.

For the stage of recession, $t \geq t_1 = T$, $r = 0$, Eq. 6.23 reduces to

$$n(i+q)q\frac{d\hat{C}}{dq} = k_e\hat{C}, \tag{6.28}$$

where $\hat{C} = 1 - C/C_s$. The integration of Eq. 6.28 leads to the solution

$$\overline{C} = 1 - \frac{1}{1+\beta}\left[\frac{q(i+q_1)}{q_1(i+q)}\right]^{k_e/ni}, t \geq t_1. \tag{6.29}$$

To get solution (6.29) it was implied that the concentration of a solute in runoff water when fallout ceases, $C(q_1)/C_s$, is defined by Eq. 6.27. If $i \to 0$, the dimensionless concentration is defined from the solution:

$$\overline{C} = 1 - \frac{1}{1+\beta}\exp\left[-\frac{k_e}{n}\left(\frac{1}{q} - \frac{1}{q_1}\right)\right], q \leq q_1. \tag{6.29a}$$

As can be seen in all variants, after the rain ceased, solute concentration, C, increases, tending to the limiting value of saturation concentration C_s.

At an arbitrary n for fixed duration of the rain event t_1, the value q_1 can be found by integrating the basic identity (6.16). The same equality, considered as transcendent, is used to find the value q, corresponding to the current process time $t > t_1$. For some particular values of n, the integration of (6.16) allows an expression for $q(t)$ to be obtained in a simple closed form. For example, with $n = 2$ at $t_0 = 0$, we have:

$$q_1 = r_e \tanh^2\left(\sqrt{\lambda r_e}\, t_1\right); \tag{6.30a}$$

$$q = i\tan^2\left[\arctan\sqrt{q_1/i} - \sqrt{\lambda i}(t - t_1)\right], \quad i > 0; \tag{6.30b}$$

$$q = \left[1/\sqrt{q_1} + \sqrt{\lambda}(t - t_1)\right]^{-2}, \quad i = 0. \tag{6.30c}$$

6.2.1.2 A General Solution for Arbitrary Rainfall Excess Function

Formula (6.25) can be generalized for multistep input function, r_e, as follows:

$$\overline{C}(q) = \frac{\beta_j}{1+\beta_j} + \left(\overline{C}_{j-1} - \frac{\beta_j}{1+\beta_j}\right)\left[\frac{q_{j-1}(r_{ej}-q)}{q(r_{ej}-q_{j-1})}\right]^{(1+\beta_j)/n\overline{r}_j}. \tag{6.31}$$

For the calculation of the concentration function $\overline{C}(q)$ in the time intervals where the condition $r_e = 0$ ($r = i$) holds, the system of Eq. (6.21) must be transformed to become:

$$\frac{n}{i}q^2\frac{d\hat{C}}{dq} = \left(1+\frac{k_e}{i}\right)\hat{C} - 1, \tag{6.32}$$

where $\hat{C}$ is as in Eq. 6.23. Equation 6.32 has a solution:

$$\overline{C}(q) = \frac{k_e}{k_e+i} - \left[\frac{k_e}{k_e+i} - \overline{C}_{j-1}\right]\exp\left[-\frac{k_e+i}{n}\left(\frac{1}{q} - \frac{1}{q_{j-1}}\right)\right], \tag{6.33}$$

which is also valid for the case $i = 0$; $\overline{C} = C/C_s$.

Given the value of t, the outflow rate, q, can be found from the integral relationship:

$$\int\limits_{q_{j-1}}^{q}\frac{q^{1/n}dq}{(r_{ej}-q)q} = \frac{n}{\tau_j^0}(t-t_{j-1}), \beta_j = \frac{k_e}{r_j}, \overline{r}_j = \frac{r_{ej}}{r_j}, \tau_j^0 = \frac{1}{r_{ej}}\left(\frac{r_{ej}}{\lambda}\right)^{1/n}. \tag{6.34}$$

As an example, we consider the results of calculation of concentration function, $C(t)$, for a hillslope hydrograph, which includes an infiltration component ($i > 0$). The hydrological conditions and the geometry of the slope are represented in Sect. 6.1.4 where Fig. 6.4 (solid curves) shows temporal variations of rainfall and rainfall excess input functions. To assess the chemical response using the approach described above, we take the transfer coefficient, k_e, to be 5 mm h^{-1}.

As can be seen from the plot (Fig. 6.5a), the behavior of function $C(t)$ is quite logical and predictable in the context of the analysis of process carried out before for the kinematic wave model (Sect. 3.2.2). A tendency toward an increase in solute concentration in surface water can be seen during the recession phase of the hydrograph. Contrary to that, the concentration decreases during the rising limb phase. The dynamics of the mass flux, $q_s = q\overline{C}$, from a lumped watershed is also controlled by the falling and rising limbs of the hydrograph (Fig. 6.5b).

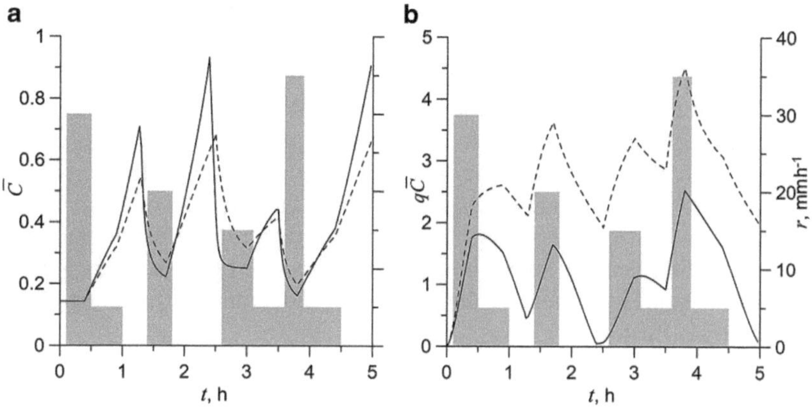

Fig. 6.5 (**a**) Solute concentration, $\overline{C}$, and (**b**) specific mass flux, $q\overline{C}$, versus time. *Solid curves* are for $i > 0$; *dashed curves* are for $i = 0$. Shaded bar is the rainfall hydrograph, $r(t)$

The described analytical model shows the main advantage of the lumped-parameter approach for coupling surface and subsurface flow study with the aim to predict the chemical response of a hillslope to a rather complex input fallout function.

6.2.2 Equilibrium Sorption-Based Models for Short-Term Washout (with Application to Radioactively Contaminated Areas)

This section discusses transient models of sorbable-solute migration in runoff water, which can be used, primarily, to forecast the removal of artificial radionuclides from contaminated surface by runoff. Historically, artificial radionuclides appeared on the land surface due to nuclear weapon tests (NWT) and atmospheric fallouts of products of gas–aerosol emissions resulting from accidents at nuclear facilities, primarily, nuclear power plants (NPP) (Chap. 7).

Many-year monitoring of radiation-related consequence of NWT and the Cheronobyl accident (1986) show that the initial washout rates are generally higher than the subsequent washout rates. The same trends are also confirmed by monitoring data collected in the impact zone of radionuclide emission from Fukushima NPP (2011). Such system behavior can be described using equilibrium models, incorporating sorption-like interactions with immobile and mobile solid phases. The appropriate models give the best description for the initial (several rainfalls, few weeks after radionuclide deposition) stage of radionuclide extraction from soil surface within a catchment or from small experimental plots and flumes under synthesized rainfalls.

The kinetics of fixation of radionuclides in the soil matrix, their downward movement into the soil column, and the heterogeneity of the contaminated landscape restricted the application of these models for long-term predictions. The removal of activity from a catchment during long periods (months, years, decades, etc.) can be described by either kinetic models with effective parameters or models explicitly incorporating the mechanisms of vertical and horizontal migration of radionuclides in the subsurface environment.

To better understand the specifics of the migration process on a lumped hillslope and the parameters controlling it, we will consider successively several models, starting from the simplest ones, which describe the release of solutes from a polluted surface into static water layer (the so-called batch conditions, when $q = 0$), to more complicated models, which take into account the unsteady-state character of water flow (when $q > 0$), forming under the effect of precipitation. In addition, in this section, taking into consideration the behavior of radionuclides at shorter time scales, we neglect the downward movement of radionuclides from the surface by diffusion.

6.2.2.1 Entrainment Coefficients

Radionuclides in surface runoff induced by rainfall or snow melting are transported in two forms, depending on the physicochemical and landscape conditions (Konoplev et al. 1992; Garcia-Sanchez and Konoplev 2009; Ueda et al. 2013): (1) dissolved ionic compounds, and (2) particulates, adsorbed onto suspended particles and colloids, which enter water flow because of soil erosion. Therefore, the transport of dissolved radionuclides is referred to as *liquid runoff* (or *liquid washout*), and the transport of radionuclides in particulate form with suspended matter, is referred to as *solid runoff* (or *solid washout*) (see also Preface to Chap. 3).

The quantitative characteristics of radionuclide removal from drainage areas can be derived from reactions that characterize the sorption equilibria of radionuclides contained in mobile water phase and retained on soil surface. The low concentrations of radionuclides allow linear equations to be used, so the equilibria can be described by two so-called entrainment coefficients (Konoplev et al. 1992; Bulgakov et al. 1999; Konoplev et al. 1992; Garcia-Sanchez et al. 2005; Garcia-Sanchez and Konoplev 2009):

(1) normalized liquid-runoff entrainment coefficient

$$K_1^* = C^d/N, \tag{6.35a}$$

which characterizes the sorption of solute ionic forms on mineral and organic particles of soil and has the dimension of $[L^{-1}]$, e.g., m^{-1};

(2) normalized solid-runoff entrainment coefficient

$$K_s^* = A_s/N, \tag{6.35b}$$

which characterizes the sorption on disperse particles and has the dimension of $[L^2M^{-1}]$, e.g., m^2g^{-1}; here N is equilibrium radionuclide activity on the polluted surface, Bqm^{-2}; C^d is the equilibrium activity of the radionuclide in dissolved form, Bqm^{-3}; A_s is the activity of suspension, Bqg^{-1}. For brevity, the coefficients K_1^* and

K_s^* are sometimes referred to as rain (or snow) washout coefficients. We remind that they are valid for short duration.

The conversion to the volumetric form of activity expression (per unit liquid volume) for a radionuclide adsorbed by particulate matter of erosion origin can be made using the following formula

$$C^s = A_s C_p, \tag{6.35c}$$

where C_p is the concentration of suspension (solid particles) in the hillslope flow, e.g., gL^{-1}.

Formally, the ratio of solid/liquid normalized entrainment coefficients determines the value of the coefficient of equilibrium sorption distribution of radionuclides in suspension:

$$K_d = K_s^* / K_1^*. \tag{6.35b}$$

Radionuclide solid washout from an eroded surface of a drainage basin is largely controlled by suspended-matter load and its size distribution. Water erosion is driven by the flow of snowmelt and rain (storm) water. Therefore, the development of water erosion within a year can be divided into two periods. The first period covers the winter thaws and spring snow melting, while the second is the period of storm rains during warm seasons.

The manifestation of the washout caused by showers significantly differs from the washout caused by intense snow melting. In the former case, a considerable amount of precipitation, falling within a short time interval, has no time to infiltrate into soil and, depending on soil properties (particle size distribution, porosity, etc.), causes the development of small groves or scours. Contrary to that, during snow melting, the coalescent, water-saturated top soil layer readily shifts over underlying frozen horizon (sheet erosion).

Analysis of rainfall and snowmelt runoff showed that radionuclide washout with snowmelt runoff is an order of magnitude less than that with rainfall runoff.

The values of C_p can be calculated based on data on the mean long-term soil washout from watersheds, G_s.

The value of G_s, as a characteristic of erosion process, varies in different natural–climatic and landscape zones within a wide range from 0.5 to 15 t/(ha year). Thus, some authors estimate the mean soil washout for sod-podzol and gray forest soils at 2–4 t/(ha year). The concentration of suspension in water can be estimated from the formula:

$$C_p \approx \gamma G_s / r_e, \tag{6.35e}$$

where r_e is the rainfall excess, γ is a coefficient introduced to account for the fact that a considerable portion of washed soil will not reach the erosion base level ($\gamma \approx 0.01$ for drainage basins of large rivers, and $\gamma \approx 0.03 - 0.05$ for those of small rivers).

6.2.2.2 Sorption in a Static Soil/Runoff System (S = const, $q > 0$)

The batch ("0-dimension") models of sorption equilibria considered below are rather useful for determining the limiting (maximal possible) degree of pollution of water layer that formed by rainfall on soil surface. Suppose that a rainfall with a rate r within time Δt results in that a *fixed* water layer with a mean depth of $S^0 = r\Delta t$ forms on the surface.

With the equilibrium character of radionuclide transfer from the surface into the solution, the following balance relationship will hold:

$$N_0 F = CS^0 F + NF, \tag{6.36}$$

where F is the area covered by water; N_0 is the initial radionuclide concentration on a unit area; N is equilibrium concentration (after a continuous water layer has formed and sorption equilibrium has been attained); C is the total concentration of radionuclide in the solution, determined by its concentration in the dissolved and "mineral" (as component of particulate matter) migration forms, such that

$$C = K_1^* N + K_s^* C_p N, \;\; K^* = K_1^* + K_s^* C_p. \tag{6.36a}$$

From here, we obtain several obvious relationships:

$$N = C/K^*, \text{ or } \overline{N} = a\overline{C}; \tag{6.37}$$

$$C = N_0/\left(S^0 + 1/K^*\right), \text{ or } \overline{C} = (1+a)^{-1}, \tag{6.37a}$$

where $\overline{C} = CS^0/N_0$, $\overline{N} = N/N_0$, $a = 1/S^0K^*$; K* is a combined sorption coefficient for radionuclide overland transport.

Water depth at the surface, S^0, is rarely greater than a few mm. Now, taking into account the variation ranges of constants K_1^* and K_s^* (see Sect. 7.1.2.1 below) and the characteristic C_p (generally, of the order of hundreds mg/L), we can show that the inequality $K^*S^0 << 1$ is true almost always. In this case, formula (6.37a) can be simplified to become $C \approx K^*N_0$ ($\overline{C} \approx a^{-1}$). This means that at equilibrium and at high sorption capacity of soil surface, the hydrodynamics of the surface flow controls the mass flux but not the absolute values of concentrations (the change in S^0 has no influence on the value of C). The observed decrease in the rate of radionuclide washout from polluted areas (see Sect. 7.2.4) here is not due to the soil surface becoming clearer because of radionuclide washout from it; it is rather due to the radionuclides losing their mobility and penetrating into deeper soil layers through infiltration and diffusion, thus becoming less available for runoff water (see also Sect. 7.2.5.2). Studying this process would require special model analysis.

6.2.2.3 Sorption in the Dynamic Soil/Runoff System (S = const, $q > 0$)

Suppose that a *mobile* liquid layer, S^0 = const, accumulating radionuclides, forms rapidly enough on the soil surface. Unlike the previous problem, the radioactive solution that forms at the first contact between the rain and the surface is supposed to be diluted by continuing precipitation. Such combination of conditions leads to a drop in radionuclide concentrations in runoff water.

The balance equation can be written as

$$S^0 \frac{dC}{dt} + \frac{dN}{dt} - r_e(C_r - C) = N_0 \delta(t); \tag{6.38}$$

here r_e is the rainfall excess; C_r is the concentration in rainfall; N_0 is the initial density of soil surface contamination; $\delta(t)$ is the delta-function mathematically formalizing a single pulse radionuclide deposition. The equilibrium state of the system (the relationship between solute concentrations in the solid and liquid phases) is described by Eq. 6.37, whence

$$K^* dN/dt = dC/dt. \tag{6.39}$$

Now Eq. 6.38 becomes

$$\left(S^0 + \frac{1}{K^*}\right)\frac{dC}{dt} - r_e(C_r - C) = N_0 \delta(t). \tag{6.40}$$

Equation (6.40) has the following dimensionless representation:

$$(1+a)\frac{d\overline{C}}{d\tau} + \overline{C} = \delta(\tau), \tag{6.41}$$

with dimensionless groups

$$\overline{C} = CS^0/N_0, \tau = r_e t/S^0, a = \left(S^0 K^*\right)^{-1}. \tag{6.41a}$$

The first-order ordinary differential Eq. (6.41) has the following general solution:

$$\overline{C} = e^{-\tau/(1+a)}\left[C(\tau = -0) + \frac{1}{R}\int_0^\tau e^{\tau/R}\delta(\tau)d\tau\right] = e^{-\tau/R}\left[0 + \frac{1}{R}e^{\tau/R}\Big|_{\tau=0}\right] = \frac{1}{R}\exp\left(-\frac{\tau}{R}\right), \tag{6.42}$$

where $R = 1 + a$ (some kind of the retardation factor). At $\tau = 0$, solution (6.42) transforms into the limiting formula (6.37a), while at $\tau \to \infty$, $\overline{C} \to 0$ $(C \to C_r)$.

6.2.2.4 Sorption in the Soil/Runoff Dynamic System ($S \neq$ const, $q > 0$)

Here, we consider a further modification of the above approach to solute uptake description for a lumped hillslope, which inherited some features of kinematic wave model (see Sects. 3.2 and 3.3). We remind that at such approach, the flow dynamics is determined by four major characteristics: rainfall excess, $r_e(t)$ (Eq. 1.1b); runoff (outflow) rate, $q(t)$ (6.1), derived from the solution of the differential Eq. (6.16); mean flow depth, $S = S(t)$ (6.5); and mean concentration, $C = C(t)$.

Now the system of differential equations describing the coupled hydrological process becomes:

$$\frac{dS}{dt} = r_e - q, \qquad \frac{dSC}{dt} + \frac{dN}{dt} - (r_e C_r - qC) = N_0 \delta(t). \tag{6.43}$$

Combining these two equations, we obtain an ordinary differential equation:

$$S\frac{dC}{dt} + \frac{dN}{dt} + r_e(C - C_r) = N_0\delta(t). \tag{6.44}$$

With Eq. (6.5), determining a relationship between the average flow depth, $S(t)$, and overland flow discharge, $q(t)$, and the equilibrium-state Eq. (6.39), the differential Eq. (6.44) can be transformed into a dimensionless form:

$$\left(a + \overline{q}^{1/n}\right)\frac{d\overline{C}}{d\tau} + \overline{C} = \delta(\tau), \tag{6.45}$$

where dimensionless groups are

$$\overline{C} = CS^0/N_0, a = \left(S^0 K^*\right)^{-1}, \tau = r_e t/S^0, \overline{q} = q/r_e; \tag{6.45a}$$

dimensionless outflow rate, $\overline{q}$, is determined from solution (6.16a); $S^0 = (r_e/\lambda)^{1/n}$ is the mean value of water depth on a hillslope at equilibrium (Fig. 6.1). The form 6.45 is valid at $r_e = \text{const}$.

Equation (6.45) is a first-order ordinary differential equation of the form

$$\frac{d\overline{C}}{d\tau} + f(\tau)\overline{C} = g(\tau), f = \frac{1}{a + \overline{q}^{1/n}}, g = \frac{\delta(\tau)}{a + \overline{q}^{1/n}}. \tag{6.46}$$

Equation (6.46) has the solution:

$$\overline{C} = e^{-F(\tau)}\left[C(\tau = -0) + \int_0^\tau g(\tau)e^{F(\tau)}d\tau\right], \tag{6.47}$$

$$F(\tau) = \int_0^\tau \frac{d\tau}{a + \overline{q}^{1/n}} = \int_0^{\overline{q}(\tau)} \frac{\overline{q}^{1/n}}{(1 - \overline{q})\overline{q}} \frac{d\overline{q}}{a + \overline{q}^{1/n}}, \int_0^\tau g(\tau)e^{F(\tau)}d\tau = \frac{1}{a}. \tag{6.47a}$$

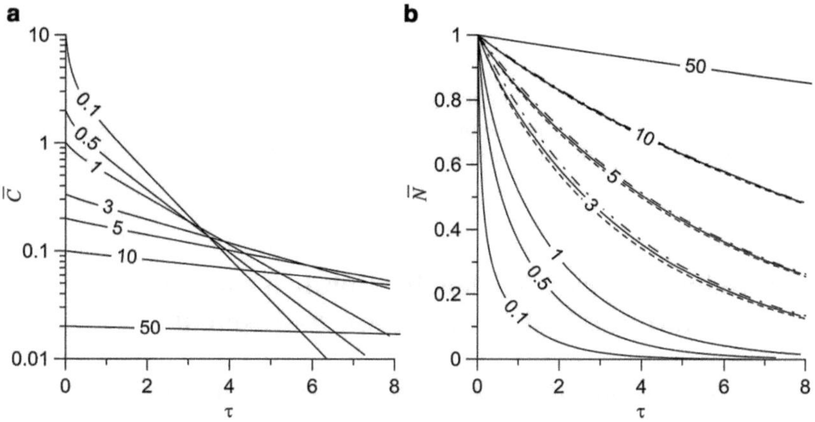

Fig. 6.6 Plots illustrating the behavior of a sorbable solute in a dynamic runoff system: **(a)** $\overline{C}(\tau)$ (Eq. 6.48) and **(b)** $a\overline{C}(\tau)$, the broken curves correspond to the approximation (6.50), the *dash*-and-*dot curves* correspond to the approximation (6.51); numbers at curves are the values of dimensionless parameter a; $n = 5/3$

Finally:

$$\overline{C} = \overline{C}(\tau) = a^{-1}\exp[-F(\tau)]. \tag{6.48}$$

Since $\overline{q}$ and τ are related through (6.16a), simple calculations allow the required function $\overline{C}(\tau)$ to be plotted at fixed a and n. Along with dimensionless concentration function $\overline{C}(\tau)$ (Fig. 6.6a), of some illustrative interest can be the normalized function $a\overline{C}(\tau) \equiv \overline{N}$ (Fig. 6.6b), which characterizes the self-purification rate of the contaminated surface (see Eq. 6.37).

The curves $\overline{C}(\tau)$ can be seen to cross the ordinate axis in points $\overline{C} = 1/a$, or $C = N_0K^*$. This concentration corresponds to the limiting concentration of solution saturation at the initial moment. At $a > 50$, in the interval of τ under consideration, $\overline{C} \approx a^{-1}$ are nearly time-independent.

An approximate solution can be obtained for a linear model of hillslope dynamics (Eq. 6.7, $n = 1$) assuming

$$\overline{q} = 1 - e^{-\tau}, \tag{6.49}$$

so that the integration of a solute balance Eq. (6.45) results in the solution:

$$\overline{C} = \frac{1}{a}\left[\frac{a}{(1 + a)\exp(\tau) - 1}\right]^{1/(a+1)}. \tag{6.50}$$

At $a \geq 2 - 3$, which corresponds to the conditions of transport with runoff water of readily absorbable solutes of which radionuclides are an example, calculations by (6.48) and (6.50) essentially coincide (Fig. 6.6b) with calculations using the approximate solution:

$$\overline{C} = \frac{1}{a}\exp\left(-\frac{\tau}{a+1}\right). \tag{6.51}$$

At the recession stage after rainfall ceases, when $t > T$ ($\tau > \tau_1$), contrary to the models considered in Sect. 6.2, which show a decrease in concentration during recession hydrograph, for the equilibrium sorption model (6.43), we have $C = C(\tau \geq \tau_1) = \text{const}$.

6.3 Catchment-Scale Models for Solute Transport in Runoff and Soil Remediation Assessment

6.3.1 A Model Based on Decoupled Hydrodynamic Formulation

6.3.1.1 Problem Setup and Basic Equations

If the catchment is regarded as a well-mixed unit (Fig. 6.7) and rainfall is steady, decoupling of water storage from discharge process can be represented as follows

$$\frac{d(\theta d_e + S)C}{dt} - J_s(t) = \left\langle \begin{array}{l} 0 \text{ when } S(t) = r_e t < S^0, \\ -r_e C \text{ when } r_e t \geq S^0 \text{ and } S = S^0, \end{array} \right. \tag{6.52}$$

where the chemical flux, $J_s(t) = -J_s(0, t)$, determines the mass exchange between surface water flow and porous (soil) domains, and S^0 is the slope storage limit. It is

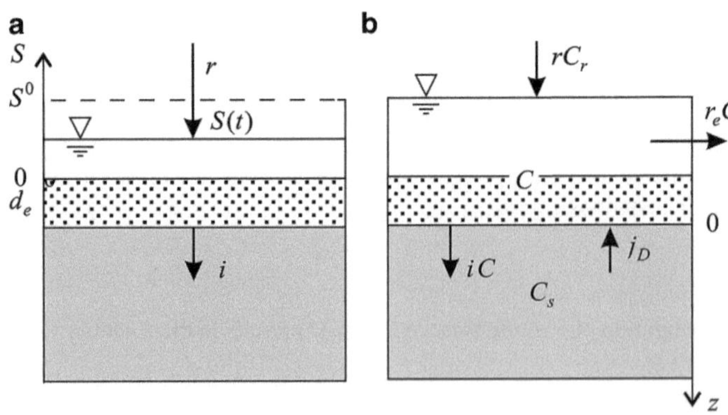

Fig. 6.7 A general concept of a lumped-parameter model: (**a**) hydrodynamic and (**b**) solute transport model components. j_D is the diffusion flux

also assumed that no solutes are present in the rainfall water. The model (6.52) includes a mixing layer with storage capacity θd_e (Sect. 3.1.2.2).

Such representation implies that the hydraulic response of the hillslope to fallout can be divided into two stages (Zhang et al. 1999). The first stage is associated with time lag, when the rainfall excess is used to build surface storage. The second stage corresponds to the runoff over the lumped slope, when flow regime is steady and can be characterized by average water depth (storage), S^0, a catchment-scale analogue of the hillslope unit (Eq. 6.11a).

The chemical flux, $J_s(t)$, can be specified from the boundary layer Eq. (3.7) coupled with the solute transport equation in the subsurface ($z > 0$). The full solution of the above mathematical problem, supplemented by a boundary condition at the lower boundary of the soil profile, consists of complex analytical constructions.

An approximate solution can be obtained if the chemical flux (from soil to runoff) is assumed independent of the concentration in water (Wallach and van Genuchten 1990), i.e. $C << C_s(0,t)$. In this case, Eq. 3.7 can be simplified:

$$J_s(t) = -\left[iC_s(0,t) - D_s\frac{\partial C_s(0,t)}{\partial z}\right] = k_e C_s(0,t); \tag{6.53}$$

obviously, due to the exclusion of the term $k_e C$, this assumption results in overestimation of the solute flux at the latest period of runoff; k_e is the transfer coefficient; $C_s(z,t)$ is the chemical content of soil water.

The general solution of the equation of solute transport for function $C_s(z,t)$, was obtained by Wallach and van Genuchten 1990. This solution yields an expression for chemical flux

$$J_s(t) = \frac{i}{2}C_0\left[(1+2\bar{k})\exp\left[4\bar{k}(1+\bar{k})\xi^2\right]\mathrm{erfc}\left[(1+2\bar{k})\xi\right] - \mathrm{erfc}(\xi)\right], \tag{6.54}$$

where $\xi^2 = i^2 t/4D_s\theta$, $\bar{k} = k_e/i$, C_0 is the initial concentration of the chemical in the soil profile; $\theta \equiv \theta_s$ is the saturated volumetric water content. At $\bar{k} >> 1$, implying infinitely large mass exchange rate between the runoff water and the pore solution at the soil surface, solution (6.54) tends to the limit

$$J_s(t) = \frac{i}{2}C_0\left[\frac{1}{\sqrt{\pi\xi^2}}\exp(-\xi^2) - \mathrm{erfc}(\xi)\right]. \tag{6.54a}$$

This result can be directly obtained using the fundamental solution of 1D advection–dispersion equation, under the conditions $C_s(z, t=0) = C_0$, $C_s(z=0,t) = 0$ (Bear 1972). Due to the initial assumption $C << C_s(0,t)$, Eqs. 6.54 and 6.54a are independent of the spatial coordinate, x.

6.3.1.2 Special Cases

For the first stage of the process ($S < S^0$), when $S = r_e t$ ($dS/dt = r_e$), Eq. 6.52 can be rewritten in a dimensionless form

$$(\varepsilon + \overline{r}_e \overline{t})\frac{d\overline{C}}{d\overline{t}} + \overline{r}_e \overline{C} = \overline{J}_s(\xi), \tag{6.55}$$

where $\overline{C} = C/C_0$, $\overline{J}_s(\xi) = J_s(\xi)/i$, $\xi^2 = a\overline{t}$, $a = i^2\tau^0/4D_s\theta$, $\overline{r}_e = r_e/i = (1-\gamma)/\gamma$, $\gamma = i/r$, $\varepsilon = d_e\theta/i\tau^0$. The time, t, is normalized by the mean residence time ($\tau^0 = S^0/r_e$): $\overline{t} = t/\tau^0$.

Equation (6.55) has a solution:

$$\overline{C} = \frac{1}{\varepsilon + \overline{r}_e\overline{t}} \int_0^{\overline{t}} \left\{ \left(\overline{k} + \frac{1}{2}\right) \exp\left[4\overline{k}(1+\overline{k})au\right] \operatorname{erfc}\left[(1+2\overline{k})\sqrt{au}\right] - \frac{1}{2}\operatorname{erfc}\left(\sqrt{au}\right) \right\} du, \quad t/\tau^0 \le 1. \tag{6.56}$$

After a steady-state regime establishes ($S = S^0$), the derivative dS/dt becomes zero and the chemical loss in runoff becomes $r_e C$ (Eq. 6.52). Under steady-state hydrodynamic conditions ($t/\tau^0 > 1$), Eq. (6.52) transforms into

$$(\varepsilon + \overline{r}_e)\frac{d\overline{C}}{d\overline{t}} + \overline{r}_e \overline{C} = \overline{J}_s(\xi). \tag{6.57}$$

Equation (6.57) has the following solution:

$$\overline{C} = \overline{C}^0 \exp\left(-\frac{\overline{r}_e(\overline{t}-1)}{\varepsilon + \overline{r}_e}\right) + \frac{1}{\varepsilon + \overline{r}_e}\exp\left(-\frac{\overline{r}_e\overline{t}}{\varepsilon + \overline{r}_e}\right) \times \left\{ \int_1^{\overline{t}} \left\{ \begin{array}{l} \left(\overline{k} + \frac{1}{2}\right)\exp\left[\dfrac{u\left(\overline{r}_e + 4a\overline{k}(\varepsilon + \overline{r}_e + \varepsilon\overline{k} + \overline{k}\overline{r}_e)\right)}{(\varepsilon + \overline{r}_e)}\right] \operatorname{erfc}\left[(1+2\overline{k})\sqrt{au}\right] \\ -\frac{1}{2}\exp\left[\dfrac{\overline{r}_e u}{\varepsilon + \overline{r}_e}\right] \operatorname{erfc}\left(\sqrt{au}\right) \end{array} \right\} du \right\}, \tag{6.58}$$

where $\overline{C}^0 = C^0/C_0$, C^0 is the initial solute concentration in the runoff water at the end of unsteady state period (when $t = \tau^0$). The concentration $\overline{C}^0$ can be found from the solution of transient problem (6.56) at $\overline{t} = 1$: $\overline{C}^0 = \overline{C}(\overline{t} = 1)$.

At $\varepsilon = 0$, solution (6.58) can be transformed into a dimensional form of the convolution integral:

$$C(t) = C^0 \exp[-(\bar{t} - 1)] + \frac{1}{r_e} \int_1^{\bar{t}} E(\bar{t} - u) J_s(u)\, du, \tag{6.59}$$

which coincides with the earlier obtained formula (Wallach 1991); here $E(\bar{t})$ is the residence time distribution of the runoff (Eq. 6.12); $\tau^0 = S^0/r_e$ is the mean residence time (Eq. 6.11b); u is a dimensionless dummy variable.

The case $J_s = 0$ (zero mass flux into the mixing layer from the underlying soil) yields a combination of partial solutions of Eq. 6.52:

$$\overline{C} = \frac{\varepsilon}{\varepsilon + \bar{r}_e \bar{t}}, \quad \bar{t} \le 1, \tag{6.60a}$$

$$\overline{C} = \frac{\varepsilon}{\varepsilon + \bar{r}_e} \exp\left(-\frac{\bar{r}_e(\bar{t} - 1)}{\varepsilon + \bar{r}_e} \right), \quad \bar{t} > 1. \tag{6.60b}$$

The obtained solutions allow a series of plots to be constructed to illustrate the behavior of the concentration function and to study runoff contamination due to a rain falling onto soil with dissolved chemicals contained in its pores. All plots (Figs. 6.8 and 6.9) predict an exponential decrease in the concentration function with $\overline{C}(\bar{t} = 0) = 1$. The dimensionless characteristics that control the process are $a = i^2\tau^0/4D_s\theta$, $\bar{k} = k_e/i$, $\bar{r}_e = (r - i)/i$, and $\varepsilon = d_e\theta/i\tau^0$. The rate of diffusion-controlled release of chemicals from the soil is determined by coefficients a and $\bar{k}$.

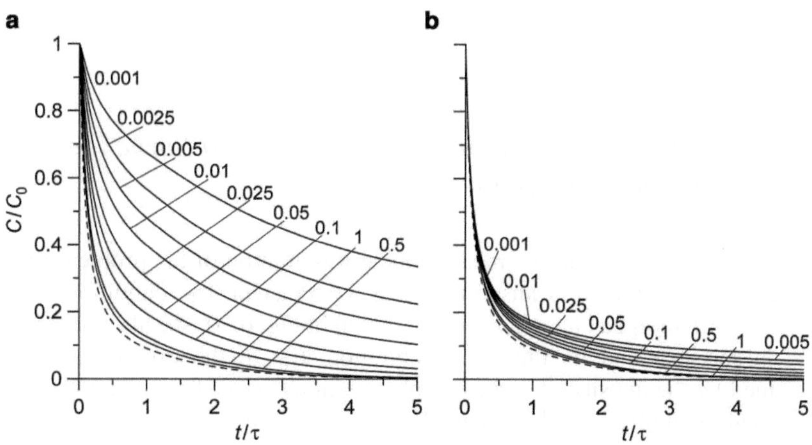

Fig. 6.8 Effect of dimensionless parameter $a = i^2\tau/4D_s\theta$ (numbers at curves) on the concentration of a solute in runoff water. (**a**) $\bar{k} = 10$, (**b**) $\bar{k} = 1$. Other values are $\gamma = 0.09$, $\varepsilon = 1$. The *dashed curve* gives the results of calculations by the limiting solution (6.60)

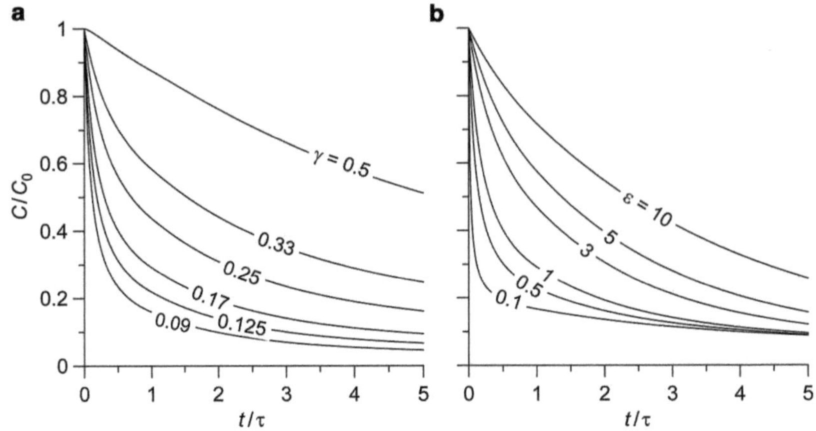

Fig. 6.9 Function $\overline{C} = \overline{C}(t/\tau)$. (**a**) At different $\gamma = i/r$ and fixed values $a = 0.01, \overline{k} = 1$ and $\varepsilon = 1$; (**b**) at different $\varepsilon = d_e\theta/i\tau^0$ and fixed values $a = 0.01$, $\overline{k} = 1$ and $\gamma = 0.17$

As follows from Fig. 6.8, the less the value of a (a high diffusion coefficient and a low infiltration), the greater diffusion flux and the saturation of runoff water by soil solutes. An increase in $\overline{k}$ amplifies this trend. At $a > 0.5-1$, the curves are well approximated by the limiting solution (6.60a), thus suggesting the flux of solutes from the soil into surface water to be negligible.

The ratio $\gamma = i/r$ with fixed i determines the rate of increase in the water layer depth on the surface and hence, the degree of solute dilution in the flow: the lesser γ, the lower solute concentration in the water flowing over the soil surface (Fig. 6.9a). The coefficient $\varepsilon = d_e\theta/i\tau^0$ characterizes the relative capacity of the active layer as a mixing zone. The greater the thickness of the mixing zone, d_e, the slower the concentration drops (Fig. 6.9b). The behavior of concentration curves depends, in a certain manner, on the mean residence time.

6.3.2 A Dynamic Catchment-Scale Model for a Sorbable Component Washout

Unlike the previous model, the mathematical formulation of the problem involves the consideration of solute transport process for transient flow conditions within a catchment with lumped characteristics as shown in Fig. 6.10. We also assume that a solute is sorbable and its initial concentration, C_0, in the mixing layer, d_e, is determined by the density of surface contamination (mass of solute applied/deposited per unit area of the land surface), N_0,

$$C(t=0) = C_0 = \frac{N_0}{(\theta + K_d^s\rho_b)d_e}, \tag{6.61}$$

where K_d^s is the equilibrium sorption distribution coefficient [L^3M^{-1}]; ρ_b is the soil bulk density [$M\,L^{-3}$].

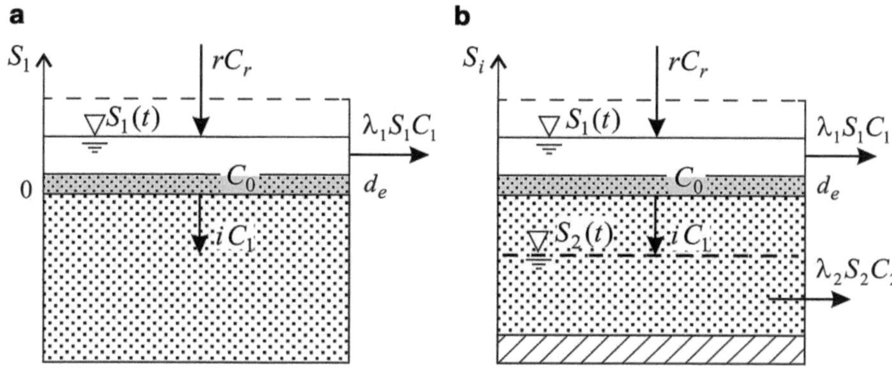

Fig. 6.10 A general concept of lumped-catchment models: (**a**) one-store catchment and (**b**) two-store catchment

Further, we will consider two conceptual models of the catchment flow structure. The first model is defined by a one-store structure, which restricts its application to surface water runoff coupled with a solute transport process in a form of mass balance equation (Fig. 6.10a). The second model, due to its two-store structure, also includes the interaction of runoff with subsurface water flow and solute transport (Fig. 6.10b). Such lumped-parameter model formulation addresses the partitioning surface–subsurface flow providing quick and slow components of the chemical hydrograph at the outlet of a catchment.

6.3.2.1 One-Store Model

Reversible Sorption Let us assume that storage function, $S_1(t) = q_1(t)/\lambda_1$, describes the behavior of a linear model of a reservoir type (Eqs. 5.3, 5.4 and 6.7 at $n = 1$), thus for the rising hydrograph ($t \leq T$) a coupled system of water dynamics and solute balance equations has the form:

$$\frac{dq_1}{dt} = \lambda_1(r_e - q_1), \tag{6.62}$$

$$\frac{1}{\lambda_1}\frac{dq_1 C_1}{dt} + \frac{d}{dt}(\theta d_e C_1 + N) = -iC_1 - q_1 C_1, \tag{6.63}$$

where q_1 is the rate of rain-induced outflow from a drainage area (catchment); λ_1 is the catchment discharge coefficient (rate constant) [T^{-1}]; $N = K_d^s \rho_b d_e C_1$ is the mass of a solute in the adsorbed state in the mixing layer per unit area [ML^{-2}]. It is implicitly assumed that sorption is reversible and no hysteresis in sorption takes place. To simplify the further analysis (Sect. 6.3.2.2), we introduce a subscript 1 for dependent and free variables.

Combining Eqs. 6.62 and 6.63, we come to a first-order ordinary differential equation:

$$\left(\theta d_e R^s + \frac{q_1}{\lambda_1}\right)\frac{dC_1}{dt} + rC_1 = 0, \tag{6.64}$$

where $R^s = 1 + K_d^s \rho_b/\theta$ is a factor of sorption retardation.

Fixing the initial condition, $q_1(t = t_0) = q_{10}$, we obtain solution of Eq. 6.62 in the form

$$q_1 = r_e + (q_{10} - r_e)e^{-\lambda_1(t-t_0)}, \tag{6.65}$$

which holds at constant rainfall excess, r_e, within time interval $[t_0, t]$. Now, the solution of Eq. 6.65 at the initial condition and $C_1(t = t_0) = C_{10} \equiv C_0$ (6.61) can be written as

$$C_1 = C_{10}\left[\frac{(\overline{R} + q_{10})\exp[-\lambda_1(t - t_0)]}{\overline{R} + r_e + (q_{10} - r_e)\exp[-\lambda_1(t - t_0)]}\right]^{\frac{r}{r_e+\overline{R}}}, \tag{6.66}$$

where $\overline{R} = \lambda_1 \theta d_e R^s$.

For rainfall event of a fixed duration, T ($t_0 = 0 \le t \le T$), when the land surface is initially dry ($q_1(0) = 0$), Eq. 6.66 becomes

- for the rising hydrograph:

$$C_1 = C_{10}\left[\frac{\overline{R}\exp(-\lambda_1 t)}{\overline{R} + r_e[1 - \exp(-\lambda_1 t)]}\right]^{\frac{r}{r_e+\overline{R}}}, \tag{6.67a}$$

- for the recession hydrograph:

$$C_1 = C_{10}\left[\frac{\overline{R}\exp(-\lambda_1 T)}{\overline{R} + r_e[1 - \exp(-\lambda_1 T)]}\right]^{\frac{r}{r_e+\overline{R}}}, \tag{6.67b}$$

where C_{10} is defined by Eq. 6.61. As seen, the concentration of a solute in the runoff after rainfall cessation remains constant and corresponds to a value which was reached at $t = T$.

If the recession lasts till all the surface water is completely discharged through catchment outlet, $t \ge T + \lambda_1^{-1}\ln[q_1(T)/i + 1]$, the catchment's response to the next (second) fallout can be predicted using the same Eq. (6.67a) where the initial concentration in the mixing zone, C_{10}, equals C_1 (6.67b) and the time counts from zero to the end of this stage.

As an example, consider the infinite series of identical storms, approximated by a periodical rainfall function when all rains have equal intensities, r, and durations, T. Basing on Eq. 6.67b, one can get a simple dimensionless formula predicting solute concentration at the end of each rainfall event due to soil decontamination:

$$\overline{C}_{1n} = \frac{1}{R^s}\left[\frac{R^s\overline{\lambda}_1 \exp(-\lambda_{r1} rT)}{R^s\overline{\lambda}_1 + \overline{r}(1 - \exp(-\lambda_{r1} rT))}\right]^{\frac{n}{\overline{r}+\overline{\lambda}_1 R^s}}, \tag{6.68}$$

where $\overline{C}_{1n} = d_e\theta C_{1n}/N_0$; n is the number of the rainfall–washout cycle; $\lambda_{r1} = \lambda_1/r$; $\overline{\lambda}_1 = \theta d_e \lambda_{r1}$; $\overline{r} = r_e/r$.

The plot in Fig. 6.11, computed using the solution (6.68) for a particular case (all curves correspond to the same value of precipitation depth in all rain cycles, $rT = 2$ cm), demonstrates that the change in solute concentration in runoff strongly depends on the retardation factor, R^s, characterizing the sorption capacity of the soil. For a well-adsorbable solute ($R^s > 50$–100), the number of washout cycles (2-h rains for this particular example) can reach hundreds, thus noticeably reducing solute concentration in runoff. Such behavior of the system is in agreement with monitoring data obtained in areas contaminated by radioactive fallouts (Chap. 7). These data show that the natural attenuation processes due to surface washout by rainfall are very slow. Besides, in Fig. 6.11, one can see that the sensitivity of the curves to kinetic coefficient, λ_1, decreases with increasing R^s.

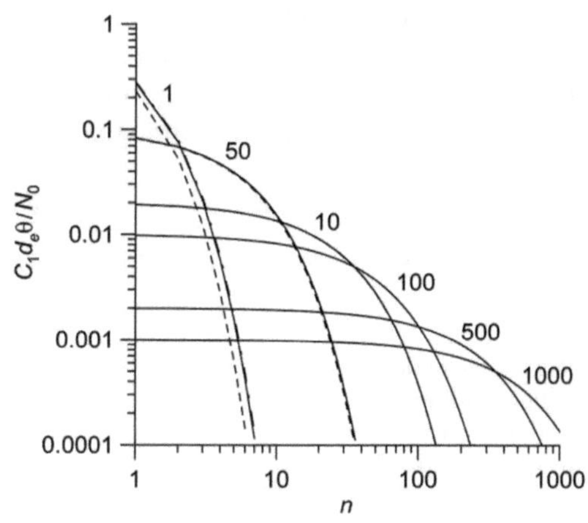

Fig. 6.11 Dependence of dimensionless solute concentration decrease on the number of washout cycles for different values of retardation factor, R^s (numbers at the curves), at catchment discharge ratio $\lambda_{r1} = \lambda_1/r$: $\lambda_{r1} = 0.1\ \text{cm}^{-1}$ (*solid curves*) and $\lambda_{r1} = 1\ \text{cm}^{-1}$ (*dashed curves*); other characteristics are $\overline{r} = 2/3$, $\theta d_e = 1$ cm, $rT = 2$ cm

The above approach can be developed to get a general solution of the problem for variable function $r_e = r_e(t)$, given its piece-wise constant representation in intervals $[t_{n-1}, t_n]$. Thus, the concentration of a solute, C_{n-1}, calculated at the end of any fallout period, $n-1$, can be used further as the initial condition for prediction of the concentration dynamics, C_n, in the following fallout period, n, using the known outflow rate q_{1n-1}:

$$C_{1n} = C_{1n-1}\left[\frac{(\overline{R} + q_{1n-1})\exp[-\lambda_1(t - t_{n-1})]}{\overline{R} + r_{en-1} + (q_{1n-1} - r_{en-1})\exp[-\lambda_1(t - t_{n-1})]}\right]^{\frac{r_{n-1}}{r_{en-1}+\overline{R}}}. \tag{6.69}$$

Also, as function $q_1(t)$ is known, the above relationships can be used to assess the specific mass flux, $q_{s1}(t)$, from the contaminated area: $q_{s1}(t) = q_1(t)C_1(t)$.

Irreversible Sorption The above approach can be extended to describe the solute removal from a contaminated soil surface, which exhibits adsorption-desorption hysteresis and irreversibility. Indeed, the *sorption* process determines the concentration of solutes in soil water within the mixing zone after land surface contamination (Eq. 6.61), e.g., due to agricultural chemical application or radioactive fallout. Once the mixing layer becomes saturated, solutes are released into runoff and into the subsoil and *desorption* becomes the dominant mechanism controlling surface and subsurface water quality (Steenhuis et al. 1994).

We suppose that adsorption–desorption process can be described in the framework of a simplified dual-site equilibrium approach (model) as shown in Fig. 6.12, where two partial sorption isotherms (*1* and *2*) for chemically heterogeneous soil matrix are responsible for linear equilibrium sorption onto sorption sites of two types (Rumynin 2011). As seen, the interaction of solutes with sorption sites of type *1* is reversible while that with sites of type *2* is irreversible.

The masses of solutes remaining in the soil matrix of the mixing zone per unit area, N, during sorption and desorption phases are

$$N^s = [fK_1C + (1-f)K_2C]d_e = K_d^s\rho_b d_e C_1, \tag{6.70a}$$

$$N^d = [fK_1C + (1-f)K_2C_0]d_e = K_d^d\rho_b d_e C_1 + N^0, \tag{6.70b}$$

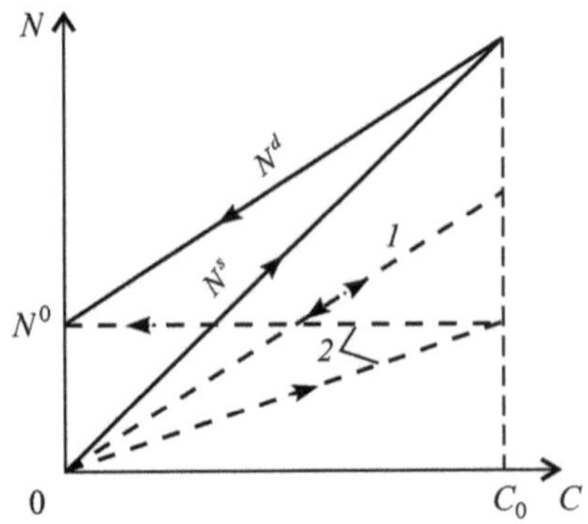

Fig. 6.12 Adsorption-desorption equilibrium in the mixing layer for a dual-site hysteresis-dependent adsorption model. The *dashed lines* show partial isotherms of sorption for two types (*1* and *2*) of sorption sites; the *solid lines* are for the total isotherms for sorption (N^s) and desorption (N^d)

where f is the mass fraction of the sites where sorption is reversible; K_1 and K_2 are partial sorption constants (dimensionless); $N^0 = (1-f)K_2 d_e C_0$ is sorption irreversibility (the mass of irreversibly adsorbed solutes per unit area); $K_d^d = fK_1\rho_b$ is equilibrium distribution coefficient for the desorption phase of the process; the equilibrium distribution coefficient for sorption in the term of the dual-site sorption model is represented by relationship: $K_d^s = [fK_1 + (1-f)K_2]/\rho_b$.

Substituting $N^d \equiv N$ from Eq. 6.70b into Eq. 6.63 and solving the system of Eqs. (6.62), (6.63) at zero initial condition, $q_{10}(t = t_0 = 0) = 0$ we come to a solution of the problem, similar to (6.66), in a dimensionless form:

$$\overline{C}_1 = \frac{1}{R^s}\left[\frac{R^d\overline{\lambda}_1 \exp(-\lambda_1 t)}{R^d\overline{\lambda}_1 + \overline{r}(1-\exp(-\lambda_1 t))}\right]^{\frac{1}{\overline{r}+R^d\overline{\lambda}_1}}, \overline{C}_1 = \frac{\theta d_e C_1}{N_0}, \qquad (6.71)$$

where $\overline{C}_1 = \theta d_e C_1/N_0$; $R^s = 1 + K_d^s\rho_b/\theta$; $R^d = 1 + K_d^d\rho_b/\theta$; $\overline{\lambda}_1 = \theta d_e \lambda_{r1}$; $\lambda_{r1} = \lambda_1/r$; $\overline{r} = r_e/r$.

As in the model of reversible sorption, solution (6.71) can be transformed to describe solute transport under periodical rainfall conditions. To do this, we are to replace $\lambda_1 t$ in (6.71) by $\lambda_1 T$ and multiply the exponent by n.

Calculations show (Fig. 6.13) that adsorption hysteresis can be a phenomenon that considerably changes the concept of the rate of natural remediation of the contaminated soils covering drainage areas.

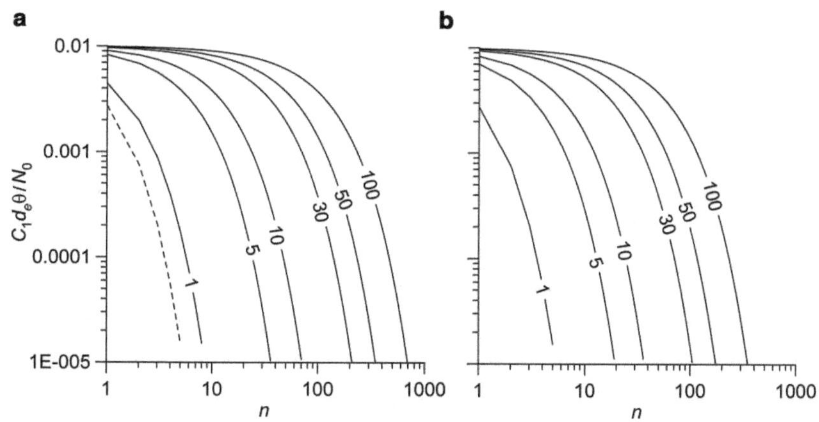

Fig. 6.13 Influence of adsorption hysteresis on washout of solutes from a contaminated catchment. Numbers at the curves are the retardation coefficients for desorption, R^d. (**a**) $\overline{\lambda}_1 = 1$; (**b**) $\overline{\lambda}_1 = 0.1$; other characteristics: $R^s = 100$, $\overline{r} = 2/3$, $\lambda_1 T = 2$

6.3.2.2 Two-Store Model

As it was discussed above (see Sect. 1.2.3), overland flow is often accompanied by lateral movement of water in the soil profile in the form of subsurface stormflow (generated in a shallow soil horizon/layer, Fig. 6.10b). Such subsurface flow is represented by a mixture of pre-event (stored, "old") soil water and current fallout ("new") water as the portion of the infiltration water, containing a solute of anthropogenic origin, quickly passes through a permeable upper part of the soil column (Sect. 3.4). The following two-store model can help determine the chemical response of the catchment to anthropogenic impact and environmental change.

The model is based on the concept that two storages define the overland flow and saturation zone in the lower part of the soil profile. The model implicitly inherited some features of the classical two-component hydrograph separation model, which is widely used to predict the isotope hydrograph separations of stream discharge and the leaching of soluble chemicals from drainage basins (Pearce et al. 1986; Richey et al. 1998; Heppell and Chapman 2006). However, unlike the conventional steady-state consideration of the problem, we propose here its unsteady-state setup basing on the consideration of a system of kinetic equations for lumped flow and solute transport.

The basic system of equations for overland flow and transport (6.62), (6.63) written for the *rising hydrograph* ($t \leq T$) is added by a coupled system of water dynamics and solute balance equations for the lower part of the soil column containing a soil water layer whose dynamics is characterized by the transient water table development and depth $S_2(t) = q_2(t)/\lambda_2$ (the second storage characteristics of the dual-storage catchment, Fig. 6.10b):

$$\frac{\phi_n}{\lambda_2}\frac{dq_2}{dt} = i - q_2, \tag{6.72}$$

$$\frac{\phi_n}{\lambda_2}\frac{dq_2C_2}{dt} + \frac{dN}{dt} = iC_1 - q_2C_2, \tag{6.73}$$

where q_2 is the outflow rate for the soil horizon; λ_2 is the catchment discharge coefficient (rate constant) of the soil horizon [T^{-1}]; ϕ_n is soil drainable porosity; C_1 is determined by the solution (6.67); $N \equiv N_2$ is the current mass of solute in the adsorbed state in the soil water horizon per unit area [ML^{-2}].

In this mathematical setup of the problem, we neglect the kinetics of solute transfer from runoff to soil water and assume that this process occurs instantaneously. Such rapid transport of solutes from surface to soil water is often recognized as a quite common phenomenon, which can be explained by a preferential flow of water and dissolved solutes through soil macropores with lengths comparable to the soil depth (Sect. 1.4.1; some experimental evidence may be found in the literature, e.g., in the paper by Heppell and Chapman 2006). This effect is most pronounced in structured (macropore, fractured) soils which are characterized by a short macropore (fracture) flow trigger time (Sect. 1.4.1). On the other hand, this may limit the usefulness of the two-store model because the quick vertical transport of solutes by the preferential flow may only be triggered by significant rainfall events (Gjettermann et al. 1997; McGrath et al. 2010; Alaoui et al. 2011).

Substituting $N \equiv K_d^s \rho_b S_2 C_2$ into Eq. 6.73, we transform this equation into

$$\frac{\phi_n R^s}{\lambda_2}\frac{dq_2 C_2}{dt} = iC_1 - q_2 C_2, \tag{6.74}$$

where $R^s = 1 + K_d^s \rho_b/\phi_n$; K_d^s is the sorption distribution coefficient for the subsurface soil material. Combining Eqs. 6.72 and 6.74, we come to an ordinary differential equation of mass balance for a reversibly adsorbed solute in a dimensionless form

$$\overline{S}_2 \frac{R^s}{\lambda_{r2}}\frac{d\overline{C}_2}{d\tau} + \left(R^s - (R^s - 1)\overline{S}_2\right)\overline{C}_2 = \overline{C}_1, \overline{C}_{1,2} = \frac{\theta d_e C_{1,2}}{N_0}, \tag{6.75}$$

where $\overline{S}_2$ is determined from solution of Eq. 6.72

$$\overline{S}_2 = \left[1 + \left(\overline{S}_{20} - 1\right)\exp(-\lambda_{r2}\tau)\right]; \tag{6.76}$$

$\overline{S}_2 = \lambda_{r2}\phi_n S_2/(1-\overline{r})$, $\overline{S}_{20} = \lambda_{r2}\phi_n S_{20}/(1-\overline{r}) \equiv q_{20}/i$, $\lambda_{r2} = \lambda_2/\phi_n r$, $\tau = rt$; S_{20} and q_{20} are the initial (pre-event) depth of the soil water layer and the corresponding pre-event outflow rate (specific soil water discharge from the catchment outlet), thus S_{20} and q_{20} characterize the antecedent conditions.

Equations 6.75 and 6.76 are valid for the recession hydrograph as well while $i > 0$. As soon as the surface runoff is over (at a certain moment, $t = t^*$) and infiltration ceases, $i = 0$, the solute transport Eq. (6.74) reduces to:

$$\frac{\phi_n R^s}{\lambda_2}\frac{dq_{s2}}{dt} = -q_{s2}, \tag{6.77}$$

where $q_{s2} = q_2 C_2 = \lambda_2 S_2 C_2$ is the specific solute flux. Solution of Eq. 6.77 is given by the formula

$$q_{s2} = q_{s2}\left(\tau = \tau^*\right)\exp\left[-\frac{\lambda_{r2}}{R^s}\left(\tau - \tau^*\right)\right]. \tag{6.77a}$$

The time characteristics τ^* can be found from a solution of Eq. 6.65: $\tau^* = rT + \lambda_{r1}{}^{-1}\ln\left[\overline{S}_1(T) + 1\right]$, $\overline{S}_1(T) = \lambda_{r1} S_1(T)/(1-\overline{r})$, $S_1(T) = S_1(t = T)$.

The ordinary differential Eq. (6.75) can be solved numerically. Some particular results are presented in Fig. 6.14 (for $\lambda_{r1} \geq \lambda_{r2}$). A superposition of two solute fluxes results in an average (total) concentration C^{tot} function that contains information on the quick and slow components of the near-surface flow:

$$\overline{C}^{tot} = \frac{q_1\overline{C}_1 + q_2\overline{C}_2}{q_1 + q_2}, q_i = \lambda_i S_i C_i. \tag{6.78}$$

Analysis of curves in Fig. 6.14 shows that an increase in dimensionless group λ_{r2} leads to earlier attainment of peak values of function $\overline{C}_2$ and an increase in the peak

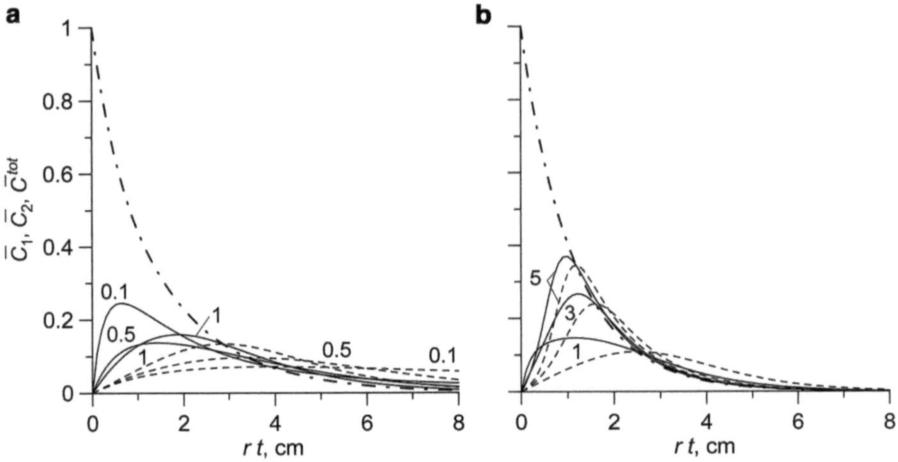

Fig. 6.14 A chemical response of a catchment to a long-term precipitation event of constant rate: solute content of overland flow ($\overline{C}_1$, *dash*-and-*dot curves*) and in subsurface water ($\overline{C}_2$, *dashed curves*); solid curves show the total concentration, $\overline{C}^{tot}$; the numbers at the curves are the values of λ_{r2}, cm^{-1}; (**a**) $\lambda_{r1} = 1$ cm^{-1}; (**b**) $\lambda_{r1} = 5$ cm^{-1}. Other parameters: $\overline{r} = 2/3$; $\theta d_e = 1$ cm; $\phi_n S_{10} = 5$ cm; $R^s = 1$

values whatever the specified values of coefficient λ_{r1}. However, the maximal values of mean concentration $\overline{C}^{tot}$, obtained at high λ_{r2}, can be lower than $\overline{C}^{tot}$, evaluated for small λ_{r2} (Fig. 6.14a).

Overall, the model can help to explore and predict the chemical response of a catchment to determine the fraction of soil and surface runoff water contributing to streamflow as a result of surface and subsurface discharge from a contaminated catchment area.

6.3.3 *Transfer Functions for Long-Term Radionuclide Washout*

The transient models discussed here consider time-dependent components controlled by the kinetics of radionuclide release from contaminated soils and overland spreading. The nature of this kinetics can be both physicochemical and flow-dynamical (induced by intrinsic properties of the distribution and interaction of watershed flowpaths).

The first steps in the development of models to be used to predict long-term changes in radionuclide runoff from catchments were due to the need to assess the environmental impact of nuclear weapon tests (NWT) in the atmosphere. The most commonly measured isotopes in fallout were Cs-137 and Sr-90. They are the main products of plutonium and uranium fission. Thus, Fig. 6.15 shows the atmospheric Cs-137 deposition flux as observed at Milford Haven site, U.K., during the period from 1954 to 1984 (He et al. 1996). Peak fallout delivery took place in the years 1963 and 1964, just prior to the making of the Atmospheric Nuclear Test Ban

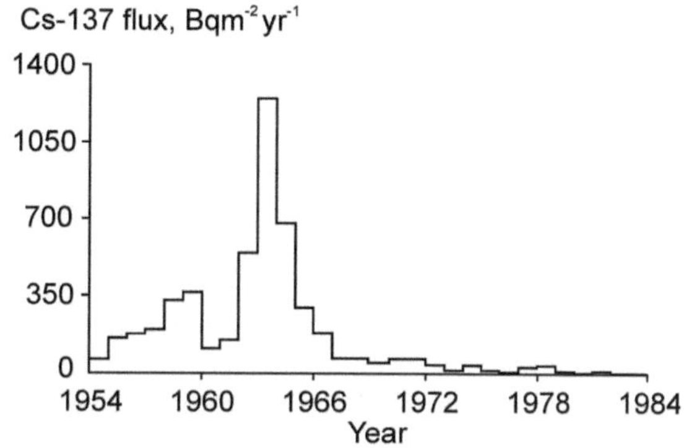

Fig. 6.15 The atmospheric Cs-137 deposition flux (1954–1984) as observed at Milford Haven, U.K. (He et al. 1996)

Treaty implemented by the United States and the former Soviet Union. No significant bomb-derived Cs-137 fallout occurred after 1984. Global data analysis indicates that the total radioactive fallout from the atmospheric NWT is greater in the northern than in the southern hemisphere, because more testing took place in the northern hemisphere (Cambray et al. 1987). The cumulative deposition of Cs-137 in the mid-1960s in the northern hemisphere reached 2–4 kBq/m^2 (Cambray et al. 1987). The radionuclide's concentration in fallout during NWT could be as high as 1–2 Bq/L (Elshamy et al. 2007). The ratio of Cs-137 to Sr-90 deposited from NWT sources has been calculated to be 1.6, allowing the annual deposition of Sr-90 from NWT sources to be estimated (Smith 2004).

Based on the long-time series measurements of radionuclides in river water (watershed outflow points), it was established that the concentration functions follow to one-, two-, or three-member exponentially decreasing relationships between the concentration and time (Helton et al. 1985). The simplest model can be derived from the principles of mass balance in a form of linear differential equation (Carlsson 1978; Monte et al. 2004; Poręba 2006)

$$\frac{dN(t)}{dt} = -(\lambda + \lambda_1)N(t) + (1 - k_1)D(t), N(0) = 0, \qquad (6.79)$$

where $D(t)$ is the rate of activity input – annual radionuclide deposition flux (Bqm^{-2} year^{-1}); λ is the physical decay constant (year^{-1}); λ_1 is the rate constant for the removal of accumulated radionuclides due to environmental effects (year^{-1}); k_1 is the fraction of radionuclides instantaneously transported to the watershed discharge area.

Integrating (6.79), one can obtain a general expression to assess the radionuclide migration following time-dependent deposition of radionuclides onto a watershed:

$$N(t) = (1 - k_1)\int_0^t D(\tau)\exp[-(\lambda + \lambda_1)(t - \tau)]d\tau. \qquad (6.80)$$

For a pulse input, $D(\tau) = N_0\delta(\tau)$, from Eq. 6.80 it follows

$$N(t) = (1 - k_1)N_0\exp[-(\lambda + \lambda_1)t]. \tag{6.81}$$

From comparison of Eqs. 6.80 and 6.81, it is seen that the exponential term plays a role of washout transfer function of a watershed which is a function of time after a unit pulse of contamination by atmospheric deposition (Garcia-Sanchez and Konoplev 2009).

The model (6.80), like many other "effective exponential decay" models, was used later for analyzing watershed radioactive contamination as a result of the widespread fallout of Chernobyl radionuclides. Most likely, such models promise future applications for analyzing the consequences of the Fukushima fallout (Evrard et al. 2013; Ueda et al. 2013), after time series of radiological measurements with appropriate duration are compiled. It should be mentioned that the global fallout was deposited over a long period compared to the Chernobyl and Fukushima fallouts, which have been depositing during short periods. Again, the distribution of Chernobyl and Fukushima radionuclides shows considerable spatial variations.

Following a single pulse of deposition of radioactivity, the concentration of radionuclide in runoff water is the sum of several exponential time functions with effective decay constants $\lambda + \lambda_i$, so called exponential model (Monte et al. 2004):

$$C(t) = \varepsilon N_0 \sum_i A_i \exp[-(\lambda + \lambda_i)t], \tag{6.82}$$

where N_0 is the radionuclide deposition onto the catchment (Bq m^{-2}); A_i is the weight of the ith component (dimensionless), such that $\sum_i A_i = 1$; λ_i are empirical parameters controlling the decay of radionuclide concentration in water due to environmental effects (year^{-1}); ε is a scaling factor (m^{-1}) which can be associated with the ratio between the initial concentration of radionuclide in water and the total deposition ($\varepsilon = C(0)/N(0) = C(0)/N_0$).

Most commonly used is the three-term modification of Eq. 6.82 (Smith et al. 2005):

$$C(t) = \varepsilon N_0\left(A_1 e^{-(\lambda+\lambda_1)t} + A_2 e^{-(\lambda+\lambda_2)t} + A_3 e^{-(\lambda+\lambda_3)t}\right). \tag{6.83}$$

The three exponential terms in Eq. 6.83 represent, respectively (Smith et al. 2004, 2005): a fast flush of activity as a result of rapid wash-off processes (first few weeks after fallout at rate λ_1); a slow decline (at rate λ_2) as a result of soil fixation and redistribution processes; and the very long-term (possibly also slowly declining with the rate λ_3) runoff fraction.

For time-dependent radionuclide deposition, $N = N(t)$, formula (6.83) becomes

$$C(t) = \varepsilon \int_{-\infty}^{t} N(\tau)\left(A_1 e^{-(\lambda+\lambda_1)(t-\tau)} + A_2 e^{-(\lambda+\lambda_2)(t-\tau)} + A_3 e^{-(\lambda+\lambda_3)(t-\tau)}\right) d\tau. \quad (6.84)$$

In many studies, the described exponential model has been widely used for assessing radionuclide behavior in different watershed environments. Based on measurements of the change in Cs-137 and Sr-90 activities as functions of time in European and Asian rivers after NWT and Chernobyl fallouts, the migration parameters, A_i, λ_i and ε, have been quantified (Cambray et al. 1987; Monte et al. 2004; Smith et al. 2004; 2005).

In particular, data analysis represented by Smith with co-workers (2004) shows that the general trend in the change in the activity of radionuclides (Cs-137 and Sr-90) over time is similar for different rivers that drain watersheds impacted by NWT and Chernobyl fallout. Thus, the following parameter values were determined from the measurements of Cs-137 from the Chernobyl accident: $A_1 = 0.905$, $A_2 = 0.09$, $A_3 = 0.005$; $\lambda_1 = 13$, $\lambda_2 = 0.41$, $\lambda_3 = 0.02$ year^{-1}. Although the time-dependent behavior of radionuclides in different rivers was similar, their activity levels varied significantly. This means that the scaling factor, ε, was different for different rivers. Estimates of ε for river-drainage basins contaminated by Chernobyl Cs-137 varied from 0.01 to 0.5 m^{-1}.

The next chapter will continue discussion of this issue in relation to the prediction of the impact of severe accidents at NPP on radionuclide contamination of the near-surface environment.

References

Agnese C, Baiamonte G, Corrao C (2001) A simple model of hillslope response for overland flow generation. Hydrol Process 15:3225–3238. doi:10.1002/hyp.182

Alaoui A, Caduff U, Gerke HH et al (2011) Preferential flow effects on infiltration and runoff in grassland and forest soils. Vadose Zone J 10:367–377

Bear J (1972) Dynamics of fluids in porous media. Dover Publication Inc., New York

Bulgakov AA, Konoplev AV, Shveikin YuV et al (1999) Experimental study and prediction of dissolved radionuclide wash-off by surface runoff from non-agricultural watersheds contaminated forests. Recent Developments in Risk Identification and Future Perspectives Part 1, NATO Science Series 2: Environ Secur 58:102–112

Cambray RS, Playford K, Lewis GN et al (1987) Radioactive fallout in air and rain: results for 1985 and 1986. United Kingdom Atomic Energy Authority. Environmental and Medical Sciences Division. Department of the Environment, London (UK)

Carlsson S (1978) A model for the movement and loss of ^{137}Cs in a small watershed. Health Phys 34:33–73

Elshamy ME, Mathias SA, Butler AP (2007) Demonstration of radionuclide transport modelling under field conditions: 50-year simulation of caesium migration in soil. Research report number: Imperial/NRP_016. Department of Civil and Environmental Engineering. Imperial College, p 47

Evrard O, Chartin C, Onda Y et al (2013) Evolution of radioactive dose rates in fresh sediment deposits along coastal rivers draining Fukushima contamination plume. Scientific Reports 3. Article number: 3079. doi:10.1038/srep03079

Garcia-Sanchez L, Konoplev A (2009) Watershed wash-off of atmospherically deposited radionuclides: a review of normalized entrainment coefficients. J Environ Radioact 100(9):774–778
Garcia-Sanchez L, Konoplev A, Bulgakov A (2005) Radionuclide entrainment coefficients by wash-off derived from plot experiments near Chernobyl. J Radioprot Suppl 40:519–524
Gjettermann B, Nielsen KL, Petersen CT et al (1997) Preferential flow in sandy loam soils as affected by irrigation intensity. Soil Technol 11:139–152
He Q, Walling DE, Owens PN (1996) Interpreting the ^{137}Cs profiles observed in several small lakes and reservoirs in southern England. Chem Geol 129:115–131
Helton JC, Iman RL, Brown JB et al (1985) Sensitivity analysis of the asymptotic behavior of a model for the environmental movement of radionuclides. Ecol Model 28:243–278
Heppell CM, Chapman AS (2006) Analysis of a two-component hydrograph separation model to predict herbicide runoff in drained soils. Agric Water Manag 79:177–207
Konoplev AV, Bulgakov A, Popov V et al (1992) Behaviour of long-lived Chernobyl radionuclides in a soil-water system. J Analy 117:1041–1047
McGrath GS, Hinz C, Sivapalan M et al (2010) Identifying a rainfall event threshold triggering herbicide leaching by preferential flow. Water Resour Res. doi:10.1029/2008WR007506
Monte L, Brittain JE, Håkanson L et al (2004) Review and assessment of models for predicting the migration of radionuclides from catchments. J Environ Radioact 75:83–103
Pearce AJ, Stewart MK, Sklash MG (1986) Storm runoff generation in humid headwater catchments: where does the water come from? Water Resour Res 22(8):1263–1272
Poręba GJ (2006) Caesium-137 as a soil erosion tracer: a review. Geochronometria 25:37–46
Richey GD, McDonnell JJ, Erbe WM et al. (1998) Hydrograph separation based on chemical and isotopic concentrations a critical appraisal of published studies from New Zealand, North America and Europe. J Hydrol (NZ) 37(2):95–111
Rose CW, Parlange JY, Sander GC et al (1983) Kinematic flow approximation to runoff on a plane: an approximate analytical solution. J Hydrol 62:363–369
Rumynin VG (2011) Subsurface solute transport models and case histories (with applications to radionuclide migration), vol 25, Series: Theory and applications of transport in porous media. Springer, Dordrecht, p 815
Singh VP (1996) Kinematic wave modeling in water resources: Surface-water hydrology. New York. Wiley-Interscience, p 1400
Smith JT, Belova NV, Bulgakov AA et al (2005) The "AQUASCOPE" simplified model for predicting 89, 90Sr, 131I, and 134,137Cs in surface waters after a large-scale radioactive fallout. Health Phys 89(6):628–44
Smith JT, Wright SM, Cross MA et al (2004) Global analysis of the riverine transport of 90Sr and 137Cs. Environ Sci Technol 38:850–857
Steenhuis TS, Boll J, Selker JS et al (1994) A simple equation for predicting preferential flow solute concentrations. J Environ Qual 23:1058–1064
Ueda S, Hasegawa H, Kakiuchi H (2013) Fluvial discharges of radiocaesium from watersheds contaminated by the Fukushima Dai-ichi Nuclear Power Plant accident, Japan. J Environ Radioact 118:96–104
Wallach R (1991) Runoff contamination by soil chemicals-time scales approach. Water Resour Res 27:215–223
Wallach R, van Genuchten MT (1990) A physically based model for predicting solute transfer from soil to rainfall-induced runoff. Water Resour Res 26(9):2119–2126
Wallach R, William AJ, William FS (1988) Transfer of chemical from soil solution to surface runoff: a diffusion-based soil model. J Soil Sci Soc Am 52:612–617
Yu B, Rose CW, Ciesiolka CCA, Cakurs U (2000) The relationship between runoff rate and lag time and the effects of surface treatments at the plot scale. Hydrol Sci J des Sci Hydrol 45(5):709–726
Zhang XC, Norton LD, Lei T et al (1999) Coupling mixinf zone concept with convection-diffusion equation to predict chemical transfer to surface runoff. Trans ASAE 42(4):987–994

Chapter 7
Prediction of the Impact of Severe Accidents at NPP on Radionuclide Contamination of the Near-Surface Environment

The fission of uranium or plutonium isotopes normally used as the fuel in nuclear reactors generates radioactive fission products, radionuclides. For nuclear reactors under normal operation and in a number of events, these radionuclides are prevented from escaping to the environment by several physical barriers (Högberg 2013). However, as experience shows, it cannot be totally excluded that at any time events occur. If all barriers fail, there is a potential substantial release of radionuclides from the damaged reactor to the environment. These aerosol-bound radionuclides being widely dispersed in the atmosphere can be removed from the atmosphere and brought to the earth surface by dry or wet deposition. The other pathway for radionuclides is connected with radioactive wastewater leak directly from the damaged reactor to the subsurface environment.

In 1990, to rate nuclear and radiological events, the International Nuclear and Radiological Event Scale (also called INES rating) was proposed by IAEA. According to this document, events are classified on the scale at seven levels: Levels 4–7 are termed “accidents” and Levels 1–3 “incidents”. Thus, the highest Level 7 is defined as follows: An event resulting in an environmental release corresponding to a quantity of radioactivity radiologically equivalent to a release to the atmosphere of more than several tens of thousands of terabecquerels of I-131. To convert this amount to a release of Cs-137, the latter has to be multiplied by a factor of 40 to be regarded radiologically equivalent to I-131 (Högberg 2013).

Many complex software systems for prediction of the consequences of such nuclear accidents at NPP are known. These systems are able to numerically model continuous or short-term releases of multiple radioactive isotopes, atmospheric transport and diffusion under changing weather conditions, and the deposition of radioactive isotopes on the surface.

For this study we rely on the simulation of *pre-defined accident scenarios* (usually worse case scenarios) for several designed and engineered NPPs in the Russian Federation. For the case studies considered here, the input for prediction of radionuclide transport in runoff and streams includes three main source term characteristics, namely, (1) specification of radionuclides of accident origin,

V.G. Rumynin, *Overland Flow Dynamics and Solute Transport*, Theory and Applications of Transport in Porous Media 26, DOI 10.1007/978-3-319-21801-4_7

(2) the amount of the released radioactive materials, and (3) radioactive fallout density. In such conceptual approach, the environmental contamination is defined mostly by the type of nuclear reactor and weather conditions, the latter assumed least favorable at the moment of hypothetical accident (heavy rain, low-level wind, etc.). An important element in the analysis and verification of model results is their comparison with the character of their environmental impact and the consequences of three severe accidents that have occurred at NPP. We will start this chapter with those severe accidents, discussed in many scientific publications and include an accident, which bears no relation to nuclear energy production, but which is related to the surface contamination through radioactive fallouts. It is worth mentioning that, in the soils contaminated with radioactive fallout components, the behavior of radionuclides is similar to that for radionuclides of global fallout caused by nuclear weapon tests (NWT, Sect. 6.3.3). In this context, data resulting from such monitoring in the second half of the twentieth century also prove to be of extreme importance for predicting the consequences of severe accidents at NPP.

The chapter ends with an example of a model with distributed parameters as applied to Pyshma watershed, within which some nuclear power units of the Beloyarsk NPP (Middle Urals, Russian Federation) are in operation or under construction. Estimates of the impact of hypothetical accidents at this NPP on radioactive contamination of groundwater are given.

7.1 Characteristics of the Severe Accidents at NPP, Radioactive Fallout Scale and Distribution

7.1.1 Accident Descriptions

Here we will focus on a brief review of three accidents that occurred at three NPPs, located in different countries, for the past nearly one third of century: USA (Three Mile Island 1979), USSR (Chernobyl 1986), and Japan (Fukushima Daiichi 2011), which had largely changed the notions of NPP safety.

The accident at the *Three Mile Island NPP* (Unit 2, equipped with a pressurized water reactor, known as TMI-1) was due to an erroneous shut down of cooling-water injection to the reactor system. As a consequence, the core boiled dry and overheated, resulting in a partial core melt. However, the nuclear fuel did not burn through the bottom of the reactor pressure vessel, so radioactive materials mostly stayed within. Although substantial amounts of gaseous and volatile fission products were released from the damaged core into the reactor containment, only a very small amount of radioactive substances was released into the environment (Högberg 2013). Released into the atmosphere were mostly radioactive inert gases, while the discharge of hazardous nuclides, such as I-131 and other fission products, was insignificant. The territory of the station was contaminated by radioactive water released from the primary coolant system. The total discharge was evaluated at $4.8 \cdot 10^{17}$ Bq.

Table 7.1 Total release of some radionuclides during the Chernobyl accident (The release, dispersion, and deposition of radionuclides... 2002)

Nuclide	Half-life	Percent of inventory	Activity, PBq[a]
I-131	8.0 d	50–60	~1760
Cs-134	2.1 y	20–40	~54
Cs-137	30.2 y	20–40	~85
Sr-90	28.8 y	4–6	~10
Pu-238	86 y	3.5	0.035
Pu-239	24 400 y	3.5	0.030
Pu-240	6 580 y	3.5	0.042
Pu-241	13.2 y	3.5	~6

[a]PBq (pectabecquerel) = 10^{15} Bq

The cause of the accident at the *Chernobyl NPP* (Unit 4), equipped with a channel-type boiling water reactor (BWR known as RBMK), was a violation of the prescribed operating limits and safety rules during a test program performed by operators. As a result of runaway fission reactions, a very strong power strike was initiated in the reactor zone, causing two subsequent thermal explosive releases of destroyed and evaporated fuel and fission products up in the atmosphere, in the form of a cloud several kilometers high, which was subsequently dispersed in the form of a plume (De Cort et al. 1998; Högberg 2013). A fire started in the remaining graphite that burned for some 10 days, causing further radioactive release. Thus, huge amounts of fission products were released into the atmosphere (Table 7.1). The Chernobyl disaster is widely considered to have been the worst NPP accident in history. The released radioactive material was widely dispersed and deposited across much of Europe.

Cs-137 was selected to characterize the magnitude of the ground deposition because (The release, dispersion and deposition of radionuclides... 2002): (a) it is easily measurable, and (b) it was the main contributor to the radiation doses received by the population once the short-lived I-131 had decayed. The three main spots of contamination resulting from the Chernobyl accident have been called the Central, Bryansk-Belarus, and Kaluga-Tula-Orel spots. The Central spot was formed during the initial, active stage of the release predominantly to the West and North-west. Ground depositions of Cs-137 of over 40 $kBqm^{-2}$ covered large areas of the Northern part of Ukraine and of the Southern part of Belarus. The most heavily contaminated area was the 30-km zone surrounding the reactor, where Cs-137 ground depositions generally exceeded 1500 kBq/m^2 (Fig. 7.1a).

A similar, though smaller scale accident, accompanied by uncontrollable release of radioactive substances, took place at the *Fukushima Daiichi NPP* because of a very strong earthquake in Japan. The earthquake and a tsunami damaged external power supply facilities and standby diesel generators, resulting in that all systems of operational and emergency cooling became inoperative and leading to the melting of the active zone of BWR-type reactors at three power units (Units 2, 3, and 4).

Large amounts of hydrogen and fission products were released to the reactor containment vessel, leading to leakage due to overpressure. The leakage of

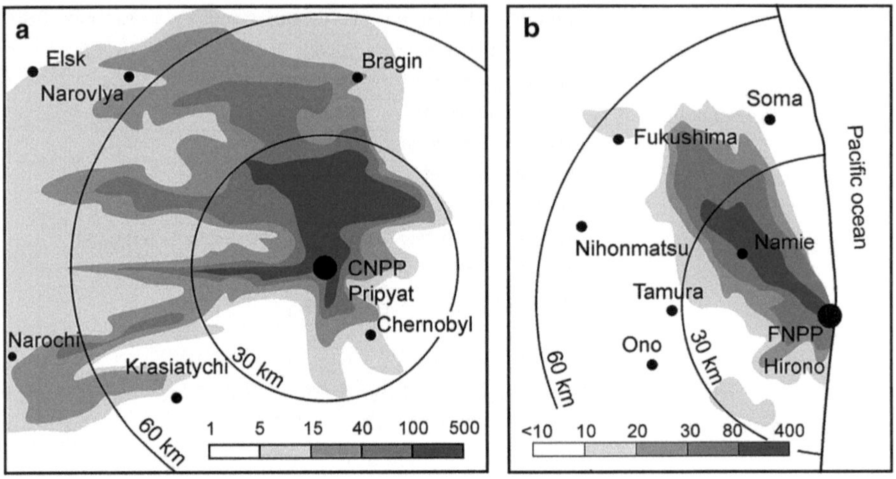

Fig. 7.1 Maps of Cs-137 deposition (grey-scale in Cikm^{-2}) after (**a**) Chernobyl (CNPP) 1986 and (**b**) Fukushima Daiichi (FNPP) 2011 accidents. 1 Cikm^{-2} = 3.7 · 10^{4} Bqm^{-2}

hydrogen caused a violent explosion destroying the upper part of the reactor buildings in three units. A substantial amount of fission products escaped to the environment (Högberg 2013).

Radioactive substance fallouts at the Chernobyl accident affected mostly the land surface, while a considerable portion of radionuclides released from the damaged reactors of the Fukushima Daiichi NPP fell onto the Pacific. The radioactive trail on the land surface (Fig. 7.1b) is mostly represented by cesium isotopes (Cs-134 and Cs-137), the deposition density of this radionuclide in the most heavily contaminated land areas reaching 3000 kBqm^{-2}. A considerable increase in I-131 concentration was also recorded in seawater and tap water, followed by its increase in food products in areas around the NPP. Water analysis in the drainage system of the second nuclear unit showed I-131 concentration of 3·10^{5} kBqL^{-1}.

The measurements made after the Chernobyl and Fukushima Daiichi reactor accidents showed that radioactive fallout is represented by a number of fission products. Some characteristics of those accidents and their consequences in terms of contamination intensity of natural landscapes are given in Table 7.2. At the very early stage of the accident, the short-lived radionuclides were, radiologically, the most important. In the following days and weeks, radionuclide I-131 was the main source of both internal and external exposure. The radiological significance of cesium (represented by isotopes Cs-137 and Cs-134) and strontium (Sr-90) radionuclides was initially small but their importance increased with time, becoming by far most important one year after the accident, especially at the larger distances (De Cort et al. 1998).

Table 7.2 also gives the characteristics of the *Kyshtym accident*, that occurred in 1957 at Mayak Amalgamated Industry (MAI), a plant, producing radioactive materials for military and civil needs (located in the South Urals region, southeast

Table 7.2 Accidents' characteristics and deposition density

Accident	Chernobyl		Fukushima Daiichi	Kyshtym	
General					
Year	1986		2011	1957	
Total release[a], TBq	$1.4 \cdot 10^7$		$7.7 \cdot 10^5$	$7.4 \cdot 10^5$	
Area, thousand km^2	208[c]		13[d]	23	
Fallout					
Radionuclide	Cs-137	Sr-90	Cs-134, 137	Cs-137	Sr-90
Fallout density[b], Bqm^{-2}	10^5–10^6	10^4–10^5	(1–3)10^6[e]	10^4	10^5–10^7

[a]The total value, including the entire spectrum of radioactive substances
[b]Mean maximal (order-of-magnitude) values
[c]Isolines $3.7 \cdot 10^4$ BqL^{-1} (for Cs-137) was used to delineate this area
[d]Insular part
[e]Evrard et al. (2013); Ueda et al. (2013)

of Kyshtym city) (Rumynin 2011). Though this radiological accident involved another source, not related to nuclear engineering, its consequences are similar to those of the accidents described above (Romanov et al. 1990; Pozolotina et al. 2012; Atlas of the EURT 2013). The cause of the accident, which resulted in a release of radioactive substances into the atmosphere, was an explosion of high-level nuclear waste stored in a tank at the MAI (after a failure of the cooling system at this tank). About 20 MCi of radioactivity was released into the air. The majority (90 %) of the nuclear waste dispersed near the tanks in the form of a liquid pulp settled out within the MAI territory. Other radioactive substances were raised by the explosion 1–2 km above the land surface to form a radioactive cloud containing liquid and solid aerosols.

The settling of liquid and solid aerosols, transported by southwestern air flow, caused a considerable land contamination with the formation of the so-called Eastern Ural Radioactive Trace (EURT). The maximal length of the EURT was 350 km, its width reaching 30–50 km (Fig. 7.2). The maximal density of the total radioactive contamination along EURT axis near the source reached $5.5 \cdot 10^{13}$ Bqm^{-2}, including $1.6 \cdot 10^8$ Bqm^{-2} due to Sr-90 (Utkin et al. 2000). The contamination of forests with total radioactivity of more than $5 \cdot 10^8$ Bqm^{-2} killed all coniferous trees. The concentration of radionuclide mixture in grass vegetation 12–18 km from the accident site several days after the accident reached $3.5 \cdot 10^{11}$ $Bqkg^{-1}$. The isoline of Sr-90 activity of $3.7 \cdot 10^6$ Bqm^{-2} embraced the area of more than 200 km^2; the same areas were 400 $km^2$for the isoline of $3.7 \cdot 10^5$ Bqm^{-2}, 1400 km^2 for the isoline of $3.7 \cdot 10^4$ Bqm^{-2}, and 23000 km^2 for the isoline of $3.7 \cdot 10^3$ Bqm^{-2} (Fig. 7.2). The Sr-90/Cs-137 ratio in Kyshtym debris was about 71 (Aarkrog et al. 1997). The safety criterion for the population was taken to be the limiting density of land contamination of $7.4 \cdot 10^4$ Bqm^{-2} in terms of strontium-90 (Utkin et al. 2000), i.e., about 10^5 Bqm^{-2}.

The relatively wide range of deposition density of the main radionuclides, represented in Table 7.2, reflects the actual situation: the spatial distribution of radionuclides (in particular, those resulting from the Chernobyl accident) is

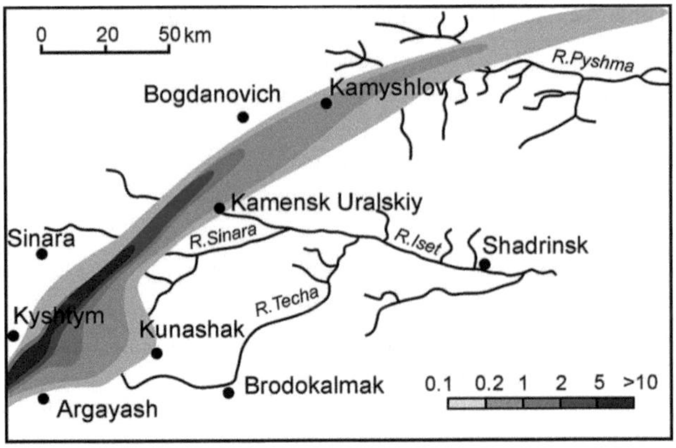

Fig. 7.2 A map of Sr-90 deposition ($Cikm^{-2}$) after the Kyshtym 1957 accident (Romanov et al. 1990)

strongly spatially variable because of both the heterogeneous character of atmospheric fallouts under unstable weather conditions and the diversity of landscapes, in which the radionuclides accumulate after their settling. The data in Table 7.2 can serve as a guide in the analysis of hypothetical emergencies at power units with different types of reactors either being designed or constructed (Sect. 7.1.3).

7.1.2 Review and Re-analysis of Historical Monitoring Data

Monitoring data described below can be used to assess both expected depositions under accidental conditions and the actual behavior of artificial radionuclides in different watershed environments.

The studying and forecasting of surface runoff contamination by artificial radionuclides has a long history. The early studies were focused on monitoring-based analysis of radionuclide export from watersheds contaminated due to *radioactivity fallout* from atmospheric NWT in the 1950s and the early 1960s (Sect. 6.3.3). The deposition of fallout radionuclides reached its maximum in the early 1960s. Since the mid-1980s, bomb-test fallout has been below detection levels (Cambray et al. 1989).

Later, studies of this type were motivated by the need to predict the consequences of the impact that can be produced on various ecosystems by severe radiation accidents, such as Chernobyl 1986 and Fukushima Daiichi 2011 accidents. Studying the fate of radionuclides within the EURT 1957 also gave important information.

In addition to observations of the radionuclide composition of river waters at river gauge stations in the watershed areas (drainage basins), special studies were carried out at experimental plots that allow controls for flow hydraulics and solute flux of Chernobyl radionuclides removing from the contaminated soil surface. The leaching of radionuclides was also determined in soil profiles at the experimental sites. The studies, carried out at both regional and experimental-plot scales under different natural–climatic and landscape conditions open ways for extrapolation of the obtained data to poorly known areas.

7.1.2.1 Radionuclide Runoff from Contaminated Watersheds

The Chernobyl fallout is characterized by two main fallout forms (Ivanov and Kashparov 2003; Smith et al. 2005): condensed components and fuel-particle components. The condensed components are represented mainly by Cs-137 and partially by radioisotopes of strontium and ruthenium. The fuel-particle component, represented by solid-phase fallout, consists of uranium oxides and chemical elements from construction materials and contains a wide spectrum of fission products (Cs-137, Cs-134, Sr-90, Ce-144, Ru-106) as well as isotopes of plutonium and transplutonium elements. It was theoretically predicted and observed that the leaching rate of these radionuclides from the particle matrix is controlled by a number of soil chemistry and landscape conditions.

The contamination of water bodies (rivers, natural and artificial water bodies) by Cs-137 and, to a lesser extent, Sr-90, after the Chernobyl accident is mostly due to the wash-out of radioactive products from the surface during rain events and snow melting, the most rapid increase in the degree of radioactive contamination of rivers being due to rain freshets. The data on 1987 and 1988 spring floods were of greatest use for studying the migration processes.

The intensity of radionuclide uptake and transport by runoff, controlling the efficiency of the natural attenuation of soil radioactive contamination in the upper parts of watersheds depends on soil type and its natural moisture content, as well as the type of vegetation cover. The mobility of radionuclides is highest in sod podzol, sandy loam, and sandy soils. Runoff from such watersheds leads to a higher radioactivity of floodplain deposits, radionuclide concentrations in which can be several times greater than those in soils in the upper part of watersheds. Soils in the upland area of watersheds represented by chernozem retain radionuclides more firmly; therefore, the runoff water contains a small amount of radionuclides, resulting in that radionuclide concentrations in the floodplains of such watersheds are commonly moderate.

The migration capacity of Sr-90 represented by dissolved forms (including organic complexes) is about ten or even more times greater than that of Cs-137. Conversely, radionuclide transport by runoff with suspended matter is most significant for Cs-137, especially, when the runoff is generated at areas represented by light arable soils, covering the steep upland area of watersheds.

Surface water and shallow groundwater carry radionuclides into rivers, where they migrate downstream. In this case, bottom sediments, as well as sediments of low floodplains during floods actively adsorb radionuclides, thus performing an important nature protection function.

An integral characteristic of the ability of soil to retain radionuclides is the so-called wash-out coefficient (or wash-out ratio), defined as (Israel et al. 1990):

$$K_c(\Delta t) = \frac{A_T}{N_0 F}, \tag{7.1}$$

where A_T is the total activity, Bq, removed from a drainage basin with runoff for observation time Δt; N_0 is the initial fallout deposition density, Bqm^{-2}; F is fallout area, m^2.

Thus, the washout of Chernobyl cesium (Cs-137) with spring flood water is characterized by the coefficient K_c, varying within (60–450)·10^{-6} at Δt = 10–20 d (Israel et al. 1990). Forecasts for the drainage basin of the Pripyat R. (which suffered most from the Chernobyl accident) were calculated with $K_c(\Delta t = 1\ \text{y}) = 0.01$ (1 %).

The discharge of Fukushima Cs-134 and Cs-137 from small catchments (Fukushima Prefecture) during 2011 was estimated to be 0.3–0.5 % of the total amount of these isotopes deposited onto the catchments (Ueda et al. 2013).

On the average, less than 1 % of the Cs-137 is transported from contaminated catchments by runoff immediately after fallout deposition, and generally less than 0.1–0.5 % moves away per year after the initial deposition; this coincides with estimates of the environmental impact of the NWT in the atmosphere (Eakins et al. 1984; Helton et al. 1985).

The interaction of radionuclides atmospherically deposited (gas–aerosol fallout) onto the catchment basins and runoff water flowing over the soil surface involves many phenomena of different nature (Monte et al. 2004; Smith et al. 2005). Adsorption–desorption has been recognized as one of the most important processes determining the migration of radionuclides in runoff water. As a first approximation, the process can be considered in an equilibrium formulation (Sect. 6.2.2).

Therefore, the special literature gives descriptions of many studies in experimental plots within the Chernobyl trace, which allowed the researchers to evaluate the entrainment coefficients K_1^* and K_s^*, characterizing the short-time retention of various radionuclides on the surface, represented by different types of soils, in the initial periods after the deposition of radionuclides (Sect. 6.2.2.1). The fullest generalization of those experimental results can be found in the studies of A.V. Konoplev, A.A. Bulgakov and L. Garcia-Sanchez with co-authors (Bulgakov et al. 1999; Konoplev et al. 1992; Garcia-Sanchez et al. 2005; Garcia-Sanchez and Konoplev 2009), focused on the behavior of Cs-137 and Sr-90 (Table 7.3).

The results given in Table 7.3 can be transformed using formula (6.35b) to evaluate equilibrium coefficients K_d, characterizing sorption equilibrium of Chernobyl 1986 radionuclides and mineral suspension. The obtained mean variation ranges of K_d are (1–3) 10^3 cm^{3}g^{-1} for Sr-90 and (5–8)10^3 cm^{3}g^{-1} for Cs-137. The latter values within the range (roughly) $1 \cdot 10^3$–$5 \cdot 10^5$ cm^{3}g^{-1} were reported by the IAEA (2010) as

Table 7.3 Variation ranges of normalized entrainment coefficients based on generalization of experimental data (Garcia-Sanchez and Konoplev 2009)

Radionuclide	K_1^*, mm^{-1}		K_s^*, m^{2}g^{-1}	
	min	max	min	max
Cs-137	$1.9 \cdot 10^{-6}$	$1.2 \cdot 10^{-4}$	$1.6 \cdot 10^{-5}$	$6.7 \cdot 10^{-4}$
Sr-90	$1.9 \cdot 10^{-6}$	$1.8 \cdot 10^{-4}$	$6.5 \cdot 10^{-6}$	$3.1 \cdot 10^{-4}$

typical values characterizing Cs-137 interaction with geosorbents in the freshwater environment.

Analysis of data collected after the Fukushima Daiichi accident has shown that the studied watershed systems are very reactive to rainfall and snowmelt events, and that export of contaminated material to the rivers, and possibly to the Pacific Ocean, has already been achieved during the year that followed the accident (Evrard et al. 2013).

The partitioning of the Fukushima cesium isotopes (Cs-137, 134) between water and suspended matter was studied in water samples from the Hiso River and Wariki River that traverse mountainous areas in Fukushima Prefecture, Japan, impacted by radioactive fallout after Fukushima Daiichi accident (Ueda et al. 2013). The obtained K_d of Cs-137 in water samples ranges roughly from 10^5 to 10^6 cm^3g^{-1}, which is one to two orders of magnitude higher than K_d for Chernobyl Cs-137.

The data of sampling in Hiso River and Wariki River showed cesium isotopes (Cs-137 and Cs-134) in river water to migrate mostly in particulate form, i.e., radionuclides are readily adsorbed onto suspended particles and colloids. The mean percentage of particulate radiocesium activity ranged from 74 % to 82 % in the Hiso River and Wariki River. Particulate matter accounted for over 90 % of the total radioactive cesium activity during periods of precipitation and flooding and for approximately 40 % during dry weather and normal water stages (Ueda et al. 2013).

In general, analysis of data collected after Chernobyl 1986, Fukushima Daiichi 2011, and Kyshtym 1957 accidents, as well as analysis of NWT historical data have shown the following tendencies with regard to non-point radioactive source contribution to runoff contamination:

- strontium and cesium in fallout particles are completely soluble elements;
- Sr-90 in overland flow water mostly occurs in ionic form and can be bound with organic ligands; a relatively small fraction of strontium is transported by runoff in adsorbed (particulate) forms;
- Cs-137 (Cs-134) has lower tendency to form complexes; this isotope is washed out from the surface along with mineral phase, mostly under the effect of erosion; the high mobility of Cs-137 on suspension is largely due to the irreversibility of sorption, as was confirmed by studies of the occurrence forms of isotope Cs-137 of Fukushima Daiichi origin (Otosaka and Kobayashi 2013); the irreversibility of cesium sorption allowed Cs-137 to be used as an erosion indicator to investigate the delivery of sediment from the hillslopes to lowland areas and water bodies (Owens et al. 1997; Poręba 2006; Konz et al. 2010);
- for fresh fallout, liquid entrainment coefficients, K_s^*, for both Sr-90 and Cs-137 do not exceed approximately 10^{-4} mm^{-1} (Garcia-Sanchez and Konoplev 2009); the coefficient K_s^* increases with increasing slope and rain intensity, because the share of fine suspension in water flow increases;
- the estimated values of coefficients K_1^* and K_s^* appreciably drop over time, thus suggesting the increasing retention ability of soils with respect to radionuclides (see Sect. 7.1.2.2). Moreover, this tendency suggests the heterogeneity of the surface of multi-site-surface sorption mineral phases, a considerable part of which includes sites irreversibly adsorbing radionuclides;
- irreversible sorption onto minerals and organic matter of soils, as well as water infiltration, limits radionuclide migration in dissolved form;

- the mobility of radionuclides adsorbed onto particulate suspended matter depends on the intensity of liquid precipitation and hence can vary widely with time;
- radionuclide losses *through runoff* and sediment transport in the first one or two years after the accident commonly do not exceed a few fractions of percent, rarely, a few percent from their total amount in the watershed. The deposited radionuclides mostly remain in watershed areas and serve as sources for long-term contamination of the environment. The abrupt slowing of the removal of radionuclides from the watersheds can be attributed to the passage of radionuclides into a less mobile state and their transport into deeper soil layers by infiltration water.

7.1.2.2 Vertical Transport of Radionuclides in Soil and Vadose Zone

The vertical migration of dissolved radionuclides in soils and their transfer to the underlying unsaturated sediments (vadose zone) mitigate the negative impact of radioactive fallout on the runoff contamination, though create a hazard of groundwater contamination.

In the years since the Chernobyl 1986 accident, the vertical distribution of fallout of Cs-137 (Cs-134) and Sr-90 in soils of different origin and different landscape conditions has been measured by many researchers. Such studies were also carried out earlier, in the period 1960–1970, as part of the analysis of NWT fallout consequences. It was established that the sorption of these radionuclides in soil depends on the mineralogical and chemical characteristics of the soil layer, such as clay minerals and clay mineral content, particle size distribution, and organic matter content.

The sorption process is usually only *partially reversible*. The process of Sr-90 and Cs-137 irreversible fixation onto soil particles has two phases – fast and slow. The latter, the so-called adsorption aging process, shows very slow kinetics, and the reversibility can last as long as tens of years. Such radionuclide behavior can be explained by the presence of several specific sorption sites in clay and hydroxide minerals (Rumynin 2011).

It is generally agreed that isotopes of cesium (Cs-137, Cs-134) migrate very slowly in most soils and landscapes. This suggests that cesium is strongly and rapidly absorbed onto soil particles (especially on clay minerals) and organic matter. The adsorption of cesium is mainly chemical and occurs through an ion exchange process. A significant fraction of the radionuclide can be retained in the soil in a non-extractable (irreversible) form. Irreversible sorption of cesium is due to the occlusion of radionuclides by soil organic matter and the diffusion into interlayer clay minerals.

In a number of publications related to the Chernobyl accident, it has been shown that Cs-137 adsorbed in the top untilled soil layer with a depth of about ten cm remains there for many years. Further downward, the transport of Cs-137 in soil by infiltration is limited.

A more mobile element is Sr-90, which can be detected at depths of up to 1–2 m from the land surface. A small but significant fraction (up to 10–20 %, on average) of Sr-90 in soil undergoes physicochemical changes making this portion of strontium non-extractable (Nilsson et al. 1985).

The distribution of radionuclides over depth in soil sections was also studied in the EURT (Sect. 7.1.1). From the vertical distribution of various radionuclides, it can be concluded that, for all sampled locations except one, essentially all activity was found in the top 30 cm soil layer (Aarkrog et al. 1997; Atlas of the EURT 2013). In case of Sr-90, the 25–30 cm layer contained on the average 1.5 % of the total inventory of Sr-90 in the 0–30 cm layer. In the case of Cs-137 and transuranics (Pu and Am), even less was found in the deeper parts of the soil column (Aarkrog et al. 1997).

Finally, first results related to vertical distribution of Cs-137 in soils were obtained for areas affected by the Fukushima Daiichi fallout (Matsunaga et al. 2013). It was determined that majority of Cs-137 is stored in the upper 5 cm of soil layer at undisturbed locations. The mean fallout contamination by Cs-137 and Cs-134 in sampling points was assessed to be about $6 \cdot 10^4$ Bqm^{-2}. The Cs-137 contamination of the top-layer soil (0–1 cm) ranged on average from 1000 to 7000 $Bqkg^{-1}$. Concentration of Cs-137 in the upper part of the soil profile (1–3 cm below the surface) varied on average from 300 to 1000 $Bqkg^{-1}$. Limited fractions of Cs-137 were extracted with water (less than 3.6 % for 0–3 cm samples) and ammonium acetate (less than 15.2 %), confirming the very low mobility of this radionuclide in response to the rainfall events (Matsunaga et al. 2013).

Thus, the soil itself acts as a shield from radiation, but a slow migration results in a slow decrease in external radiation, and near-surface accumulation of accidental radionuclides restricts land use, e.g., for food production, in highly contaminated areas for a long time.

Special mention should be made of the studies aimed at revealing the contribution of preferential flow paths in soils and deeper vadose zone to the rapid vertical transport of water and solutes. The causes of the formation of such anomalous zone where solutes migrate with velocities far in excess of the mean velocity of water flow in near-surface sediments include the weathering processes and the activity of organisms resulting in soil structure disintegration (Flury 1996; Steenhuis et al. 1997; Gerke et al. 2007). Macropores and sporadically distributed cracks and fissures become active carriers of contaminants from the surface to groundwater table, especially, during storm precipitation events, accompanied by soil surface ponding and runoff generation. As it was mentioned many times, not only high-mobility components, but also substances with a high aptitude to sorption, rapidly reach aquifers in such periods.

The role of macropores and fractures increases in the case of morphological heterogeneity of the relief, providing conditions for focusing surface runoff (Nieber 2001; Bixio et al. 2002; Shestopalov et al. 2006, 2007). Mostoften, the focused-flow patterns form when the relief contains drainless or partially drained local depressions (Fig. 7.3). Their distribution on land surface can be random or can have some trend determined by the confinement of morphological depressions to persistent deep-seated geological structures or specific stress–strain zones of the earth crust (Shestopalov et al. 2006).

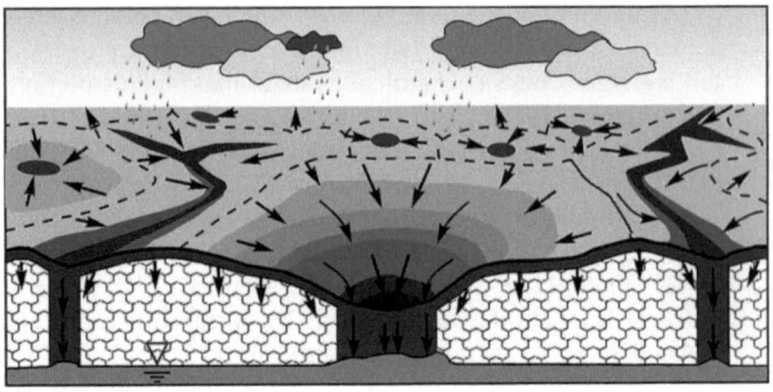

Fig. 7.3 Conceptual scheme of groundwater recharge and contamination by subvertical preferential flow zones (Shestopalov et al. 2006)

The presence of morphological depressions in a watershed determines a high degree of spatial heterogeneity and time variations of groundwater infiltration recharge via the aeration zone. In dry seasons, the rate of infiltration shows a uniform horizontal distribution and the motion of moisture front is uniform throughout the ambient soil matrix. During rainfall events or active melting of snow cover, the generation of overland flow leads to partial filling of morphological depressions, because of which the soil moisture content in the depression bed increases and at a certain moment may exceed the field capacity, resulting in downward water percolation to recharge groundwater. In addition, the central part of a depression, as better washed, shows higher permeability, making the infiltration process even less homogeneous. Thus, microrelief forms, focusing precipitation, provide rapid solute transport in the vadose zone. Groundwater contamination in this case is of impulsive character.

This mechanism of water and solute transport has been studied in experimental sites within the Chernobyl exclusion zone (Shestopalov et al. 2006, 2007; Bublias and Shestopalov 2001). Those studies showed the penetration depth of radionuclides into soil profiles, whose position in plan coincides with the central lines of depressions, can be many times greater than the penetration depths of radionuclides in infiltration water in reference areas, where there are no such morphological relief forms (Bublias and Shestopalov 2001). The presence of anomalous zones of this type can be the cause of the rapid (a few years after the accident) appearance of Sr-90 and Sc-137 in phreatic waters and confined aquifers within the Chernobyl trace.

A complete description of the process requires the application of coupled models describing the formation of surface runoff and soil moisture and solute transport in the vertical direction for unsaturated groundwater recharge (Bixio et al. 2002). Approximate analytical estimates can be based on simplified conceptual models of tank type (Steenhuis et al. 1997). Overall, the construction of models, which would explain the behavior of the distributions of moisture content and concentration in the soil and aeration zone in the cases where such behavior radically differs from its classical description (satisfying Richards' and Fick equations) is an urgent research line in the modern hydrogeology and a focus of active studies.

7.1.3 *Pre-defined (Worse-Case) Scenarios for the Release and Deposition of Accidental Radionuclides*

The prediction of radionuclide specification, the amount and levels of the radioactive release and fallout is difficult because of several factors: the type of nuclear reactor and its thermal power output, the type of accident scenario, the meteorological conditions, such as wind direction and speed, and the nature of the surface that will be affected by fallout. Therefore, in this section, basing on accident simulation, we only give tentative characteristics of the accidental radioactive releases (so-called "source term") for two types of reactors (Table 7.4), which are used to evaluate fission products fallout characteristics. For comparison, some data on the Chernobyl 1986 (Table 7.1) and Fukushima Daiichi 2011 (Nagai et al. 2014) accidents were introduced as well.

The source term data (Table 7.4) were estimated based on 'catastrophic' and 'beyond design basis' accident scenarios for VVER (Nalbandyan et al. 2012) and BN (Beloyarsk NPP... 2011; see also Sect. 7.2.3) types of nuclear reactors. These data correspond to accidents associated mainly with situations known as *station blockout,* during which losing both the grid connection (offsite power) and standby power generators (inside power) occurs. Under normal conditions, they are sources of electricity providing the cooling system in operation to reduce *decay heat* after reactor shut-down and mitigate negative effects during an accident. It is postulated that, unless grid or standby power is restored, the reactor will overheat. In this period, which may last for as long as a few hours, reactor fuel reaches melt-down, releasing highly radioactive fission products.

The worst-case scenarios postulate the loss of containment of more than half of fuel elements and melting of a noticeable amount (up to several hundred kg) of fuel in the reactor core. Numerical simulation shows that the explosion that would breach the final containment, releasing fission products to the environment is not possible. The reactor vessel at this emergency remains intact, and radioactivity release into the environment takes place because of the actuation of a hydraulic hitch in the gas system of reactor.

Table 7.4 The major source term characteristics, M (TBq)[a]

		Radionuclide			
Reactor type	Scenario	I-131	Cs-134	Cs-137	Sr-90
VVER-1000[b]	Catastrophic release	$2.7 \cdot 10^3$	$4.4 \cdot 10^3$	$2.8 \cdot 10^3$	44
	Beyond design basis	31.5	5.2	3.3	42
BN-800[c]	Beyond design basis	3–30	3–35	5–50	–
RBMK[d]	Catastrophic (Chernobyl)	$1.8 \cdot 10^6$	$5.4 \cdot 10^4$	$8.5 \cdot 10^4$	$1 \cdot 10^4$
BWR[e]	Catastrophic (Fukushima)	$1.2 \cdot 10^5$	n.d.	$8.8 \cdot 10^3$	n.d.

[a]$\text{TBq} = 10^{12}$ Bq
[b]Pressurized water reactor, thermal neutron reactor design
[c]Fast neutron reactor, sodium-cooled design
[d]Channel-type boiling water reactor; [e]Boiling water reactor

Modeling results show that the fission products that release from a damaged reactor are initially a complex mixture of hundreds of different radionuclides, however most of these nuclides have short half-lives and decay to longer living daughter elements at very rapid rates.

Of greatest interest in terms of radiation contamination hazard of the surface are four radionuclides: I-131, Cs-134, Cs-137, and Sr-90 (Table 7.4). The first of them, short-living as it is, can be of hazard, because it is inert and very mobile in the natural environments. Of greatest importance here are short-range (a few days or weeks) forecasts correlated with hydrological and landscape features of the region that has suffered from radiation accident. The other three radionuclides (Cs-134, Cs-137, and Sr-90), because of their higher sorption capacity, are less mobile; however, their half-life is relatively long, so the analysis of their migration in natural environments is of interest for long-range forecasts of the consequences of emergency releases.

The fission products escaped to the atmosphere become rapidly incorporated into aerosol particles and raindrops that are subject to the fallout or remain in the gas phase. In these forms, fission products are carried by the wind until they fall out. The physical, chemical, and radiochemical properties of fallout particles and the frequency distribution of particle sizes and types vary significantly as well. These characteristics, along with the meteorological and landscape conditions, control the rates and locations of radioactive products deposition on the ground after having been injected into the atmosphere by a nuclear accident.

The meteorological prediction of the radionuclide deposition involves two steps: (a) dispersion of the fission products in the atmosphere, and (b) estimation of the amount of radionuclides deposited on the ground. In other words, the deposition forecasts are calculated with a deposition model using results of atmospheric dispersion modeling. The deposition of radionuclides on the ground results from two processes: the impaction of aerosols on the ground surface (dry deposition) and precipitation (wet deposition).

Many mathematical models have been developed to describe those processes. However, their applicability to the quantification of the accident consequences largely lose its significance because of the unpredictability of the accident moment and the accompanying weather conditions, controlling the location of the radioactive fallout trace, and other poorly known factors like physicochemical form of the radionuclide deposition. Therefore, the formation of input data for hydrological models to describe the consequences of radioactive contamination of soil surface can be based on alternative approaches, leading to maximally conservative estimates, which do not depend on the meteorological and other factors.

Thus, we can assume that all radionuclides having been released to the atmosphere in the amount of M (Table 7.4), deposit in gas–aerosol or precipitation forms on the ground surface, forming near the NPP an area-distributed source with an intensity of $N_0 = M/F$, where F is an area, which can be established empirically, e.g., it can be associated with drainage basins, lying within NPP-site boundary or pre-described sanitary protection zone (Sect. 7.2.3). It can be assumed a priori that the maximal allowable levels of surface contamination can be reached within a

5-km zone around the damaged reactor, with the emergency plume assumed low and dispersion from it assumed uniform. At distances far in excess of the size of the NPP-site (up to 30 км in the case of a high plume), the settling of fission products will most likely be of a "sectoral character," determined by wind direction (Sect. 7.2.3).

The area of radioactive contamination is considered as a non-point source. The subsequent contamination of streams is due to radionuclides transported by overland flow. Groundwater contamination takes place because of vertical infiltration of precipitation through the soil and aeration zone.

The comparison of data that characterize the consequences of hypothetical accidents at power generation units VVER-1000 and BN-800 with estimates of Chernobyl release show the values of M given in Table 7.4 to be several orders of magnitude less than the radionuclide release during the Chernobyl accident.

7.2 A Case Study: Assessment of Watershed Contamination after a Hypothetical Accident at the Beloyarsk NPP (the Middle Urals, the Russian Federation)

The main priority in the operation of nuclear power plants (NPP), as well as in the designing and construction of new NPP is accident prevention and the minimization of the environmental effect of the plants. The construction and operation of NPP have various impacts, primarily, on the ecosystems of surface water bodies and the geological medium, including groundwater. Integral assessment and prediction of the state of natural waters in the zone of influence of nuclear-industry plants involves the consideration, in addition to the normal operation regime of facilities, emergencies at power generation units, leading to the release of radionuclides into the environment. This thesis, supported by present-day IAEA requirements, is still relevant in the context of large-scale radiation accidents at NPP in operation (Chernobyl 1986 and Fukushima Daiichi 2011 accidents).

These requirements are also relevant to the Beloyarsk NPP (BNPP) constructed in 1964 in the Middle Urals, Russia (Fig. 7.4). The fifty-year history of the NPP operation has seen several generations of nuclear power units (Table 7.5). The first two units were provided with boiling water reactors (Unit 1 and 2). Their era ended after decommissioning Unit 2, AMB 200, in 1990. The slow-neutron boiling-water reactors were replaced with a third generation of fast-neutron power generating units of higher capacity (BN 600 – Unit 3), being in operation until now (BN is a Russian abbreviation for the name of Fast Neutron type of nuclear reactors).

In compliance with the governmental program of developing nuclear-power engineering, two more nuclear power units BN 800 (Unit 4) and BN 1200 (Unit 5) are being designed in the immediate vicinity of the currently operating nuclear power unit.

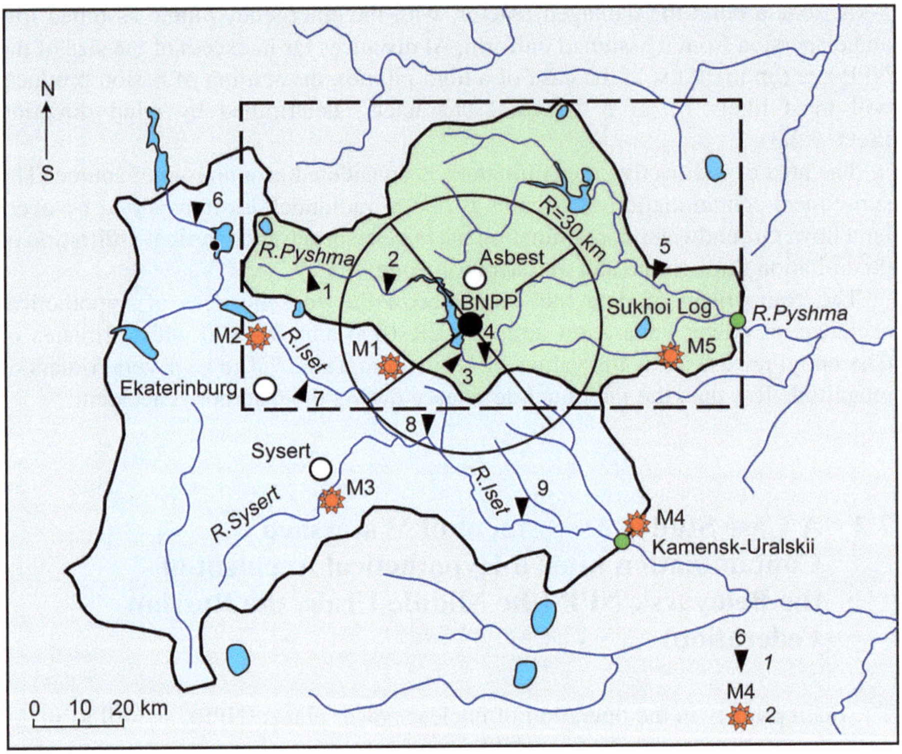

Fig. 7.4 Main watersheds and location of gauge (*1*) and meteorological (*2*) stations. The circle divided into segments is the 30-km area of potential NPP influence (Sect. 7.2.3). The dash rectangle delineates the model area for the Pyshma watershed

Table 7.5 History and prospects of the NPP development

Unit	Name	Commission	Decommission
1	AMB 100	1964	1983
2	AMB 200	1969	1990
3	BN 600	1981	2025
4	BN 800	2016	–
5	BN 1200	2020	–

On-site environmental engineering survey following the design of the Beloyarsk NPP provides a good basis for implementing new approaches to forecasting the effect of NPP on surface water and groundwater systems. First, field material was obtained for the parametric support of models; second, a considerable progress was observed in recent years in the development of computing facilities adjusted to solving forecast problems of this type. A combination of those factors was the main motive for writing this section of the book.

Table 7.6 River discharge characteristics

No. in the map	River–gauge station	Catchment area, km^2	Discharge, m^3s^{-1}	
			Avr.	Max
1	Pyshma R.–«Berezit»	197	1.24	10,9
2	Pyshma R.–«Sarapulka»	663	2.49	24.1
5	Pyshma R.–«Sukhoi Log»	3180	10.8	70.0
6	Chernaya R.–«Sagra»	220	1.16	36.6
7	Iset R.–«Mill No.3»	1470	8.03	39.6
8	Iset R.–«Kolyutkino»	3500	13.3	114

7.2.1 Description of the Study Area

7.2.1.1 Relief and the Hydrological and Hydrometeorological Conditions in the Area of the Beloyarsk NPP

BNPP is situated on the eastern branch of the Middle Urals mountain ridge. The relief in the area is hilly with alternation of plateaus and depressions, the latter often being swampy. Relief elevations vary from 169 to 329 m.

The geomorphological features of the region are determined by its geological structure. Vast water divide areas represent denudation hilly–knap peneplanation plane. The relative elevations of individual relief forms reach 30–50 m. A hilly–knap surface of near-valley slopes and a modern floodplain terrace with 2–3 peneplanation planes extend along well-developed river network.

BNPP area is situated on Pyshma R. watershed on the shore of the Beloyarsk Reservoir; Iset River watershed neighbors the area (Fig. 7.4). The hydrographic network of the area belongs to Kara Sea water system. The Pyshma R., flowing into the Tura R., is the third-order tributary of the Tobol R. (Pyshma–Tura–Tobol). The rivers of the region belong to the class of rivers with pronounced spring flood, summer–autumn rain floods, and long and stable winter dry period. The rivers are mostly fed by snowmelt water. Its share in the total river runoff reaches 50 % and that of rainwater, 22 %; groundwater runoff accounts for 28 %. Surface-water regime observations are based on a gauge network (Fig. 7.4, Table 7.6).

By its chemistry, river water is hydrocarbonate–calcium; HCO_3^- ions dominate in its anion composition (20.0–183 mgL^{-1}), and Ca^{2+} ions dominate in the cation composition (16.4–74.1 mgL^{-1}). pH varies from 6.6 to 8.8. The water is moderately hard (1.10–3.90 $meqL^{-1}$).

According to observation data on sediment discharge of the Pyshma R. (Beloyarka V.), collected during surveys for the construction of the Beloyarsk Reservoir (1955–1958), the mean annual suspended-sediment discharge in the Pyshma R. under natural conditions is 0.05 kgs^{-1}, which corresponds to the mean annual sediment input of 1.6 thousand t (about 1 thous. m^3y^{-1}). The maximal value was 2.3 thous. ty^{-1}. The within-year distribution is uneven, varying from 0.4–0.5 kgs^{-1} during spring flood to a few hundredths kg/s in autumn and winter (Fig. 7.5). Those data can be of use for the assessment of slope erosion – a process that can contribute to the transport of accident-related radionuclides in runoff (Sect. 7.2.5).

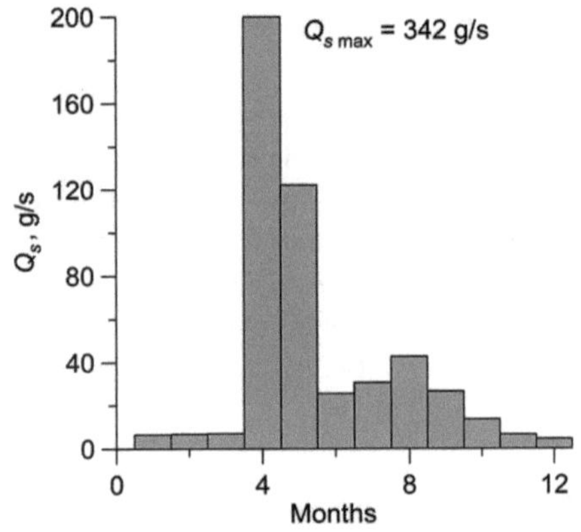

Fig. 7.5 Annual distribution of sediment discharge (Pyshma R.)

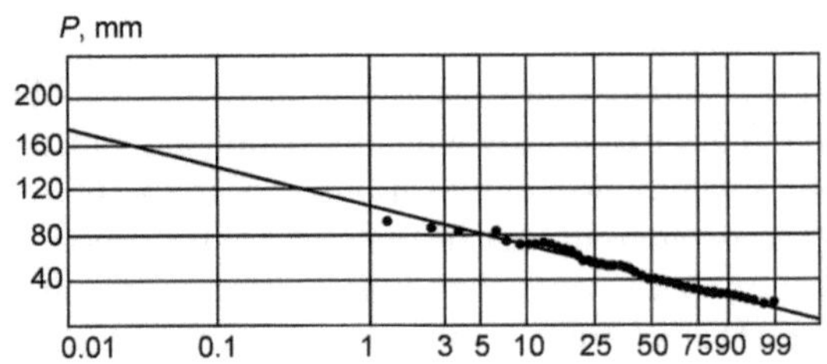

Fig. 7.6 The frequency distribution of maximum daily precipitation

Analysis of many-year observation data (weather station M1 "Dubovo V.," Fig. 7.4) shows the mean annual precipitation to be 540 mm (the area lies in the zone of sufficient moistening), the largest total monthly precipitation (274.5 mm) was recorded in September 1987, and the least one (0.8 mm) was recorded in March 1976). The average monthly precipitation is 49.6 mm. The total precipitation is maximal in the warm season (76 % of the annual precipitation). The total duration of rain events within a year averages about 1800 h. Summer precipitation is mostly of storm character. The daily maximum of precipitation observed over many years was 94 mm, which approximately corresponds to 1 % occurrence probability (Fig. 7.6).

The obtained time series provide a substantiation of the forecasted synoptic scenario. The need to obtain conservative estimates of the environmental impact of BNPP emergency releases implies the consideration of the worst weather conditions, corresponding to a rainy period of warm season, when the generated runoff leads to intense river floods. A typical example is the flood in the summer of 1971, which can be seen, in particular, in the hydrograph measured at the gauge station 5 "Sukhoi log" (see Fig. 7.11 below). The mean river discharge in this section is 10.8 m^3s^{-1}. The runoff water transports radionuclides from the upper parts of watersheds to river valleys. A description of this process will be given in the final sections of this work.

7.2.1.2 Analysis of Monitoring Data

As mentioned above, BNPP is a firstling of Russian power engineering. Several power units with accompanying facilities of nuclear power cycle were successively commissioned since the mid-1960s. Some radioactive materials involved in those cycles enter natural and artificial water bodies (rivers, marshes, hydrotechnical reservoirs etc.) as low-level radioactive waste (wastewater).

Until 1979, the Russian water and nature-protection legislation imposed relatively weak requirements to the quality of treatment of industrial wastes discharged into water bodies, resulting in their appreciable radionuclide contamination. Wastewater discharge standards did not limit the total volume (amount) of discharge, the main parameter at the discharge of those waters being the maximum allowable concentration of specified contaminating substances in the wastewater discharged into water bodies. It should be mentioned that, although radionuclide concentrations in surface water bodies in the considered area were many times greater than the global background concentrations, this has not led to any appreciable radiation impact on the population. Later, starting from about 1980, the procedure of wastewater treatment changed and the radiation impact on natural water bodies dropped significantly.

There are two main natural units for wastewater discharge from the plant (Fig. 7.7): the Beloyarsk Reservoir and Olkhovskaya marsh-river system. The damping effect of the reservoir is due to the dilution of discharged wastewater by

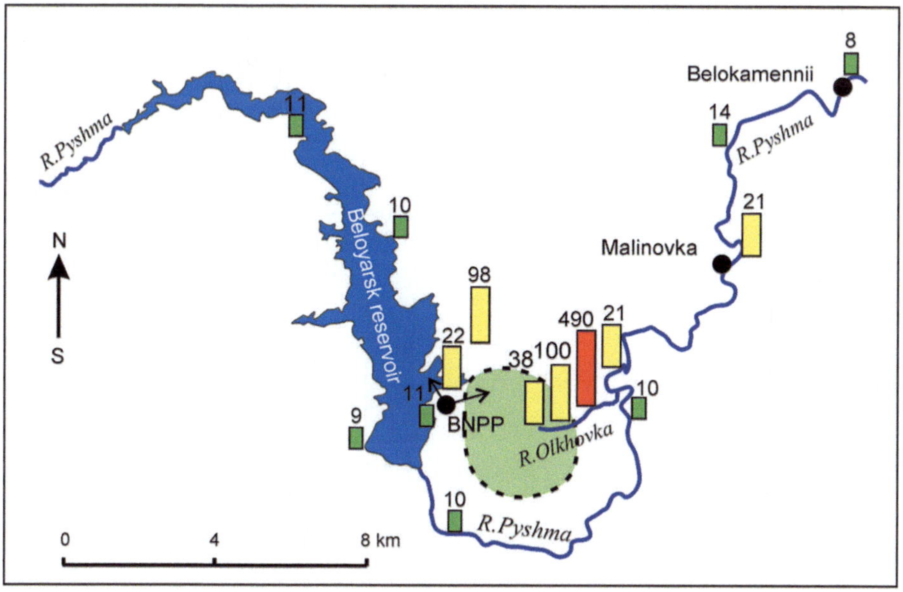

Fig. 7.7 A schematic map of the main water bodies located within the area of potential BNPP influence. *Arrows* show the location of the waste discharge zones. The bars show the radiation intensity, microrentgen/hour, near water bodies (2012)

river water (Pyshma R.), flowing through the reservoir, and sorption on bottom sediments. The dissipated wastewater flow within the marsh is deprived of the contaminants it carries during seepage through the marsh vegetation cover and sorption on bottom formations saturated with organic matter.

Monitoring data collected during the accumulation of radionuclides in the environment and the subsequent natural attenuation of the environment provide useful information about some migration processes and mechanisms. This information is of interest in the context of problems discussed in this chapter. However, considering that not all data characterizing radioactive waste discharge are available, the analysis given below is mostly of qualitative character.

Beloyarsk Reservoir Contamination The maximal level of reservoir water contamination by radionuclides was recorded in the mid-1970s, during the parallel operation of the first reactors AMB-100 (1964–1981) and AMB-200 (1967–1989) (Table 7.5, Fig. 7.8). Near the discharge site of surplus hot water, the concentrations reached 100 BqL^{-1} for H-3, 1.3 BqL^{-1} for Co-60, 1.2 BqL^{-1} for Cs-137, and 0.1 BqL^{-1} for Sr-90. The concentration curves have similar shapes and differ only in the absolute values.

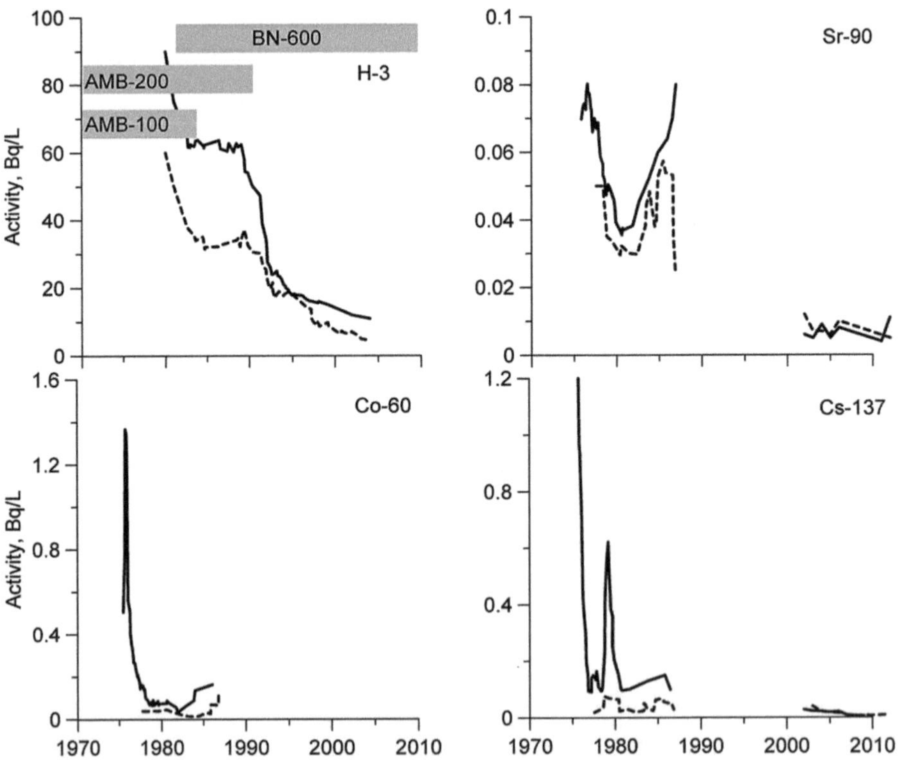

Fig. 7.8 Dynamics of radionuclide activity in the Beloyarsk reservoir. The *solid line* is for the middle reach; the *dashed line* is for the upper reach

The abrupt drop in H-3 concentration in water correlates with decommissioning of Unit 2 (AMB-200) in 1989. The second peak may be due to the commissioning of Unit 3 (BN-600). In this period, the total annual discharge of radionuclides into the water body was about $4.9 \cdot 10^9$ Bq for Co-60, $8.6 \cdot 10^9$ Bq for Cs-90, and $8.4 \cdot 10^9$ Bq for Cs-90 (data of 1985).

Radiation water monitoring in 1976–1987 gave significant materials, reflecting a decrease in radionuclide concentrations in surface water with the distance from NPP (Table. 7.7). The zone of influence of the plant includes the upper part of the reservoir lying more than 15 km away from the discharge point. Radionuclide activities in the upper part of the Beloyarsk Reservoir show a statistically significant difference from the respective concentrations in the water of the Reftinskoe Reservoir, which can be considered as an indicator of regional background level (Table 7.7).

Bottom sediments are a major deposition place of radionuclides, which determine the radiation dose for the population. According to 2007–2011 monitoring data, the concentration of Cs-137 in bottom sediments of the Beloyarsk Reservoir generally varies from 10 to 300 Bqkg^{-1}. This intervals approximately corresponds to the initial (1970–1980) contamination level of bottom sediments.

Nowadays, Beloyarsk Reservoir water also contains technogenic radionuclides (H-3, Sr-90, and Cs-137); however, their detected activities are hundreds and thousands times less than the established intervention level (Safety Standard, NRB-99/2009). Variations of the activities of Sr-90 and Cs-137 in Beloyarsk Reservoir water are of the order of a few mBqL^{-1}, while the activity of tritium never exceeds 20 BqL^{-1} (Table 7.8). The specific activity of Pu-238,239 varies from 0.16 to 0.3 mBqL^{-1}.

Table 7.7 Many-year mean monthly radionuclide concentrations in the water of the Beloyarsk and Reftinskoe reservoirs, BqL^{-1} (1976–1987)

Sampling site	Co-60	Sr-90	Cs-137	Distance from BNPP, km
Beloyarsk Reservoir				
Lower reach	0.264 ± 0.075 (54)	0.061 ± 0.022 (59)	0.307 ± 0.060 (49)	0.5
Middle reach	0.090 ± 0.019 (21)	0.044 ± 0.003 (17)	0.107 ± 0.029 (21)	8.4
Upper reach	0.037 ± 0.009 (23)	0.048 ± 0.007 (21)	0.044 ± 0.009 (22)	16
Reftinskoe Reservoir (Background)				
Mid-part	Not detected (9)[a]	0.034 ± 001 (9)	0.011 ± 0.003 (9)	36

[a]Parenthesized are the numbers of determinations

Table 7.8 Variations of the specific activity of artificial radionuclides in Beloyarsk Reservoir water (2012)

Radionuclides	Units	Lower reach	Upper reach	Safety Standard
H-3	Bq/L	7–11	5–8	7 600
Sr-90	mBq/L	6–11	4–6	4 900
Cs-137	mBq/L	10–20	8–10	11 000

Overall, the obtained data suggest efficient natural attenuation of the radionuclide-contaminated water system.

Contamination of the Olkhovskaya Marsh–River System The Olkhovskoe marsh is situated about 5 km southeast of the NPP (Fig. 7.7). The marsh–river system is the main reservoir accumulating radionuclides that entered the environment with liquid wastes from the BNPP. The system includes a low peat moor about 30 ha in area surrounded by swampy areas. The moor is the source of the Olkhovka R., 3.5 km in length, which empties into the Pyshma R.

The main anthropogenic contamination of the Olkhovskoe march was taking place before 1980 during the operation of Units 1 and 2 of the BNPP. As mentioned above, the then sanitary standards and regulations did not limit the volume of liquid radioactive waste discharge. Many-year (from 1964 to 1980) waste discharges caused the accumulation of radionuclides in the marsh in the amount estimated at 3.0 TBq (80 Ci) to 7.4 TBq (200 Ci) (Utkin et al. 2000). After the standards on the admissible waste discharges became stricter, the input of radioactivity into the Olkhovskoe marsh decreased considerably.

Observations at Olkhovskaya marsh–river system have been carried out since 1978 (Table 7.9). Peaks of specific activity of technogenic radionuclides in all years were recorded in the upper reach of the marsh – in the vicinity of the discharge point. The activity of radionuclides in water in the marsh–river system gradually decreases with the distance from the contamination source because of dilution by pure surface water and groundwater. Physicochemical transformations, including

Table 7.9 Specific activity of radionuclides in water of the Olkhovskaya marsh–river system (Molchanova et al. 2009)

Sampling point	Sr-90, BqL^{-1}			Cs-137, BqL^{-1}		
	1978–1988	1989–1991	1999–2009	1978–1988	1989–1991	1999–2009
Olkkhovskoe marsh						
Upper reach	0.8 (0.1–1.6)[a]	1.3 (1.0–1.6)	0.2 (0.1–0.3)	16.6 (2.3–49.2)	8.4 (2.0–10.6)	0.05 (0.03–0.08)
Mid-reach	0.9 (0.1–1.7)	1.2 (0.7–1.2)	0.3 (0.2–0.4)	20.7 (4.1–43.4)	17.0 (9.9–24.1)	0.2 (0.1–0.3)
Olkhovka R.						
Upper reach	0.8 (0.2–0.5)	1.0 (0.8–1.1)	0.5 (0.2–1.4)	16.0 (9.7–35.0)	11.6 (4.7–20.2)	0.3 (0.1–0.5)
Mid-reach	0.4 (0.2–1.0)	0.5 (0.2–1.0)	0.3 (0.1–0.7)	10.0 (2.4–16.0)	7.9 (4.1–17.9)	0.2 (0.1–0.6)
Pyshma R. (downstream of Olkhovka mouth)						
0.5 km downstream	0.2 (0.1–0.3)	0.35 (0.04–0.5)	0.1 (0.06–0.1)	1.3 (0.1–4.4)	0.5 (0.16–0.9)	0.2 (0.1–0.2)
0.5 km upstream	0.1 (0.1–0.2)	0.08 (0.03–0.1)	0.03 (0.01–0.04)	0.2 (0.1–0.4)	0.1 (0.08–0.15)	0.05 (0.03–0.08)

[a]Parenthesized are variation ranges

radionuclide passage from solution into solid phase because of adsorption on the rock matrix, bottom sediment, and suspension, also play an important role in the natural removal of radionuclides from water.

Data of many-year studies show that the specific activities of Cs-137 in water in the Olkhovskaya marsh–river system gradually dropped throughout the observation period. For example, Cs-137 activity at the source of the Olkhovka R. varied from 9.7 to 35 BqL^{-1}, while in 2012, it was 0.17 BqL^{-1}, the Safety Standard (NRB-2009) being 11 BqL^{-1}. Sr-90 showed an increase in activity in the Olkhovka R. up to 1 BqL^{-1} in period 1989–1991, which is commonly associated (Molchanova et al. 2009) with the stage of decommissioning of the first stage of the BNPP (Units 1 and 2). In 2012, the activity of Sr-90 at the source of the Olkhovka R. was 0.075 BqL^{-1} (the Safety Standards is 4.9 BqL^{-1}). Starting from 1980, the activity of tritium in water dropped by almost two orders of magnitude from 8000 to 130 BqL^{-1}.

In observation period 1999–2010, Cs-137 activity in bottom sediments of the marsh–river system varied from 0.02 to 6.2 $kBqkg^{-1}$. The activity of Sr-90 varied within much narrower limits (from <0.01 to 0.36 $kBqkg^{-1}$) because of the lesser adsorption of Sr-90 on solid phase. Interestingly, the maximal radionuclide contamination of bottom sediments in 1978–1988 was recorded in the upper reach of the marsh near the waste discharge point from the plant. After the discharge of radioactive wastes into the Olkhovskoe marsh decreased, the accumulated radionuclides started redistributing in the natural ecosystem. Radionuclide desorption, decay, and removal from the marsh through river systems on suspended matter caused a decrease in the activities of Cs-137 and Co-60 by factors from 5 to 50 compared with those in the 1980s. The contamination front gradually migrates so that activity peaks are now recorded in the middle of the marsh and in the Olkhovka R., rather than in the upper part of the marsh.

Special studies (Molchanova et al. 2009) were carried out to assess radionuclide release from the Olkhovskoe marsh into open hydrographic network in 2007–2008. Water samples were taken at the source of the Olkhovka R., where water discharges were also measured. Ultra-filtration experiments were carried out with the samples. The obtained data were used to evaluate the proportions of radionuclides' export from the marsh in ionic and particulate (with solid matter) forms (Table 7.10).

Olkhovka R. water discharge is 0.3 m^3/s during spring flood and 0.04 m^3s^{-1} in dry season; the sediment load in the same period is 0.023 and 0.001 kgs^{-1}, respectively. High concentrations of Cs-137 were recorded in both liquid (water) phase and solid (suspended) phase of river flow. The concentrations of radionuclides sorbed onto suspended particles are 3–4 orders of magnitude greater than those in ambient water (converted to 1 kg of sample weight).

The contribution of the liquid fraction to the "two-component transport" of Sr-90 amounts to 94–96 % in all seasons. Cs-137 release from the contaminated site in the ionic form is significant (77 %) only during dry season, while in other seasons, the contributions of liquid and sediment flows to runoff transport of Cs-137 are similar.

Table 7.10 Radionuclide concentrations in runoff components in the Olkhovka R. (Molchanova et al. 2009)

Radionuclide	Units	Liquid river runoff		Flow of solid matter	
		Spring flood period	Low-water season	Spring flood period	Low-water season
Sr-90	$Bqkg^{-1}$	0.25	0.87	206	787
	Bqs^{-1}	87	35	5	1
	%	94.6	96	5.4	4
Cs-137	$Bqkg^{-1}$	0.60	0.09	5460	1420
	Bqs^{-1}	160	4	120	1
	%	56.3	77.3	43.7	22.7

Thus, the data obtained in this study suggest the significance of *particulate forms* of radionuclides in runoff, at least for cesium isotopes, a fact to be taken into account in the simulation of the consequences of accidents accompanied by a release of artificial radionuclides in BNPP. This will be discussed below.

7.2.2 *The Study of Soil Properties*

The radionuclides that reach landscape surface after accidents at NPP and form dispersion flux in runoff interact with different environmental components (soil, vegetation cover, and surface and subsurface water systems). The spatial structure and the velocity of radionuclide transport in different natural landscapes (Fig. 7.9a) are largely determined by the physicochemical and mechanical properties of soils and various near-surface deposits. Therefore, we paid special attention to studying these properties in the zone of BNPP impact.

BNPP is situated in the zone where soils show relatively high fertility and wide diversity. Out of the agricultural lands, 36.3 % are used as plough lands; 6.5 %, as hey-fields; and 4.35, as pastures. Overall, agricultural lands account for 30 % of the area under study. The most widespread soil types (Fig. 7.9b) are gray forest, sod-podzolic, and soddy-gley (in wetlands etc.).

To study soil sections within the 30-km zone around the BNPP, shallow soil (prospecting) pits (Fig. 7.9b) were dug, outcropping different types of soils all over their depth from the surface to underlying deposits represented by bedrock weathering crust. Soil samples were taken with intervals of 10–15 cm, so that each soil section was characterized by 3–5 samples, which, after their natural moisture content was determined, were sent to the laboratory. Overall, 64 soil samples were taken. The characteristics determined in the laboratory included (Tables 7.11 and 7.12): particle size distribution, moisture content (initial, θ_i, residual, θ_r, maximum water content, θ_s, and the wilting point, θ_{wp}), capillary rise height h_c, the coefficients of sorption distribution for isotopes of cesium and strontium, K_d.

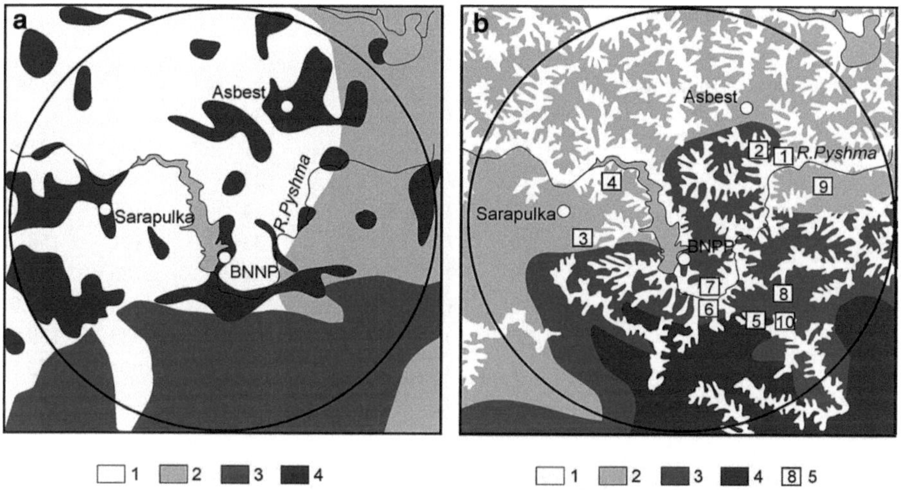

Fig. 7.9 (**a**) Schematic landscape and (**b**) soil type distribution maps within the 30-km zone around the BNPP. (**a**): *1* – mixed forest, *2* – pine forest, *3* – ploughlands and meadows, *4* – urban areas; (**b**): *1* – alluvial soils, *2* – sod-podzolic soils, *3* – grey forest soils, *4* – meadow soils and chernozems, *5* – the number of the experimental site (soil pit)

Table 7.11 Characteristics of landscapes at the examined soil sections and soil particle size distribution

No.	Experimental site	Landscape unit	Particle size (mm) distribution, % 1.00–0.25	0.25–0.05	0.05–0.01	0.01–0.005	0.005–0.001
1	Malinovka V.	Field	40	23	27	5	5
2	Malinovka V.	Mixed forest	30	41	19	5	5
3	Berezovskii T.	Pine forest	33	31	29	3	4
4	Reservoir shore	Mixed forest	60	20	12	2	5
5	Bazheno Settl.	Ploughland	20	22	35	17	5
6	Beloyarskii Settl.	Pine forest	33	31	30	3	3
7	Beloyarskii Settl.	Meadow	20	12	36	15	17
8	Shepilovo r/w st.	Farm field	20	12	36	15	17
9	Asbest T.	Mixed forest	60	20	12	3	5
10	Brusnyatskoe V.	Field	20	12	36	15	17
–	Mean values	–	34	23	27	8	8

Summary tables (Tables 7.11 and 7.12) give depth-averaged characteristics (down to 40–60 cm from the surface) of soil sections (commonly based on data of 3–5 samples). One can see that the soils of field landscapes, in the overwhelming majority of cases, are loamy soils (the share of fraction $d_{<0.05\ \mathrm{mm}}$ exceeds 30 %), while the soils of forest landscapes are sandy-loam ($d_{<0.05\ \mathrm{mm}}$, commonly less than 10 %). The soils show moderate moistening (on the average, 20–40 %). Full-saturation moisture content is in excess of 40 %.

Table 7.12 Physical and sorption properties of soils in different landscape units

		Moisture content, %					K_d, cm^{3}g^{-1}	
No. of profile	Landscape index	θ_i	θ_r	θ_s	θ_{wp}	h_c, cm	Cs-137	Sr-90
1	1	17	8	51	12	34	9384	151.1
2	1	25	9	55	13	21	2094	121.8
3	1	27	9	51	13	23	3637	159.8
4	1	17	9	54	13	24	1708	192.1
5	3	29	7	45	11	33	1191	158.3
6	1	40	9	56	14	24	2185	157.1
7	1	39	8	48	12	34	14684	249.9
8	3	36	10	65	16	35	18323	139.2
9	2	24	9	53	13	21	4064	153.5
10	3	28	10	54	14	33	10256	123.0

7.2.3 *Emergency Scenario and the General Concept of Model Analysis*

BNPP development programs includes the designing and commissioning of new Units 4 and 5 (based on fast-neutron reactors BN-800 and BN-1200). To meet the new regulatory requirements regarding environmental impact assessment, including the requirement contained in IAEA documents (Accidental Analysis . . ., 2002), implies assessing the possible consequences for the environment from a hypothetical accident at the BNPP. The assessment includes: (1) simulating an emergency release of radionuclides into the environment and their precipitation onto soil layer and (2) simulating the short- and long-time dynamics of radioactive contamination of natural (surface and subsurface) waters.

The radiation consequences are worst for the accident scenario with total blockout described earlier (Sect. 7.1.3). For this scenario, the source term corresponds to the following amounts (M) of the released radionuclides (TBq): 2.9 for I-131, 3.5 for Cs-134, 5.1 for Cs-137, $1.5 \cdot 10^5$ for Xe-133, and 0.7 for Na-24. The release-time type is instantaneous. Of greatest interest in terms of radiation contamination hazard of the surface are three radionuclides: I-131, Cs-134, and Cs-137 (Sect. 7.1.3 and Table 7.4).

The precipitation of radionuclides from a gas–aerosol release forms the so-called non-point contamination source on the land surface. As can be seen from Fig. 7.4, the zone of possible impact of the emergency release within the 30-km zone around the BNPP includes mostly the Pyshma R. watershed; the contaminated area in the nearby Iset R. watershed is several times less. The initial characteristic for forecasting is surface contamination density (N_0). It can be estimated by expert method with the assumption that all radionuclides of the emergency gas–aerosol release, depending on weather conditions at the moment of accident, will precipitate within one of the four sectors of the 30-km zone of influence of the plant (Fig. 7.4): $N_0 = M/F$ (Sect. 7.1.3). The axes are oriented with the wind diagram taken into

Table 7.13 Predicted deposition density, N_0 (Bqm^{-2})

Radionuclide	I-131	Cs-134	Cs-137	Sr-90[a]
M, TBq	2.9	3.5	5.1	–
N_0	$4.1 \cdot 10^3$	$4.9 \cdot 10^3$	$7.2 \cdot 10^3$	–
$N_0 \times 10$	$4.1 \cdot 10^4$	$4.9 \cdot 10^4$	$7.2 \cdot 10^4$	$(1 \cdot 10^5)$

[a]Reference radionuclide

account. The area of the zone sector is $F = \pi R^2/4 = 7.1 \cdot 10^8\ m^2$. The obtained estimates of N_0 are given in Table 7.13.

The forecasted values of M (Table 7.13) correspond to the lower boundary of the range of values given in Table 7.4 for this type of reactor. The estimated values of N_0 are much less than the deposition density for Cs-137 and Cs-134, recorded in the areas affected by Chernobyl 1986 and Fukushima Daiichi 2011 accidents (Table 7.2). Considering the conservative character of the forecasts, this allows us to increase the estimates of N_0 by a factor of 10, which is reflected in Table 7.13. Moreover, taking into account the methodological character of this study, we considered also a strontium isotope (Sr-90), taking N_0 equal to the value typical of the Cherbobyl trace in its periphery domain.

Preliminary analysis of the general situation shows the basin of the Iset R. to be only slightly affected by the emergency release; therefore, of main interest is a description of the processes of radionuclide contamination of soils and natural waters in the Pyshma R. basin. The initial contamination area is limited to two sectors – the northern (N) and eastern (E). Those two sectors are supplemented by a circular zone of initial contamination with a radius of 5 km with the plant as the center.

The general concept of model analysis proceeds from a simplifying assumption that the conditions in the Pyshma R. drainage basin are homogeneous, i.e., it is proposed to neglect the spatial variations of parameters that control hydrological processes. With this in view, the data of experimental studies of soil samples taken under different conditions (Sect. 7.2.2) were averaged in a special manner. Additionally, they were supplemented by a number of lacking parameters based on the analysis of published results and expert estimates.

The forecasting of all hydrological processes was carried out for the time interval corresponding to the system's response to summer flood period, which in this region commonly takes place in August (the so-called worse-case weather scenario): radionuclide washout from contaminated areas in the period of intense rains is the most hazardous process in terms of radioactive impact on the environment. As mentioned above (Sect. 7.2.1.1), this type of reference weather scenario can be associated with the moist summer of 1971, when a high flood was recorded in the Pyshma R. In August of this year, river discharge was 10 times greater than its dry-season value.

Studying the long-term consequences of the accident, namely, the forecasting of rehabilitation of the area affected by radioactive contamination (such rehabilitation commonly lasts for many years) is an independent problem, whose solution goes beyond the scope of this study.

7.2.4 *The Choice of Numerical Simulator and Parametric Support of the Model*

To forecast the space and time dynamics of radionuclide washout from the hypothetically contaminated (accident scenario) Pyshma watershed, we chose GSSHA software package, which creates a distributed-parameter physically sound dynamic model of the area (Downer and Ogden 2006; GSSHA Wiki 2014). The description of the process of radioactive contamination of soils and water bodies after the precipitation of radionuclides from the postulated accident onto land surface implies the interaction between all major modules of this software package, responsible for the simulation of

- runoff generation due to infiltration excess;
- water flow over the surface as slope sheet flow and gully flow, and river channel flow;
- soil erosion, leading to the appearance in water of disperse particles – potential radionuclide transporters;
- transport of radionuclides in dissolved form and adsorbed on suspended particles in surface water contamination transfer to subsoil water table.

A digital model of ground surface topography (Digital Elevation Model, DEM) of the Pyshma watershed was constructed with the use of TOPAZ (TOpographic PArameteriZation program) module. DEM data were used for watershed delineation, river network definition, and watershed characteristics extraction. All computation procedures were performed on a rectangular grid domain (Fig. 7.4), consisting of square blocks 500 × 500 m. The total amount of blocks was 30912 (38 × 224 blocks). The model domain is 112 km long in the east–west direction and 19 km long in the north–east direction.

The Pyshma watershed shows four types of landscapes (Fig. 7.9a): (1) mixed forest, (2) pine forest, (3) ploughland, (4) urban areas (parenthesized is a conventional landscape index). They occupy about 35, 20, 30, and 15 %, respectively (ω_i is the percentage of the total area of Pyshma R. drainage basin), (Table 7.14).

To account for the interception of precipitation by vegetation, the so-called rainfall (canopy) interception coefficient, γ, was introduced, its numerical value depending on vegetation types and landscape conditions. As seen from Table 7.14, where coefficients γ are given (Shestakov and Pozdniakov 2003), the weighted area-averaged value of γ_e is 34 %, i.e., on the average, not more than two thirds of total precipitation reach the land surface. This fact was taken into account in the input data for the model.

GSSHA software allow the user to implement several sub-models of infiltration. We chose Green and Ampt model (Sect. 1.3.2.1). The parameters determining the infiltration velocity in Green and Ampt model, are hydraulic conductivity at saturation (k_s), suction head at the wetting front (which is identified here with capillary height rise, h_c), initial saturation deficit, $\Delta\theta$ (the difference between the total moisture capacity θ_s and natural moisture content θ_i). All those characteristics, except k_s, were determined in the field (Table 7.14). The effective value k_s will be determined below during hydrological model calibration.

After the formation of a water layer on the surface of the model domain, the runoff is computed. The overland flow module in the GSSHA employs a simplified set of the Saint Venant equations. The diffusive wave approximation describing the water movement on the surface (solved by the finite difference method) seemed to be a good option to get a

Table 7.14 Watershed properties used for surface water flow and solute transport simulation

i	ω_i	γ	m, s/m$^{1/3}$	θ_i	θ_r	θ_{wp}	θ_s	h_c, cm	K_d (Cs), cm^{3}g^{-1}	K_d (Sr), cm^{3}g^{-1}
1	0.35	0.25	0.20	27	8.7	12.8	52.5	26.7	5615	172
2	0.20	0.30	0.18	24	9.0	13.0	53.0	21.0	4064	154
3	0.30	0.15	0.15	31	9.0	13.7	54.7	33.7	9923	140
4	0.15	1.0	0.02	21	7.0	12.0	45.0	25.0	–	–
K_e	–	0.34	0.15	27	8.6	13	52.1	27.4	6770	157

reliable solution. In addition to relief topography, the obtained solution depends on surface roughness, described by Manning coefficient (m) – a characteristic that controls, along with surface slope, the relationship between flow discharge and water depth on the surface. The value of coefficient m is determined by several natural factors, reflecting the landscape conditions of the area. In our example, the initial value of m for different landscapes was chosen in accordance with the study (Downer and Ogden 2006) (Table 7.14). Next, it was corrected based on simulation results.

As mentioned above, the landscapes were characterized by data of studying several soil profiles (Tables 7.11 and 7.12), allowing the evaluation of the mean values of appropriate characteristics for each landscape unit K_i, (conventional index of the parameter) and next, the effective (weighted mean) values of parameters for the entire drainage area (Table 7.14):

$$K_e = \sum K_i\omega_i / \sum \omega_i$$

Soil erosion is a factor governing the contamination rate of surface water due to the transport of radionuclides adsorbed on suspended particles. The description of this process in the GSSHA modeling complex takes into account the main physical mechanisms responsible for soil destruction, transport of soil matter, and its re-deposition on the surface. The rates of the processes is determined by the sediment transport capacity of flow and the physical properties of the soil material. The model reflects two types of soil erosion (Sect. 4.1.1): the erosion resulting from the detachment of soil particles by surface runoff (flow/hydraulic erosion) and the erosion caused by detachment by raindrop impact.

The detachment rate by surface runoff (Eqs. 4.7 and 4.8), e_h, is calculated for the case when hydraulic shear stress, τ_s, exceeds the critical shear stress, τ_c, of the soil and when sediment load is less than the sediment transport capacity, T_c (Foster et al. 1995; GSSHA Wiki 2014):

$$e_h = K_r(\tau_s - \tau_c)(1 - q_s/T_c), \tag{7.2}$$

where $q_s = qS$ is the sediment load, K_r is an empirical coefficient (a rill erodibility parameter) (s/m). The dimensions of all characteristics are given in Sect. 4.1.1. This form of detachment rate representation (Eq. 7.2) assumes that the rill erosion and the flow are uniformly distributed within the model grid cell.

When simulating hydraulic erosion, the user of GSSHA software complex is (1) to choose (substantiate) the type of equation (model) for sediment transport capacity and (2) to determine (specify) the parameters that control this type of erosion.

In the opinion of GSSHA designers, the sediment transport capacity can be adequately described by Kilinc–Richardson equation, establishing a power functional dependence between, T_c (kgm^{-1}s^{-1}), specific water flow discharge on the hillslope, q (m^2/s), and the

Table 7.15 Soil erosion parameters

Parameter	Value
General erodibility, K	0.001
Critical detachment shear stress, τ_c	3.5 Pa
Raindrop soil erodibility coefficient, K_I	32 J^{-1}
Rill erodibility parameter, K_r	0.0115 cm^{-1}

surface slope S_0, (mm^{-1}) (Eq. 4.14). In the original version of GSSHA package (Nelson et al. 2012),

$$T_c = 25.5\, S_0^{1.664} q^{2.035} \frac{K}{0.15}, \tag{7.3}$$

where K is an overland transport capacity erosion coefficient (a combined factor in a range 0–1), which reflects the joint effect of soil erodibility, vegetation, and land-use.

The description of this type of erosion takes into account the fact that the process begins when the hydraulic shear stress on the surface due to water flow exceeds the shear strength of the sediment (Sect. 4.1.1). Therefore, the critical shear stress, τ_c (Pa), is an additional parameter to be specified. Moreover, an erosion coefficient is to be specified to control the possibility of formation of rill and gully flow.

The rate of raindrop erosion of soil surface, e_r ($kgm^{-2}s^{-1}$), is proportional to the rainfall (Eqs. 4.2 and 4.3)

$$e_r = K_I C_w C_G C_i r^{\beta}, \tag{7.4}$$

where r is rainfall rate; K_I is the soil erodibility for detachment by raindrop impact (J^{-1}); C_w, C_G and C_i are coefficients, accounting for the effects of water layer and vegetation on soil surface and land use.

The coefficients in formulas (7.3) and (7.4), which determine the rate of soil layer erosion in the territory (Table 7.15), were specified basing on published materials. Thus, in accordance with the results of studies generalized in (Wicks and Bathurst 1996), raindrop erodibility coefficient K_I, characterizing rain erosion of sandy–loamy soils, varies from 23.4 to 39.8 J^{-1}. We took the mean value of 32 J^{-1}. The values of the coefficients C_w, C_G, and C_i were taken equal to unit and $\beta = 1$. The choice of the value of τ_c was based on recommendations in user's guide (GSSHA Wiki 2014).

The sensitivity analysis given in the study (Nelson et al. 2012), shows the highest sensitivity of model results to variations in the erodibility coefficient, K. For a surface represented by loamy sand and silt loam soils with undisturbed structure, not protected by vegetation and not subject to agricultural treatment, the values of K vary from 0.12 to 0.48 (loamy sand–silt loam) (Downer and Ogden 2006). The presence of a vegetation cover, e.g., forest massifs, can lead to a decrease in this coefficient by a factor of tens or hundreds. Strictly speaking, the true value of this coefficient K can be derived from model calibration based on data on solid runoff from the area under study. For bare soils, a medium value of $K = 0.2$ was chosen. For areas covered with forest or grass, $K = 0.2 \times 0.005 = 0.001$ (Downer and Ogden 2006). This is the value we will use below as an effective initial approximation for the entire Pyshma watershed.

Finally, the transport of disperse products of soil destruction and their redeposition on the surface depends on the grain size distribution of the soil that experiences the impact of this type. The size distribution of fractions of cover deposits (Table 7.16) was specified based on data of laboratory studies of soil samples (Table 7.11).

All processes described above are of hydrodynamic and geomechanic nature. They determine the conditions of water layer generation on soil, the appearance on soil surface of mobile products of erosion and the transport capacity of flow. However, the behavior of

Table 7.16 Soil particle size distribution

Fraction	Particle size, mm	Fraction content, %
Medium sand	1.00–0.25	33.6
Fine-grained sand	0.25–0.05	22.4
Silt	0.05–0.005	35.7
Clay	0.005–0.001	8.3

dissolved components in surface runoff is also governed by mass transfer processes, taking place on the boundary between the contaminated soil and water flow, as well as processes between those components and suspension (of erosion origin) in the flow. Such exchange processes are also simulated by GSSHA software complex.

The computation scheme is based on the concept of a near-surface mixing layer, where interaction between the materials settled onto the surface, rainwater, and water flow takes place (Sect. 3.1.2.2). This interaction is characterized by (1) mass transfer coefficient, k_e (md^{-1}); (2) distribution coefficient for equilibrium sorption, K_d (m^{3}kg^{-1}); (3) characteristics of the mixing layer, i.e., effective thickness, d_e (m), hydraulic conductivity, k_s (md^{-1}), porosity ($n=\theta_s$), initial moisture content (θ_i), and matrix density (taken as $\rho_s=2650$ kgm^{-3}).

Numerous studies with tracers at experimental plots, as well as field observations show the effective thickness of the mixing layer to be several mm, rarely exceeding a few cm. Therefore, assuming a characteristic value $d_e = 2$ cm, we will not introduce a large error into the subsequent calculations, considering the even greater uncertainty associated with the choice of the value of mass transfer coefficient between the soil surface and the surface runoff, k_e.

Formally, coefficient k_e could be determined from the well-known expression from boundary layer theory (Wallach et al. 1989):

$$k_e = D_e/\delta, \tag{7.5}$$

where δ is boundary layer thickness, D_e is effective diffusion coefficient. Under static conditions, the coefficient D_e is close to molecular diffusion coefficient in porous medium D_p. However, when water starts moving over the surface (in this case, over the surface of deposits in drainage areas), the transport velocity of dissolved components from pore solution increases many times due to the advective component, induced by water flow. The effective coefficient D_e now depends on a number of characteristics (Wallach et al. 1989; Grant et al. 2012), including the kinetic energy of raindrops, flow velocity, the roughness of the slope surface, shear stress of soil material, and many others. In fact, D_e can be tens, hundreds, or even thousands times greater than the characteristic values of D_p. Therefore, the determination of true values of the kinetic exchange coefficient k_e is unlikely. Only some order-of-magnitude estimates can be discussed. For example, assuming $\delta = 0.1d_e$, $D_e = (10-100)D_p$ ($D_p \approx 10^{-4}$ m^{2}d^{-1}), we obtain the likely variation range of k_e from 0.5 to 5 m/d. As a first approximation, in further calculations we take $k_e = 1$ md^{-1}.

Finally, the list of input parameters includes the radionuclide activity in soil, M_0 (Bqkg^{-1}), expressed in terms of deposition density, N_0 (Bqm^{-2}). At the uniform distribution of radionuclide in the mixing layer, its weight concentration at the initial time moment is $M_0 = N_0/\left[d_e(1-\theta_s)\rho_s\right]$ (Bqkg^{-1}). Now the concentration in pore water is calculated by the second formula (4.125), or

$$C_{we} = \frac{(1-\theta_s)\rho_s M_0}{\theta_i}, C_{we} = \frac{(1-\theta_s)\rho_s M_0}{\theta_i + K_d(1-\theta_s)\rho_s} \left(\mathrm{Bqm}^{-3}\right), \tag{7.6}$$

for unsorbable and sorbable radionuclides, respectively (Table 7.17).

Table 7.17 Radionuclides and characteristics of contamination impact

Radionuclide	N_0, Bqm^{-2}	M_0 $Bqkg^{-1}$	C_{we} BqL^{-1}	$T_{1/2}$	K_d, cm^3g^{-1}	SS[b], BqL^{-1}
I-131	$4.1 \cdot 10^4$	1216	7593	8.0 day	0	6.2
Cs-137	$7.2 \cdot 10^4$	2830	0.42	30.2 year	6770	11.0
Cs-134	$4.9 \cdot 10^4$	1926	0.28	2.1 year	6770	7.2
Cs-137 + 134[a]	$1.2 \cdot 10^5$	4717	0.70	–	6770	–
Sr-90	$1.0 \cdot 10^5$	3931	25.0	28.8 year	157	5.0

[a]The ratio Cs-137:Cs-134 ~ 0.6:0.4
[b]the Safety Standard

7.2.5 *Modeling Results and Their Discussion*

7.2.5.1 Calibration of the GSSHA Pyshma Watershed Model and Water Balance

The goal of calibration is to obtain good estimates for the actual parameters of the watershed. The summer hydrograph based on historic data of 35-day observations conducted in 1971 at the "Sukhoi log" gauge station No. 5 (Fig. 7.4) was used as a sample for model calibration (Fig. 7.10). This period featured abundant rains; hence, it meets the requirements for the subsequent estimates to be conservative. The plot was corrected taking into account the partial interception of precipitation by vegetation ($\gamma = 0.34$, see Table 7.14). The simulation of precipitation flow over the surface yielded a hydrograph for watershed outlet (Fig. 7.10). The plot shows that the abrupt increase in the discharge at the Pyshma R. gage begins almost since the beginning of storm precipitation and reaches its maximum after about 2 days. Next, the discharge gradually decreases to the mean discharge of the Pyshma R. at this outlet section (6.9 m^3s^{-1}). The maximal value of discharge at this section is 84 m^3s^{-1}.

The model was calibrated by varying two parameters: Manning coefficient, m, and hydraulic conductivity, k_s. Other parameters, which characterize soil moisture content and capillary rise remained unchanged (Table 7.14). The observed and simulated hydrographs were found to agree well at $m = 0.12$ $sm^{-1/3}$ and $k_s = 0.016$ cmh^{-1} ($3.8 \cdot 10^{-3}$ md^{-1}). The main distinction was due to the presence of a second local peak in the model plot, which can be attributed to the model generation of inflow into the Pyshma R. from a model subdomain, in which surface runoff does not form under natural conditions.

The obtained value of m is very close to the mean value given in Table 7.14 and derived from generalized literary data. The value of k_s is typical of covering deposits with sand–loam composition.

The overall water balance of the model (Table 7.18) shows that only about 25 % of precipitation that reaches soil surface form surface runoff, while the rest 75 % increase soil saturation.

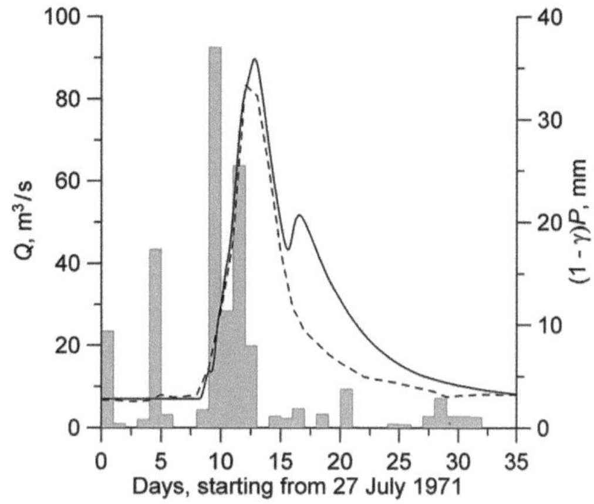

Fig. 7.10 Observed (*dashed curve*) and simulated (*solid curve*) hydrographs, combined with precipitation plot (*bar chart*) onto the surface of the Pyshma R. drainage area (27.07–31.08.1971)

Table 7.18 Volumetric water balance of the Pyshma watershed (35 d), mln. m^3

Total precipitation	Vegetation intersection	Infiltration loss	Water rest on the surface	River flow outlet
0.75	0.25	0.38	0.053	0.071

7.2.5.2 Forecast Calculations (Migration of Radionuclides)

Migration calculations were carried out successively for the eastern and northern sectors within the Pyshma R. basin (Fig. 7.4). The contaminated area in both cases was about 740 km^2. The behavior of four radionuclides (Cs-137, Cs-134, Sr-90, and I-131) was studied with cesium isotopes considered jointly, because they have similar sorption properties, while a difference in their half-lives for short-term forecasts has no effect on the final result. When necessary, the relationship Cs-137: Cs-134 ~ 0.6:0.4 can be used to evaluate the contribution of each cesium isotope to radioactive contamination.

Table 7.19 gives the main characteristics of the post-accident radiation balance of the area. The most impressive is the abrupt difference between the rates of export of radionuclides settled with aerosols in the eastern (E) and northern (N) sectors: the impact is maximal when the eastern sector is contaminated. Therefore, our further analysis will be limited to this emergency scenario.

Analysis of the balance of activities shows that the major portion of adsorbed radionuclides remains in the soil, notwithstanding the considerable surface and river runoff (about 130 mm during the period under consideration). The decrease in soil radioactivity due to the drop in the amount of cesium isotopes (Cs-137 and Cs-134) is about 0.2 %, and that for strontium isotope (Sr-90) is some hundredths of percent (about 0.01 %). Due to exchange processes during rain events, the major

Table 7.19 Radiation balance (35 d), Bq

Radio-nuclide	Scenario	Initial activity	Final activity	Loss of activity			$K_c(\Delta t)$, %
				River[a]	Infiltration[b]	Decay	
Sector E							
Cs-137 + 134	Erosion	$8.8716 \cdot 10^{13}$	$8.8562 \cdot 10^{13}$	$157 \cdot 10^{9}$	$2.41 \cdot 10^{9}$	0	0.180
	No erosion		$8.8713 \cdot 10^{13}$	$2.79 \cdot 10^{9}$	$0.07 \cdot 10^{9}$	0	
Sr-90	Erosion	$8.0818 \cdot 10^{13}$	$8.0807 \cdot 10^{13}$	$9.90 \cdot 10^{9}$	$1.15 \cdot 10^{9}$	0	0.012
	No erosion		$8.0813 \cdot 10^{13}$	$3.80 \cdot 10^{9}$	$0.92 \cdot 10^{9}$	0	
I-131	Erosion	$3.0123 \cdot 10^{13}$	$0.1485 \cdot 10^{13}$	$2.16 \cdot 10^{9}$	$13.82 \cdot 10^{9}$	$2.86 \cdot 10^{13}$	0.007
	No erosion		$0.1485 \cdot 10^{13}$	$2.16 \cdot 10^{9}$	$13.82 \cdot 10^{9}$	$2.86 \cdot 10^{13}$	
Sector N							
Cs-137 + 134	Erosion	$9.39995 \cdot 10^{13}$	$9.39994 \cdot 10^{13}$	$1.43 \cdot 10^{3}$	$0.15 \cdot 10^{9}$	0	$<10^{-10}$

[a]Activity removal from the drainage area by river flow; [b] input into groundwater due to infiltration

portion of those radionuclides will be involved in slope flow to reach the river; only a small portion of them will enter groundwater with infiltrating water. As can be seen from Table 7.19, the amount of radionuclides exported from the watershed by river water, is largely determined by soil erosion; this effect will be discussed below.

The behavior of I-131 is radically different: by the end of the model period, the soil is almost completely decontaminated, containing not more than 5 % of the initial activity of I-131. This is due to the short half-life of this radionuclide ($T_{1/2} = 8$ d). In this case, however, the total amount of radionuclides that enter groundwater is greater than their export out of the basin by river waters.

Analysis of the dynamics of radioactive contamination of river water was based on time-series plots, obtained for model gauge at the main outlet of the watershed (Fig. 7.11). The comparison of the record moments of isotope peak concentrations suggests that the sorption proper does not increase significantly the migration times of radionuclides such as Cs-137, Cs-134, and Sr-90: their peak concentrations (Fig. 7.11a, b) in river water were recorded 1.5–2 days after the detection of the nonreactive I-131 (Fig. 7.11c). On the other hand, it is worth noting that the beginning of radioactive contamination of river water (by all radionuclides) shows a considerable time lag (up to 4–6 days) behind the beginning of flood, such that peak activities coincide with the descending branch of hydrograph (Fig. 7.11).

To assess the role of the transport of radionuclides adsorbed onto suspended matter, the data of simulation with erosion processes taken into account were compared with those obtained with such processes being neglected (Fig. 7.11a, b). We can see that the disregard of erosion, which generates the appearance of mineral particles in overland water flow leads to a considerable underestimation of river water activity at the watershed outlet.

Interestingly, the maximal concentration of cesium isotopes, whose sorption is high, is about 12 BqL^{-1}, while the peak concentrations of radioactive strontium, whose sorption is lower, are lower by nearly an order of magnitude (about 1 BqL^{-1}). Such behavior of radionuclides, which seems paradoxical at the first sight, may be attributed to the effect of infiltration on the distribution profile of radionuclides in the top soil layer. Indeed, the drainage basin loses a considerable amount of water through infiltration saturation of soil (as shown in Sect. 7.2.5.1). At the same time, the descending (vertical) flow of water particles is accompanied by vertical transport of radionuclides, resulting in a decrease in their concentration at the contact between the soil and surface water layer, the magnitude of this decrease being the greater the lesser the sorption of the radionuclide, in this case, Sr-90. This means that its potential for accumulation in this water layer due to erosion decreases. Conversely, the very high sorption of other radionuclides, Cs-134 and Cs-137, contributes to their retention at the soil–water contact, and, accordingly, those radionuclides become much more involved in the process of sorption on suspension that forms due to surface erosion of watershed slopes at precipitation. Where there was no erosion, one should not expect such anomalous processes to occur. Indeed, in the absence of erosion and at the nearly equal density of initial

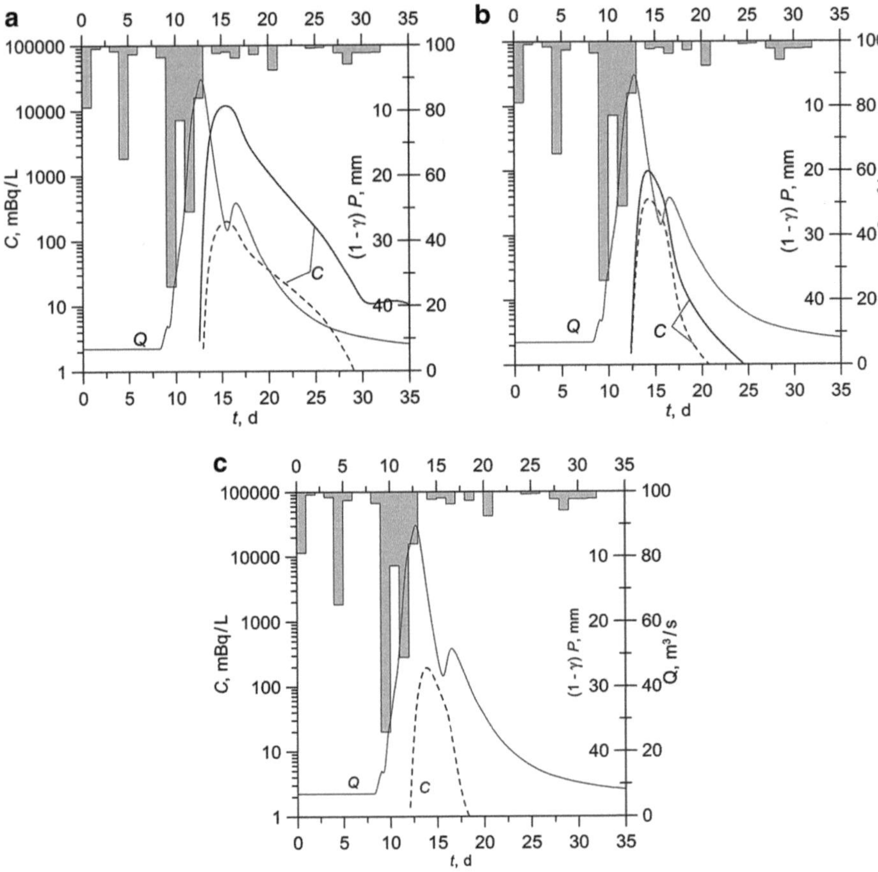

Fig. 7.11 Dynamics of the activity (excess above the background level, *solid curves*) of (**a**) Cs-137 + Cs-134, (**b**) Sr-90, and (**c**) I-131 in the Pyshma R. («Sukhoi log» gauge). The *dashed curves* show *C* calculated with no account for radionuclides in particulate form

contamination of the surface by cesium and strontium (Table 7.19), the concentration of Sr-90 in water is almost twice as large as the total concentration of Cs-134 and Cs-137 (dashed curves in Fig. 7.11).

The identified inversion is yet another example of anomalous behavior of artificial radionuclides, determined by the superposition of different solute transport mechanisms (Rumynin 2011).

Chemically inactive radionuclide I-131 with short half-life has the least effect on the radioactive contamination of river water (Fig. 7.11c): its maximal activity is about 0.2 BqL^{-1}. Soil erosion has no effect on the radioactive contamination of water with this radionuclide (Table 7.19).

The character and rate of soil erosion is illustrated by plots in Fig. 7.12. The first plot (Fig. 7.12a) characterizes variations of the total suspended sediments in river water during flood period. Unlike the previous plots, the input of first portions of

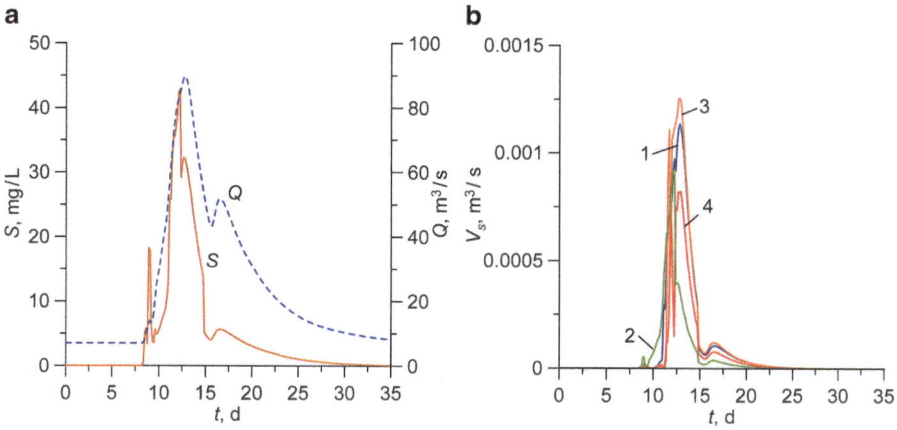

Fig. 7.12 TSS (total suspended sediment) plots: (**a**) total concentration of suspended particles in the river water, and (**b**) volumetric distribution of particle size (1 and 2 are coarse and fine sand fractions, 3 and 4 are silt and clay fractions)

suspended particles into river flow coincides with the beginning of the flood. The concentration of sediments varies from a few to several tens of mg/L, a fact that does not contradict the current notions regarding river water chemistry during floods. At the same time, conversion to mass flux yields a value of about 3000 gs^{-1}, which is appreciably higher than the actual measured values (Fig. 7.5). The distribution of *fractions of suspended* matter (Fig. 7.12b) is approximately proportional to the distribution between size-fractions in soil.

Data given in Table 7.19 can be used to evaluate the removal of radionuclides from the Pyshma watershed, characterized by *wash-out coefficient*, $K_c(\Delta t)$ (Eq. 7.1). Calculations for Cs-137/Cs-134 yield a change of about 0.2 % in $K_c(\Delta t = 35\ \text{d})$. The respective values for Sr-90 and I-131 are even less. Thus, simulation data are in general agreement with the current notions regarding the impact of such accidents on natural waters (Sect. 7.1.2.1).

The results given here were obtained for a relatively short period (1 month). The later rains and snow melting will also contribute to the export of activity (Cs-137, Cs-134, and Sr-90) from the watershed; however, the maximal radionuclide concentrations in river water will gradually decrease due to the decay of radionuclides and their export into deeper soil layers, which are less exposed to washout and water erosion.

In addition to the presented analysis, Fig. 7.13 gives examples of maps illustrating the dynamics of contamination of the watershed slopes and river water by Cs-137 + Cs-134 for a case of the initial deposition of radionuclides in the eastern sector (Fig. 7.4). The major contamination flow can be seen to be carried by surface runoff towards the watershed outlet. Since the *wash-out coefficient* is of the order of hundredths of percent, the concentration of adsorbed radionuclides on the surface after a storm rain remains nearly the same. Clearly, contaminated soils are a source of long-term contamination of surface and subsurface waters.

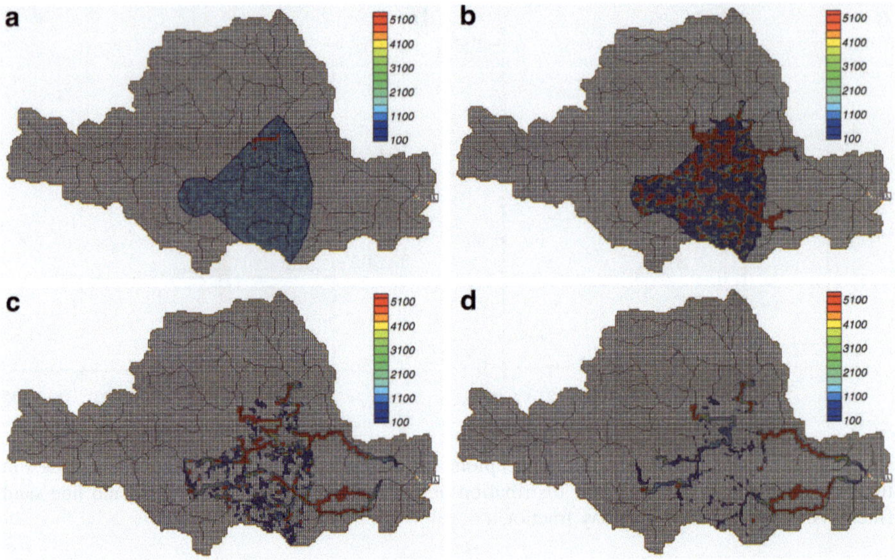

Fig. 7.13 Predicted dynamics of areal contamination of the Pyshma watershed by isotopes of cesium (Cs-137 and Cs-134) after fallout in the eastern segment (Bqm^{-3}). The maps are given for moments 8, 10, 12, and 14 days after the fallout of accident radionuclides.

Overall, the obtained simulation results regarding the role of different forms of radionuclides, are in qualitative agreement with the data of field observations of surface runoff water originating from the radioactivity-contaminated Olkchovkaya marsh-river area (Sect. 7.2.1.2). According to the study, the monitoring data prove that the more significant the sorption (in this context, of Cs-137 in comparison with Sr-90), the more important the role of radionuclide transport in runoff facilitated by soil erosion.

References

Aarkrog A, Dahlgaard H, Nielsen SP (1997) Radioactive inventories from the Kyshtym and Karachay accidents: estimates based on soil samples collected in the South Urals (1990–1995). Sci Total Environ 201:137–154

Accident Analysis for Nuclear Power Plants. Safety Reports Series 23 (2020) IAEA, Vienna, p 121

Atlas of the EURT, East Ural Radioactive Trace (2013) Izrael YA. IGCE Roshydromet and RAS, «Infosphere» Foundation, Moscow, p 140

Beloyarsk NPP (2011) Unit 4. Final report on safety assessment. Book 4. Sect. 15.3. Analysis of the design beyond accidents. "OKBM Afrikantov". BL4-0-0-OOOB-001/15.4

Bixio AC, Gambolati G, Panikoni C et al (2002) Modeling groundwater–surface water interactions including effects of morphogenetic depressions in the Chenobyl exclusion zone. Environ Geol 42(2–3):162–177

Bublias VN, Shestopalov VM (2001) Anomaly zones and their role in redistribution of radionuclides between soils and aquifers. Water exchange in hydrogeological structures and Chernobyl catastrophe. Institute of Geological Sciences. Ukrainian Acad Sci Kiev 1:251–356

Bulgakov AA, Konoplev AV, Shveikin YV et al (1999) Experimental study and prediction of dissolved radionuclide wash-off by surface runoff from non-agricultural watersheds/contaminated Forests. Recent developments in risk identification and future perspectives Part 1, NATO science series 2. Environ Secur 58:102–112

Cambray RS, Playford K, Lewis GN et al (1989) Radioactive fallout in air and rain: results to the end of 1988. AERE-R-13575. Atomic energy authority report, Harwell, UK

De Cort M, Dubois G, Fridman SD et al (1998) Atlas of caesium deposition on Europe after the Chernobyl accident. EUR report N 16733, EC, Official Publication of the European Communities. Luxembourg, p 65

Downer CW, Ogden FL (2006) Gridded surface subsurface hydrologic analysis (GSSHA). User's Manual. Version 1.43 for Watershed Modeling System 6.1, p 208

Eakins JD, Cambray RS, Chambers KC (1984) The transfer of natural andartificial radionuclides to Brotherswater from its catchment. In: Haworth EY, Lund JWG (eds) Lake sediments and environmental history: studies in palaeolimnology and palaeoecology in honour of Winifred Tutin. Leicester University Press, Leicester, pp 125–144

Evrard O, Chartin C, Onda Y et al (2013) Evolution of radioactive dose rates in fresh sediment deposits along coastal rivers draining Fukushima contamination plume. Scientific reports 3. Article number: 3079. doi:10.1038/srep03079

Flury M (1996) Experimental evidence of transport of pestcides through field soils. J Environ Qual 25:25–45

Foster GR, Flanagan DC, Nearing MA (1995) Chapter 11. Hillslope erosion component. In: Flanagan DC, Nearing MA (eds) Technical documentation. USDA – water erosion prediction project (WEPP). NSERL. Report N10. National Soil Erosion Research Laboratory, West Lafayette, Indiana, USA

Garcia-Sanchez L, Konoplev A (2009) Watershed wash-off of atmospherically deposited radionuclides: a review of normalized entrainment coefficients. J Environ Radioact 100(9):774–778

Garcia-Sanchez L, Konoplev A, Bulgakov A (2005) Radionuclide entrainment coefficients by wash-off derived from plot experiments near Chernobyl. J Radioprot Suppl 40:519–524

Gerke HH, Dusek J, Vogel TJ et al (2007) Two-dimensional dual-permeability analyses of a bromide tracer experiment on a tile-drained field. Vadose Zone J 6:651–667

Grant SB, Stewardson MJ, Marusic I (2012) Effective diffusivity and mass flux across the sediment-water interface in streams. Water Resour Res. doi:10.1029/2011WR011148

GSSHA Wiki (2014) Gridded surface subsurface hydrologic analysis. http://www.gsshawiki.com

Helton J, Muller A, Bayer A (1985) Contamination of surface-water bodies after reactor accidents by the erosion of atmospherically deposited radionuclides. Health Phys 48(6):757–771

Högberg L (2013) Root cases and impacts of severe accidents at large nuclear power plants. Ambio 42:267–284

Israel YA, Vakulovskii SM, Vetrov VA et al (1990) Chernobyl: radioactive contamination of the environment. Hydrometeoizdat, Moscow, p 223 (in Russian)

Ivanov YA, Kashparov VA (2003) Long-Term dynamics of the radioecological situation in Terrestrial ecosystems of the Chernobyl exclusion zone. Environ Soil Pollut Res 1(Special Issue):13–20

Konoplev AV, Bulgakov A, Popov V et al (1992) Behaviour of long-lived Chernobyl radionuclides in a soil-water system. J Analyst 117:1041–1047

Konz N, Baenninger D, Konz M (2010) Process identification of soil erosion in steep mountain regions. Hydrol Earth Syst Sci 14:675–686

Matsunaga T, Koarashi J, Atarashi-Andoh M et al (2013) Comparison of the vertical distributions of Fukushima nuclear accident radiocesium in soil before and after the first rainy season, with physicochemical and mineralogical interpretations. Sci Total Environ 447:301–314

Molchanova IV, Karavaeva EN, Mikchailovskaya LN (2009) Results of long-term radio-ecological investigations of the natural ecosystems in zones of liquid waste discharge from Beloyarskaya NPP. Prob Radiat Safety 4:20–27
Monte L, Brittain JE, Håkanson L et al (2004) Review and assessment of models for predicting the migration of radionuclides from catchments. J Environ Radioact 75:83–103
Nagai H, Katata G, Terada H (2014) Source term estimation of I-131 and Cs-137 discharged from the Fukushima Daiichi Nuclear Power Plant into the atmosphere. In: Takahashi S (ed) Radiation monitoring and dose estimation of the Fukushima nuclear accident. Springer, Tokyo, pp 155–173
Nalbandyan A, Ytre-Eide MA, Thørring H (2012) Potential consequences in Norway after a hypothetical accident at Leningrad nuclear power plant. Potential release, fallout and impacts on the environment. Norwegian Radiation Protection Authority, Østerás
Nelson EJ, McCarthy JE, Paudel M et al (2012) Watershed erosion evaluation of empirical and physical models at Aguacate Reservoir. In: Munoz RM (ed) River Flow. pp 889–896
Nieber JL (2001) The relation of preferential flow to water quality, and its theoretical and experimental quantification. Preferential flow. Water movement and chemical transport in the environment. In: Proceedings of the 2nd international symposium. Honolulu, Hawaii, pp 1–10
Nilsson K, Jensen SB, Carlsen L (1985) The migration chemistry of strontium. Eur Appl Res Rept Nucl Sci Technol 7(1):149–200
Otosaka S, Kobayashi T (2013) Sedimentation and remobilization of radiocesium in the coastal area of Ibaraki, 70 km south of the Fukushima Dai-ichi Nuclear Power Plant. Environ Monit Assess 185:5419–5433
Owens PN, Walling DE, He Q et al (1997) The use of caesium-137 measurements to establish a sediment budget for the Start catchment, Devon, UK. Hydrol Sci 42(3):405–423
Poręba GJ (2006) Caesium-137 as a soil erosion tracer: a review. Geochronometria 25:37–46
Pozolotina VN, Molchanova IV, Mikhaylovskaya LN et al (2012) The current state of terrestrial ecosystems in the Eastern Ural Radioactive Trace. In: Gerada JG (ed) Radionuclides: sources, properties and hazards. Nova Science Publishers, Huntington, pp 1–21
Romanov GN, Nikipelov BV, Drozhko EG (1990) The Kyshtym accident: causes, scale and radiation characteristics. Proceedings of Seminar on Comparative Assessment of the Environmental Impact of Radionuclides Released during Three Major Nuclear Accidents: Kyshtym Windscale, Chernobyl, Commission of the European Communities, EUR 13574, 1–5 October, Luxembourg, pp 25–40
Rumynin VG (2011) Subsurface solute transport models and case histories (with applications to radionuclide migration), vol 25, Theory and applications of transport in porous media. Springer, Dordrecht, p 815
Smith JT, Voitsekhovitch OV, Konoplev AV et al (2005) Radioactivity in aquatic systems. In: Smith JT, Beresford NA (eds) Chernobyl catastrophe and consequences. Praxis Publishing Ltd, Chcester, pp 139–181
Shestakov VM, Pozdniakov SP (2003) Geohydrology. IKC “Academkniga”, Moscow, p 176 (in Russian)
Shestopalov VM, Bohuslavsky AS, Bublias VN (2007) Assessment of groundwater protection with respect to preferential flow zones. Institute of Geological Sciences. Ukrainian Academy of Sciences, Kiev, p 120
Shestopalov VM, Rudenko YF, Bohuslavsky AS et al (2006) Chernobyl-born radionuclides: groundwater protectability with respect to preferential flow zones. In: Vereecken H, Binley A, Revil A, Titov K (eds) Applied hydrogeophysics. NATO Science Series, Springer, Dordrecht, pp 341–383
Steenhuis TS, Bodnar M, Geohring LD (1997) A simple model for predicting solute concentration in agricultural tile lines shortly after application. Hydrol Earth Syst Sci 4:823–833
The release, dispersion and deposition of radionuclides. Chernobyl: Assessment of Radiological and Health Impact Update of Chernobyl: Ten Years On (2002)

Ueda S, Hasegawa H, Kakiuchi H (2013) Fluvial discharges of radiocaesium from watersheds contaminated by the Fukushima Dai-ichi Nuclear Power Plant accident, Japan. J Environ Radioact 118:96–104

Utkin VI, Chebotina MY, Evstigneev AV (2000) Radioactive disasters of the Ural. Ural Branch of the Russian Academy of Sciences, Ekaterinburg, p 94

Wallach R, Jury WA, Spencer WF (1989) The concept of convective mass transfer for prediction of surface-runoff pollution by soil surface applied chemicals. J Trans ASAE 32:906–912

Wicks JM, Bathurst JC (1996) SHESED: a physically based, distributed erosion and sediment component for the SHE Hydrological Modeling System. J Hydrol 175:213–238

Conclusion

In this conclusive message we would like to underline again that the book does not exhaust the wide diversity of hydrological problems related to water quality but deals adequately with the basic concepts concerning the recognition of (1) the effect of transient rainfall–runoff–infiltration partitioning on the chemical response of drainage areas to excess precipitation under certain field conditions determined by soil and hillslope characteristics and contaminant properties; (2) soil erosion as a key factor, which enhances the potential of adsorbed chemical transport in runoff; (3) common tendencies in radionuclide behavior in the near-surface environment contaminated by radioactive fallout from the sadly remembered accidents at nuclear units, as well as the consequences of the nuclear weapon tests in the atmosphere since 1952.

On the other hand, from the book one can get a sense that despite the ongoing exponential growth in modeling approaches to study the mention problems, the selection of an appropriate mathematical model for a specific area is still a matter of uncertainty and the various models exhibit a large spread. Much of this uncertainty results from the lack of specificity in characterizing the structure of a watershed hydrologic system. The heterogeneity of soil structure and input random variables, which are usually assumed uniformly distributed, are among the main obstacles to the implementation of different models to actual catchments. Many of these issues out of scope of this study, and therefore a fundamental question whether the mathematical model used is appropriate can be addressed in other specific publications.

Meanwhile, the author hopes that the book's goal to provide a conceptual foundation to enable readers to apply scientific knowledge for solving practical problems in environmental hydrology and radiology has been reached.

V.G. Rumynin, *Overland Flow Dynamics and Solute Transport*, Theory and Applications of Transport in Porous Media 26, DOI 10.1007/978-3-319-21801-4

Index

V.G. Rumynin, *Overland Flow Dynamics and Solute Transport*, Theory and Applications of Transport in Porous Media 26, DOI 10.1007/978-3-319-21801-4

MIX
Papier aus verantwortungsvollen Quellen
Paper from responsible sources
FSC® C105338

If you have any concerns about our products,
you can contact us on
ProductSafety@springernature.com

In case Publisher is established outside the EU,
the EU authorized representative is:
Springer Nature Customer Service Center GmbH
Europaplatz 3, 69115 Heidelberg, Germany

Printed by Libri Plureos GmbH
in Hamburg, Germany